Learning Techniques for the Internet of Things

Praveen Kumar Donta • Abhishek Hazra •
Lauri Lovén
Editors

Learning Techniques for the Internet of Things

Springer

Editors
Praveen Kumar Donta
Distributed Systems Group
TU Wein
Vienna, Austria

Abhishek Hazra
Indian Institute of Information Technology
Sri City, India

Lauri Lovén
Center for Ubiquitous Computing
University of Oulu
Oulu, Finland

ISBN 978-3-031-50513-3 ISBN 978-3-031-50514-0 (eBook)
https://doi.org/10.1007/978-3-031-50514-0

This Springer imprint is published by the registered company Springer Nature Switzerland AG
The registered company address is: Gewerbestrasse 11, 6330 Cham, Switzerland

Paper in this product is recyclable.

To My Family, Dear Friends, My Teachers, Colleagues: You have been my pillars of strength, my source of inspiration, and my trusted companions on this literary voyage. It is your enduring support, affection, and the strong bond we share that have brought this book into being.
– Praveen Kumar Donta

For all computer science students aspiring to create a positive impact on society through the application of technology.
– Abhishek Hazra

To my loving family, whose unwavering support and understanding have been my anchor, and to my dedicated colleagues, whose collaboration and insights have enriched this work.
– Lauri Lovén

Preface

Learning for the Internet of Things is a combination of advanced learning techniques for the Internet of Things (IoT) encompassing a range of cutting-edge approaches, including deep learning with CNNs, RNNs, and transformers, federated learning, edge AI for local data processing, reinforcement learning for autonomous decision-making, and their applications in real time. With this aim, we invited renewed professors and researchers around the world, and received around 29 chapters. After careful evaluation we confirmed 13 chapters, which are included in this book. This book can be a reference book suitable for lecturing the most relevant topic of IoT in intelligent environments through advanced learning techniques. It is suitable for Bachelor's or Master's students as well as young researchers. Throughout this book, the author provides basic insights into IoT using traditional learning algorithms, such as machine learning, federated learning, and deep learning, in addition to multiobjective reinforcement learning and inference strategies.

The book is structured into 13 chapters; each comes with its own dedicated contributions and future research directions. Chapter 1 provides an introduction to IoT and the use of Edge computing, particularly cloud computing, and mobile edge computing. This chapter also mentions the use of edge computing in various real-time applications such as healthcare, manufacturing, agriculture, and transportation. Chapter 2 motivates mathematical modeling for federated learning systems with respect to IoT and its applications. Further, Chap. 3 extends the discussion of federated learning for IoT, which has emerged as a privacy-preserving distributed machine learning approach. Chapter 4 provides various machine learning techniques in Industrial IoT to deliver rapid and accurate data analysis, essential for enhancing production quality, sustainability, and safety. Chapter 5 discusses the potential role of data-driven technologies, such as Artificial Intelligence, Machine Learning, and Deep Learning, focusing on their integration with IoT communication technologies. Chapter 6 presents the requirements and challenges to realize IoT deployments in smart cities, including sensing infrastructure, Artificial Intelligence, computing platforms, and enabling communications technologies such as 5G networks. To highlight these challenges in practice, the chapter also presents a real-world case study of a city-scale deployment of IoT air quality monitoring within Helsinki city.

Chapter 7 uses digital twins within smart cities to enhance economic progress and facilitate prompt decision-making regarding situational awareness.

Chapter 8 provides insights into using multiobjective reinforcement learning in future IoT networks, specially for efficient decision-making systems. Chapter 9 offers a comprehensive review of intelligent inference approaches, with a specific emphasis on reducing inference time and minimizing transmitted bandwidth between IoT devices and the cloud. Chapter 10 summarizes the applications of deep learning models in various IoT fields. This chapter also presents an in-depth study of these techniques to examine new horizons of applications of deep learning models in different areas of IoT. Chapter 11 explores the integration of Quantum Key Distribution (QKD) into IoT systems. It delves into the potential benefits, challenges, and practical considerations of incorporating QKD into IoT networks. In Chap. 12, a comprehensive overview regarding the current state of quantum IoT in the context of smart healthcare is presented, along with its applications, benefits, challenges, and prospects for the future. Chapter 13 proposes a blockchain-based architecture for securing and managing IoT data in intelligent transport systems, offering advantages like immutability, decentralization, and enhanced security.

The book is suitable for a wide range of disciplines, including Computer Science, Artificial Intelligence, Mechanical or Automation, Robotics, and so on. This book's primary audience is Bachelor's or Master's level students. It may be appropriate to consider this book for courses to motivate students in these areas since multiple subdomains/branches are being opened up by many universities. The book provides simplified approaches and real-time applications, so readers without background knowledge of Artificial Intelligence or the Internet of Things can easily understand them. Furthermore, a few chapters (3, 6, 8, 9, and 13) in the book are extensive and useful for PhD students, where they can use them as basic reference material for advancing technologies. As we keep undergraduates in mind, we try to simplify the text, so basic math and a brief knowledge of communication and networking skills are enough to understand this book.

Vienna, Austria — Praveen Kumar Donta
Sri City, India — Abhishek Hazra
Oulu, Finland — Lauri Lovén

Acknowledgements

We would like to express our sincere gratitude to all the esteemed authors who contributed their invaluable expertise and insights to this edited book. Your dedication, arduous work, and commitment to your respective chapters have made this book a reality. Each of you enriched the content with your unique perspectives and knowledge, and we deeply appreciate your time and efforts to contribute. We are honored to have had the opportunity to work with such a talented group of individuals. We thank you for your collaborative spirit and excellent work.

This book is supported by the Academy of Finland through the 6G Flagship program (Grant 318927); by the European Commission through the ECSEL JU FRACTAL project (Grant 877056), receiving support from the EU Horizon 2020 program and Spain, Italy, Austria, Germany, France, Finland, Switzerland; and finally, by Business Finland through the Neural pub/sub research project (diary number 8754/31/2022). We also thank Center for Ubiquitous Computing, University of Oulu, Oulu, Finland, for providing the necessary support to conduct this book.

This book also received support from the European Commission through the TEADAL project (Grant 101070186), and AIoTwin (Grant 101079214), by EU Horizon 2020 program and partners from different countries like Spain, Italy, Greece, Germany, Israel, Portugal, Switzerland. We also thank Distributed Systems Group, Technische Universität Wien, Vienna, Austria, for providing the necessary support to conduct this book.

We also thank the Networks and Communications Lab, Department of Electrical and Computer Engineering, National University of Singapore and Department of Computer Science and Engineering, Indian Institute of Information Technology, Sri City, Andhra Pradesh, India, for providing the necessary support to conduct this book.

Vienna, Austria — Praveen Kumar Donta
Sri City, India — Abhishek Hazra
Oulu, Finland — Lauri Lovén

Contents

Editors and Contributors

About the Editors

Dr. Praveen Kumar Donta (Senior Member IEEE and Professional Member ACM) is currently working as Postdoctoral Researcher at Distributed Systems Group, TU Wien (Vienna University of Technology), Vienna, Austria. He received his PhD from the Indian Institute of Technology (Indian School of Mines), Dhanbad, in the field of machine learning-based algorithms for wireless sensor networks in the year of 2021. From July 2019 to Jan 2020, he is a visiting PhD fellow at Mobile & Cloud Lab, Institute of Computer Science, University of Tartu, Estonia, under the Dora plus grant provided by the Archimedes Foundation, Estonia. He received his Master in Technology and Bachelor in Technology from the Department of Computer Science and Engineering at JNTUA, Ananthapur, with distinction in 2014 and 2012. Currently, he is a Technical Editor and Guest Editor for *Computer Communications*, Elsevier, Editorial Board member for *International Journal of Digital Transformation*, Inderscience, and *Transactions on Emerging Telecommunications Technologies* (ETT), Wiley. He is also serving as Early Career Advisory Board in *Measurement and Measurement: Sensors*, Elsevier journals. He served as IEEE Computer Society Young Professional Representative for Kolkata section. His current research includes Learning-driven Distributed Computing Continuum Systems, Edge Intelligence, and Causal Inference for Edge. Contact him at pdonta@dsg.tuwien.ac.at.

Dr. Abhishek Hazra currently works as an Assistant Professor in the Department of Computer Science and Engineering, Indian Institute of Information Technology Sri City, Chittoor, Andhra Pradesh, India. He was a Postdoctoral Research Fellow at the Communications and Networks Lab, Department of Electrical and Computer Engineering, National University of Singapore. He has completed his PhD at the Indian Institute of Technology (Indian School of Mines) Dhanbad, India. He received his MTech in Computer Science and Engineering from the National Institutes of Technology Manipur, India, and his BTech from the National Institutes of Technology Agartala, India. He currently serves as an Editor/Guest

Editor for *Physical Communication*, *Computer Communications*, *Contemporary Mathematics*, *IET Networks*, *SN Computer Science*, and *Measurement: Sensors*. He is also a conference general chair for IEEE PICom 2023. His research area of interest includes IoT, Fog/Edge Computing, Machine Learning, and Industry 5.0. Contact him at abhishek.hazra1@gmail.com.

Dr. Lauri Lovén DSc(Tech), is a Senior Member of IEEE and the coordinator of the Distributed Intelligence strategic research area in the 6G Flagship research program, at the Center for Ubiquitous Computing (UBICOMP), University of Oulu, in Finland. He received his DSc at the University of Oulu in 2021, was with the Distributed Systems Group, TU Wien in 2022, and visited the Integrated Systems Laboratory at the ETH Zürich in 2023. His current research concentrates on edge intelligence, and on the orchestration of resources as well as distributed learning and decision-making in the computing continuum. He has co-authored 2 patents and ca. 50 research articles. Contact him at lauri.loven@oulu.fi.

Contributors

Martha Arbayani Zaidan Department of Computer Science, University of Helsinki, Helsinki, Finland

Lalit Kumar Awasthi National Institute of Technology Uttarakhand, Srinagar, India

Quentin De La Cruz Department of Mathematics & Computer Science, Brandon University, Brandon, MB, Canada

Chinmaya Kumar Dehury Institute of Computer Science, University of Tartu, Tartu, Estonia

Schahram Dustdar Distributed Systems Group, TU Wien, Vienna, Austria

Iwada Eja Cloud Platform Team, Finnair, Estonia

Balqees Talal Hasan Department of Computer and Information Engineering, Nineveh University, Mosul, Iraq

Min Huang College of Information Science and Engineering, Northeastern University, Shenyang, China

Abhishek Hazra Indian Institute of Information Technology, Sri City, India

Naser Hossein Motlagh Department of Computer Science, University of Helsinki, Helsinki, Finland

Ali Kadhum Idrees Department of Information Networks, University of Babylon, Babylon, Iraq

Surbhi Bhatia Khan Department of Data Science, School of Science, Engineering and Environment, University of Salford, Manchester, UK

Haodong Li College of Computer Science and Engineering, Northeastern University, Shenyang, China

Ying Li College of Computer Science and Engineering, Northeastern University, Shenyang, China
Distributed Systems Group, TU Wien, Vienna, Austria

Sindri Magnússon Department of Computer and Systems Science, Stockholm University, Stockholm, Sweden

Hariprasath Manoharan Department of Electronics and Communication Engineering, Panimalar Engineering College, Chennai, Tamil Nadu, India

Poonam Maurya Aalborg University, Aalborg, Denmark

Roberto Morabito Department of Computer Science, University of Helsinki, Helsinki, Finland

Ilir Murturi Distributed Systems Group, TU Wien, Vienna, Austria

Petteri Nurmi Department of Computer Science, University of Helsinki, Helsinki, Finland

Victor Casamayor Pujol Distributed Systems Group, TU Wien, Vienna, Austria

Somya Rathee Informatics, HTL Spengergasse, Vienna, Austria

Haider Daniel Ali Rizvi Yogananda School of AI, Shoolini University, Bajhol, India

Shitharth Selvarajan School of Built Environment, Engineering and Computing, Leeds Beckett University, Leeds, UK

Megha Sharma Netaji Subhas University of Technology, Delhi, India

Aditya Srivastava Department of Computer Science & Engineering, Amity School of Engineering and Technology, Amity University, Lucknow, India

Atul Srivastava Department of Computer Science & Engineering, Amity School of Engineering and Technology, Amity University, Lucknow, India

Gautam Srivastava Department of Mathematics & Computer Science, Brandon University, Brandon, MB, Canada

Department of Computer Science and Math, Lebanese American University, Beirut, Lebanon

B. Sundaravadivazhagan Department of Information and Technology, University of Technology and Applied Sciences, Al Mussana, Muladdah, Oman

Kartick Sutradhar Indian Institute of Information Technology, Sri City, India

Sasu Tarkoma Department of Computer Science, University of Helsinki, Helsinki, Finland

Abhinav Tomar Netaji Subhas University of Technology, Delhi, India

Shubham Vaishnav Department of Computer and Systems Science, Stockholm University, Stockholm, Sweden

Priyanka Venkatesh Presidency University, Bengaluru, India

Ranjitha Venkatesh Gandhi Institute of Technology and Management, Bengaluru, India

Shangguang Wang State Key Laboratory of Network and Switching, Beijing University of Posts and Telecommunications, Beijing, China

Xingwei Wang College of Computer Science and Engineering, Northeastern University, Shenyang, China

Rongfei Zeng College of Software, Northeastern University, Shenyang, China

Dingge Zhang State Key Laboratory of Network and Switching, Beijing University of Posts and Telecommunications, Beijing, China

Qiyang Zhang State Key Laboratory of Network and Switching, Beijing University of Posts and Telecommunications, Beijing, China
Distributed Systems Group, TU Wien, Vienna, Austria

Chapter 1
Edge Computing for IoT

Balqees Talal Hasan and Ali Kadhum Idrees

1.1 Introduction

In as early as 1966, an insightful prediction emerged from Karl Steinbuch, a pioneer in German computer science. He predicted that within a few decades, computers would be a necessary component of almost every industrial product. The term "pervasive computing" was first introduced by W. Mark in 1999. It means integrating computers into everyday objects so seamlessly that they become a natural and unnoticed part of the environment, with people interacting with them effortlessly. In the same year (1999), the term "Internet of Things" was coined by Kevin Ashton at a presentation at Procter & Gamble (P&G) (Elazhary 2019).

IoT is a new paradigm for attaching various physical objects to the Internet so they can interact and make informed decisions. The technologies that fall under this paradigm include pervasive computing, RFID, communication technologies, sensor networks, and Internet protocols. In IoT, physical things have the ability to intelligently collaborate and establish connections with the Internet, operating autonomously and introducing innovative applications. These applications span a variety of industries, such as manufacturing, transportation, healthcare, industrial automation, and emergency response (Lorenzo et al. 2018; Idrees et al. 2020).

IoT has become permeated our daily lives, providing crucial measuring and data-gathering capabilities that influence our decision-making. Numerous sensors and gadgets run continuously, producing data and enabling vital communication over complex networks. It is challenging to execute complicated computations on most

B. T. Hasan
Department of Computer and Information Engineering, Nineveh University, Mosul, Iraq
e-mail: balqees.hasan@uoninevah.edu.iq

A. K. Idrees (✉)
Department of Information Networks, University of Babylon, Babylon, Iraq
e-mail: ali.idrees@uobabylon.edu.iq

P. K. Donta et al. (eds.), *Learning Techniques for the Internet of Things*,
https://doi.org/10.1007/978-3-031-50514-0_1

of the IoT devices due to their limited CPU and energy resources. In general, IoT devices collect data and transmit it to more robust processing centers for analysis (Alhussaini et al. 2018). The data is subjected to extra processing and analysis at these centers (Idrees et al. 2020). In order to lighten the load on resources and avoid overuse of them, edge computing has become prevalent as a novel way to address IoT and local computing requirements (Yu et al. 2017). In edge computing, small servers that are located closer to users are used to process data. In close proximity to the devices of the consumers, these edge servers are capable of doing complex tasks and storing enormous volumes of data. As a result of their proximity to users, processing and storage at the network edge becomes faster and more efficient (Hassan et al. 2019). In 2022, the worldwide edge computing market was worth USD 11.24 billion. Experts predict that it will experience significant expansion, with an expected yearly from 2023 to 2030, the growth rate is 37.9% (Edge Computing Market Size, Share & Trends Analysis Report By Component (Hardware, Software, Services, Edge-managed Platforms) 2023).

Edge computing is different from the usual cloud computing approach. Instead of processing and storing data in centralized data centers far from users, edge computing involves positioning resources closer to users, specifically at the "edge" of the network. This means there are multiple computing nodes spread throughout the network, which reduces the burden on the central data center and makes data exchange much faster, as there is less delay in sending and receiving messages (Yu et al. 2017). Edge computing allows for the intelligent collection, analysis, computation, and processing of data at every IoT network edge. This implies that data can be filtered, processed, and used close to the devices or data sources, where it is generated. Edge computing makes everything faster and more effective by pushing smart services to the edge of the network. Making decisions and processing data locally can also help deal with significant limitations in networks and resources, and it can address concerns related to security and privacy too (Zhang et al. 2020; Shawqi Jaber & Kadhum Idrees 2020).

Here is how this chapter is organized: Sect. 1.2 provides a comprehensive explanation of computing paradigms for IoT. Moving on to Sect. 1.3, a detailed introduction to edge computing paradigms is presented. Section 1.4 outlines the architecture of edge computing-based IoT. In Sect. 1.5, the focus shifts to illustrate the advantages of edge computing-based IoT. The enabling technologies for edge computing-based IoT are introduced in Sect. 1.6. In Sect. 1.7, the chapter reviews edge computing in IoT-based intelligent systems. Section 1.8 illustrates the challenges and future research directions for edge computing-based IoT. Finally, Sect. 1.9 concludes the chapter.

1.2 Computing Paradigms for IoT

This section describes the fundamental concepts underlying the three major computing paradigms and how they are integrated with IoT: cloud computing, edge

Fig. 1.1 Three-tier architecture of computing paradigms

computing, and fog computing Srirama (n.d.); Fig. 1.1 shows the architecture of the 3 tiers computing paradigms.

1.2.1 Cloud Computing

As mobile hardware evolves and improves, it will always face limitations in terms of available resources compared to stationary hardware. Regarding devices that people wear or carry for extended periods, prioritizing improvements in weight, size, and battery life takes precedence over enhancing computational power. This is a fundamental aspect of mobility rather than a transient restriction imposed by modern mobile electronics. Therefore, there will always be trade-offs when using computational power on mobile devices. The resource limitations of mobile devices can be solved simply and effectively by using cloud computing. With this approach, a mobile device can execute an application that requires a lot of resources on a robust remote server or a cluster of servers, allowing users to interact with the application through a lightweight client interface over the Internet (Satyanarayanan et al. 2009). The cloud computing paradigm gives end users on-demand services by utilizing a pool of computing resources. These resources include computing power, storage, and more, and they are all immediately available at any time (Khan et al. 2019).

IoT and the cloud have had separate evolutionary processes. However, its integration has produced a number of benefits for both parties. On the one hand, IoT can greatly profit from the boundless capabilities of cloud computing to overcome its own technological limitations, such as storage, processing power, and energy requirements. The cloud can take advantage of IoT through expanding its range of applications to handle real-world objects in a more distributed and adaptable fashion, thus providing new services in various real-life situations (Alessio et al. 2014).

1.2.2 Edge Computing

Typically, the architecture of the cloud is used to manage the massive amount of data that IoT devices produce. However, cloud computing encounters various challenges such as lengthy transmission time, increased bandwidth requirements, and latency between IoT devices and the cloud. The concept of edge computing has emerged to overcome these difficulties. This approach enhances scalability, latency, and privacy factors while enabling real-time predictions by processing data at the source (Naveen et al. 2021; Idrees et al. 2022).

In an extension of cloud computing, edge computing places computer services at the edge of the network, where they are more accessible to end users. Edge computing shifts services, computational data, and applications out from cloud servers and toward the edge of a network. This enables content providers and application developers to provide users with services that are located nearby. Edge computing is unique in that it may be used for a variety of applications, thanks to its high bandwidth, very low latency, and fast access to network data (Khan et al. 2019; Idrees & Jawad 2023).

In the world of IoT, both edge computing and cloud computing offer major advantages due to their distinct characteristics, such as their capacity to execute complex computations and store large amounts of data. However, when it comes to IoT, edge computing outperforms cloud computing, despite having somewhat limited compute capability and storage capabilities. In particular, IoT demands fast responses rather than powerful computing and massive storage. Edge computing fulfills the requirements of IoT applications by offering satisfactory computing capability, sufficient storage, and fast response times. Edge computing, on the other hand, can also leverage IoT to expand its framework and adapt to the dynamic and distributed nature of edge computing nodes. These edge nodes can serve as providers and may consist of either IoT devices or devices with some residual computational capabilities (Yu et al. 2017).

1.2.3 Fog Computing

Cisco introduced the concept of fog computing in January 2014 (Delfin et al. 2019). This computing paradigm offers numerous advantages across various fields, especially the IoT (Atlam et al. 2018; Idrees & Khlief 2023b). According to Antunes, a senior official in charge of advancing corporate strategy at Cisco, edge computing is a division of fog computing. He explains that fog computing primarily focuses on managing the location of data generation and storage. In essence, edge computing involves processing data in proximity to its source of origin (Kadhum & Saieed Khlief n.d.). Fog computing, on the other hand, leverages edge processing and the necessary network connections to transfer data from the edge to the endpoint (Delfin et al. 2019). The fog computing system was not designed to replace cloud computing; instead, its development aimed to fill any service gaps present in cloud computing. Fog computing emphasizes on bringing abilities of cloud computing closer to the edge of the network so that users can access communication and software services faster. This approach works well for offering cloud solutions for highly mobile technologies like vehicular ad hoc networks (VANET) (Ravi et al. 2023) and the IoT (Alwakeel 2021). Fog computing serves the endpoints or edges of the network of interconnected devices. It prioritizes the analysis of time-sensitive data near the sources, sending only the selected and abridged data to the cloud (Delfin et al. 2019; Idrees & Khlief 2023a).

The concept of "fog as a service" (FaaS) is a new service possibility brought about by the integration of fog computing and IoT. According to this concept, a service provider creates a network of fog nodes throughout the area covered by its service, operating as a landlord to numerous tenants from diverse businesses. Each fog node provides storage, networking, and local computing capabilities. Through FaaS, customers can access services using innovative business models. Unlike clouds, which are often managed by huge businesses with sizable data centers, FaaS enables both small and large businesses to create and manage public or private computing, storage, and control services at different scales, meeting the requirements of various clients (Atlam et al. 2018; Idrees et al. 2022).

1.3 Edge Computing Paradigms

Edge computing emerged due to the evolution of cloud computing, and it provides different computing advantages. Several paradigms for operating at the edge of the network have been established throughout the growth of edge computing, including cloudlet and mobile edge computing. The two primary edge computing paradigms are introduced in this section.

1.3.1 Cloudlet

In 2009, Satyanarayanan and his team first proposed cloudlet computing as a remedy for the problems that can arise while using traditional cloud computing. These restrictions cover things like delay, jitter, and congestion (Satyanarayanan et al. 2009). Cloudlets, as shown in Fig. 1.2, are essentially small data centers, often referred to as miniature clouds, which are frequently just a hop away from a user device (Yousefpour et al. 2019). Instead of relying on a far-off "cloud," a nearby cloudlet with abundant resources can be used to alleviate the limited resources of a mobile device. To fulfill the requirement for real-time interactive responses, a solution is to establish a wireless connection with a nearby cloudlet that provides high-bandwidth, one-hop, and low-latency wireless access. In this situation, the mobile device serves as a lightweight client, with the cloudlet in close proximity

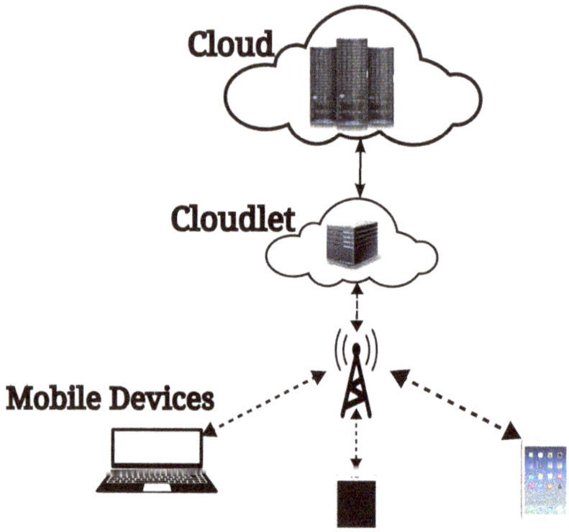

Fig. 1.2 Cloudlet architecture

handling the majority of the complex computational operations (Satyanarayanan et al. 2009). Cloudlet relies on technologies like Wi-Fi, making it reliant on a strong Internet connection (Abbas et al. 2018; Donta et al. 2023).

Owners of the network infrastructure, such as Nokia and AT&T, can enable cloudlets to be positioned in closer proximity to mobile devices, in hardware with smaller footprints than the cloud computing's massive data centers. As a result of their smaller footprint, cloudlets have less computational power than typical clouds, but they still have advantages over them including lower latency and energy usage (Yousefpour et al. 2019).

A "data center in a box" is how cloudlets are like. It operates autonomously, with minimal power consumption, and only needs an Internet connection and access control for setup. This administrative simplicity correlates with a device-based computing architecture, enabling easy implementation in a variety of business premises such as doctor's offices or coffee shops. From an internal viewpoint, a cloudlet can be perceived as a cluster of computers equipped with multiple cores, fast internal connections, and a high-bandwidth wireless LAN. For safe implementation in unsupervised areas, the cloudlet can be housed in a protective casing designed to resist tampering with third-party remote monitoring of hardware integrity (Satyanarayanan et al. 2009). To avoid any serious implications in the event of loss or malfunction, it is crucial to emphasize that the cloudlet should only store transient data and code, such as cached copies. In critical circumstances like military operations, disaster-affected areas, and even cyberattacks, cloudlets are essential. In these circumstances, the cloudlet, which is the middle layer in a three-tier architecture consisting of the mobile, cloudlet, and cloud, is required to fill in and provide vital services, making up for the cloud's unavailability. Cloudlets also have the benefit of reducing the dangers linked to multi-hop networks, such as the possibility of DoS attacks (Elazhary 2019).

1.3.2 Mobile Edge Computing

The idea behind a cloudlet is to strategically place powerful computers so that they can provide neighboring user equipment (UE) with computing and storage capabilities. Similar to Wi-Fi hotspots, cloudlets deliver cloud services to mobile customers as opposed to offering Internet connectivity. The fact that mobile UEs predominantly use Wi-Fi connections to access cloudlets has a potential disadvantage because it requires users to alternate between the mobile network and Wi-Fi anytime they need cloudlet services (Mach & Becvar 2017).

Another option that enables cloud computing at the edge is mobile edge computing (MEC), which first announced in 2014 by the European Telecommunications Standards Institute (ETSI). The MEC platform is characterized as a system that provides IT and cloud computing functionalities within the radio access network (RAN), positioned in close proximity to mobile subscribers (Mao et al. 2017).

MEC is defined by the ETSI as having low latency, local processing and storage resources, network awareness, and better service quality supplied by mobile carriers. MEC makes mobile end-user services more accessible by providing computing and storage resources. These resources are intended to be deployed on mobile networks near end users. MEC resources can be used in a variety of locations, including radio access networks (RANs), indoor and outdoor base stations, and access points, which connect user equipment to mobile network operators' (MNOs') core networks (Haibeh et al. 2022).

Within the radio access network, MEC provides cloud computing capabilities, as depicted in Fig. 1.3. Rather than routing mobile traffic from the core network to end users, MEC creates a direct link between users and the nearest edge network empowered with cloud services. By deploying MEC at the base station, it enhances

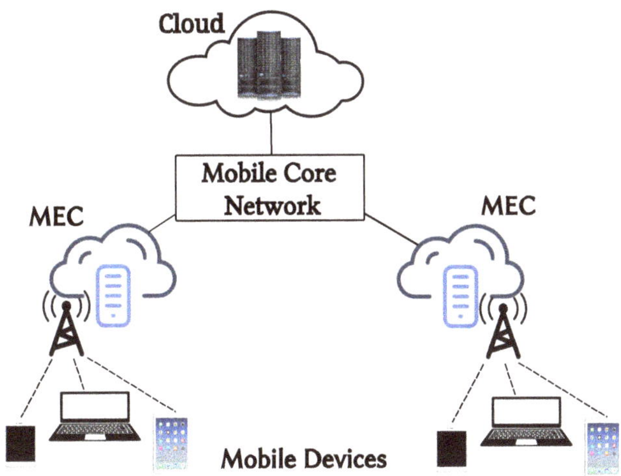

Fig. 1.3 Mobile edge computing architecture

computation, mitigates bottlenecks, and reduces the risk of system failure (Abbas et al. 2018). MEC is implemented on a virtualized platform that takes advantage of the most recent advancements in information-centric networks (ICN), network function virtualization (NFV), and software-defined networks (SDN). A single-edge device with NFV at its core can provide computational services to numerous mobile devices by producing several virtual machines (VMs). These VMs can handle several tasks or perform diverse network operations at the same time (Mao et al. 2017).

1.4 Architecture of Edge Computing-Based IoT

In the context of IoT, edge computing is primarily focused on its implementation across various IoT scenarios, aiming to minimize decision-making latency and network traffic (Fazeldehkordi & Grønli 2022). The edge computing-based IoT architecture, as depicted in Fig. 1.4, consists of three distinct layers, IoT, edge, and cloud, all of these are built on top of existing edge computing reference designs. Our primary focus is to define the specific functions allocated to each layer and explore the communication mechanisms established among these layers (Fazeldehkordi & Grønli 2022; Qiu et al. 2020):

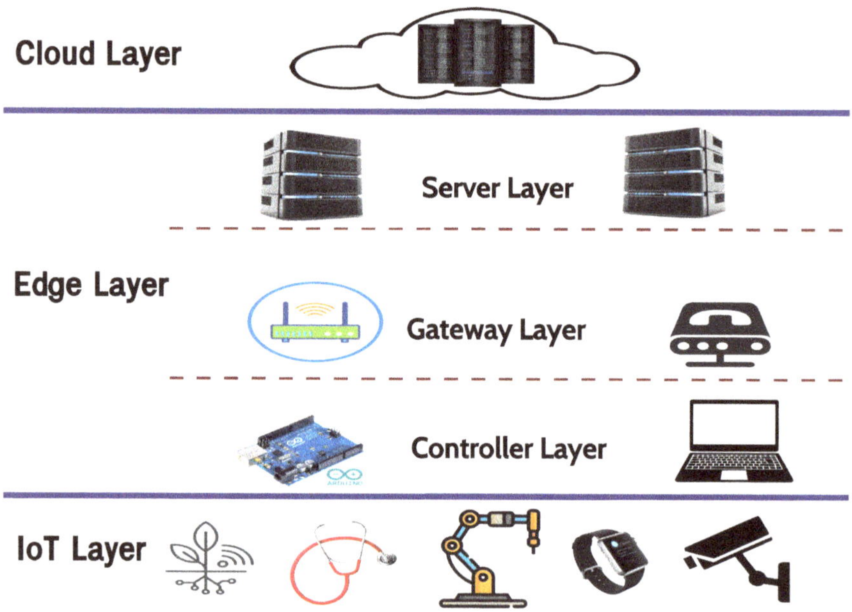

Fig. 1.4 Edge computing-based IoT architecture

1. **IoT layer**: The IoT layer encompasses a broad spectrum of devices and equipment, such as smart cars, robots, smart machinery, handheld terminals, instruments and meters, and other physical goods. These objects are tasked with overseeing the functioning of services, activities, or equipment. Furthermore, the IoT layer consists of actuators, sensors, controllers, and gateways constructed expressly for IoT contexts, which enable the administration of computational resources within IoT devices (Fazeldehkordi & Grønli 2022).
2. **Edge layer**: The main purpose of this layer is to receive, process, and send streams of data from the device layer. It offers real-time services like intelligent computing, security and privacy protection, and data analysis. Based on the equipment's ability to process data, three further sub-layers are separated from the edge layer: the near-edge layer, the mid-edge layer, and the far-edge layer (Qiu et al. 2020):

    (a) Far-Edge Layer (Edge controller layer): In this layer, data is collected from the IoT layer by edge controllers and subsequently undergoes initial threshold assessment or data filtering. After that, the edge layer or cloud layer directs the control flow back to the IoT layer. After IoT device data has been collected, it is preprocessed to determine thresholds or perform data filtering. Consequently, the edge controllers in this layer must incorporate algorithm libraries tailored to the environment's configuration to consistently improve the strategy's efficiency. Additionally, these edge controllers should convey the control flow back to the IoT layer via the programmable logic controller (PLC) control or action control module after receiving decisions from the edge controller layer or upper layers (Fazeldehkordi & Grønli 2022).
    (b) Mid-Edge Layer (Edge gateway layer): This layer is often made up of edge gateways, which can connect to wired networks like industrial ethernet or wireless networks like 5G to receive data from the edge controller layer. Furthermore, the layer enables diverse processing capabilities and caches the accumulated data. Moreover, the edge gateways in this layer play a crucial role in shifting control from the upper layers, such as the cloud layer or edge server layer, to the edge controller layer. Simultaneously, they monitor the equipment in both the edge gateway layer and the edge controller layer (Fazeldehkordi & Grønli 2022). The mid-edge layer has more storage and processing power than the far-edge layer, which can only carry out basic threshold judgment or data filtering. As a result, it can handle IoT layer data in a more thorough manner (Qiu et al. 2020).
    (c) Near-Edge Layer (Edge server layer): The edge server layer is equipped with robust edge servers. Within this layer, advanced and crucial data processing takes place. The edge servers leverage dedicated networks to gather data from the edge gateway layer and generate directional decision instructions based on this collected information. Additionally, platform administration and business application management features are anticipated for the edge servers in the edge server layer (Fazeldehkordi & Grønli 2022).

3. **Cloud layer**: This layer primarily focuses on in-depth data mining and seeks to allocate resources optimally on a big scale, across a whole organization, a region, or even the entire country. Data from the edge layer is sent to the cloud layer through the use of the public network. Additionally, the edge layer has the ability to receive feedback from cloud layer-provided business applications, services, and model implementations (Fazeldehkordi & Grønli 2022).

1.5 Advantages of Edge Computing-Based IoT

Edge computing plays a vital role as a computing paradigm for IoT devices, involving the utilization of cloud centers located near the IoT devices for tasks such as filtering, preprocessing, and aggregating IoT data (Fazeldehkordi & Grønli 2022). The primary advantages of edge computing include:

(A) **Low Latency**: The close proximity and low latency of edge computing provide a solution to the response delay faced by user equipments (UEs) while accessing typical cloud services. Edge computing can drastically reduce response time, which includes communication, processing, and propagation delays. Cloud computing typically results in an end-to-end latency of more than 80ms (or 160ms for response delay), making it unsuitable for time-sensitive applications such as remote surgery and virtual reality (VR), which require near-instantaneous replies within 1ms. Edge computing, on the other hand, benefits UEs by reducing total end-to-end delay and reaction delay due to their close proximity to edge servers. This enhancement enables faster and more efficient interactions for time-critical applications, meeting the requirements for tactile speed and responsiveness (Hassan et al. 2019).

(B) **Energy Saving**: IoT devices often have limited energy supply due to their size and intended usage scenarios, yet they are expected to conduct complicated activities that are frequently power-intensive. It is difficult to design a cost-effective system to properly power numerous distributed IoT devices since regular battery charging or discharging is not always practicable or possible. However, edge computing offers a solution by enabling IoT devices to offload power-consuming computation tasks to edge servers. This not only substantially lowers energy use but also enhances processing efficiency, enabling billions of IoT devices to function optimally (Wang et al. 2020).

(C) **Security and Privacy**: Among the most important features of cloud platform services is enhancing data security and privacy. Customers of these services can obtain centralized data security solutions from these providers, but any compromise of the centralizedly held data may have severe consequences. In contrast, edge computing has the benefit of allowing local deployment of customized security solutions. With this approach, less data transport is necessary because the majority of processing can be done at the network edge. As a result, there is a lower chance of data leakage during transmission, and less data is stored on the cloud platform, lowering the security and privacy risks (Fazeldehkordi & Grønli 2022).

(D) **Location Awareness**: Edge servers with location awareness can acquire and handle data generated by user equipments (UEs) based on their geographical locations. As a result, personalized and location-specific services can be offered to UEs, allowing edge servers to collect data directly from nearby sources without sending it to the cloud. This allows for more efficient and targeted service provisioning customized to specific UE needs (Hassan et al. 2019).

(E) **Reduce Operational Expenses**: Transmitting data directly to the cloud platform incurs substantial operational expenses due to the demands for data transmission, sufficient bandwidth, and low latency. Edge computing, on the other hand, has the advantage of minimizing data uploading volume, resulting in less data transmission, lower bandwidth consumption, and lower latency. As a result, edge computing reduces operational costs when compared to direct data transfer to the cloud platform (Fazeldehkordi & Grønli 2022).

(F) **Network Context Awareness**: Edge servers are able to understand the network context through network context awareness. This includes user equipment (UE) information, such as allocated bandwidth and user locations, as well as real-time network conditions, such as traffic load in a network cell and radio access network specifics. With this invaluable knowledge, edge servers are better equipped to adapt and accommodate to the various UEs and network conditions, which leads to an optimum use of network resources. As a result, edge servers can effectively handle a large amount of traffic, improving network performance. Additionally, the availability of fine-grained information enables the development of services that are specifically customized to the needs of various traffic flows and individual users (Hassan et al. 2019).

1.6 Enabling Edge Computing-Based IoT Technologies

Edge computing-based IoT can be implemented with the integration of several enabling technologies. This section illustrates the relevant enabling technologies by using artificial intelligence and lightweight virtualization as examples.

1.6.1 Edge Intelligence

As the need for intelligent edge devices has grown, the industry has responded with innovation and the adoption of intelligent edge architectures. These innovative architectures support real-time, mission-critical applications that work with a wide variety of devices. Any machine can qualify as intelligent if it mimics human behaviors and skills including perception, attention, thinking, and decision-making. Machine learning has gained a lot of traction as a field of advancement in artificial intelligence. This has led to a surge in the presence of intelligent devices, fueled primarily by advancements in deep learning techniques (Naveen et al. 2021).

Deep neural networks (DNNs) have received substantial attention in the machine learning era because of their unrivaled performance across different use cases such as computer vision, natural language processing, and image processing (Marchisio et al. 2019). Notably, deep learning has even outperformed human players in complex games like Atari Games and the game of Go. The integration of deep learning and edge computing holds promise for addressing challenges and opening up new possibilities for applications. On one hand, edge computing applications greatly benefit from the powerful processing capabilities of deep learning, enabling them to handle intricate scenarios like video analytics and transportation control. On the other hand, edge computing offers specialized hardware foundations and platforms, such as the lightweight Nvidia Jetson TX2 development kit, to effectively support deep learning operations at the edge (Wang et al. 2020). Many techniques have been introduced to improve the performance of deep learning when performed on edge computing devices, such as:

(A) **Model design**: When machine learning researchers design DNN models for resource-constrained devices, they commonly emphasize creating models with fewer parameters in order to minimize memory usage and execution latency while still maintaining high accuracy. Several techniques are employed to achieve this, including MobileNets, SSD, YOLO, and SqueezeNet. These methods are aimed at optimizing DNN models for efficient performance on such devices (Chen & Ran 2019).

(B) **Run-Time Optimizations**: Depending on the particular requirements of the application, suitable run-time optimizations can be employed to minimize the quantity of samples that need to undergo processing. For instance, in object detection applications, a high-resolution image can be divided into smaller images (tiling), and a selection criterion can be used to choose images with high activity regions. This approach allows the design of DNNs that can handle smaller inputs, resulting in improved computational and latency efficiency.

(C) **Hardware**: In the pursuit of accelerating deep learning inference, hardware manufacturers are adopting various strategies. These include utilizing already existing hardware like CPUs and GPUs, as well as developing custom application-specific integrated circuits (ASICs) dedicated to deep learning tasks, like Google's Tensor Processing Unit (TPU). Additionally, there are novel custom ASICs like ShiDianNao, which prioritize efficient memory access to minimize latency and energy consumption. FPGA-based DNN accelerators also show promise, as FPGAs can deliver fast computation while remaining reconfigurable (Chen & Ran 2019).

1.6.2 Lightweight Virtualization

Virtualization technologies are widely employed in cloud computing due to their effective method of harnessing the cloud's capabilities by partitioning a physical host into smaller, more manageable virtual components. By leveraging these technologies, cloud computing services become more user-friendly and economically efficient. Hypervisors such as VirtualBox and VMware are frequently used in cloud computing hardware virtualization. However, this approach has limitations such as increased resource cost, longer startup times, and larger attack surfaces. To solve these limitations, lightweight virtualization technologies such as Unikernels and Containers have evolved and are currently used in both cloud and edge computing. These lightweight virtualization technologies offer fast deployment and high efficiency, effectively overcoming the limitations posed by traditional hypervisor-based virtualization (Chen & Zhou 2021). Considering that the computational capabilities of edge computing devices are less potent than data centers, the adoption of emerging lightweight virtualization technologies offers numerous advantages. These benefits encompass swift initialization, minimal overhead, high instance density, and commendable energy efficiency, making them well-suited for the edge computing environment (Morabito & Beijar 2016).

Lightweight virtualization technology is critical in edge computing because it allows the deployment of resource management, orchestration, and isolation services without the need to account for different hardware configurations. This technology has brought about a significant transformation in software development and deployment practices. Container-based virtualization can be regarded as a lightweight alternative to the traditional hypervisor-based virtualization (Chen & Zhou 2021). Container-based virtualization offers a different level of abstraction in terms of isolation and virtualization when compared to hypervisors, as illustrated in Fig. 1.5. Hypervisors virtualize hardware and device drivers, resulting in increased

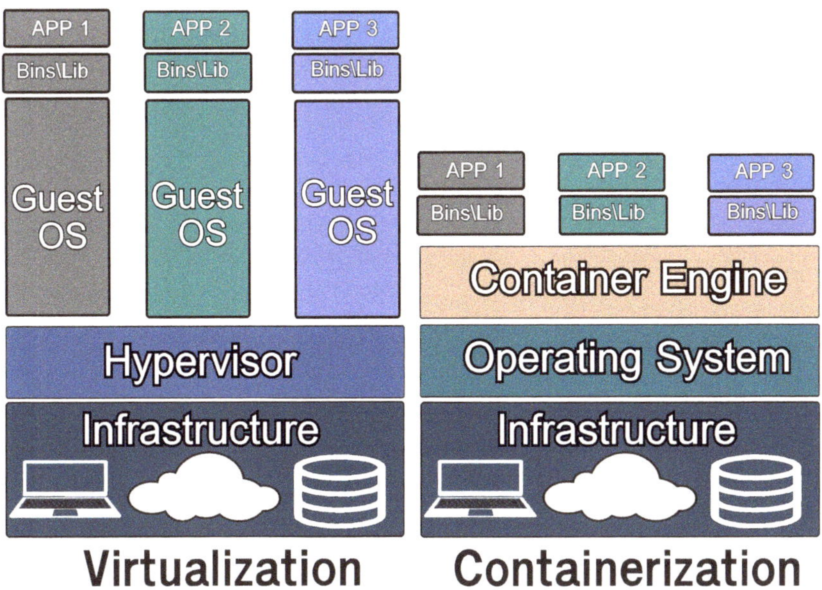

Fig. 1.5 Virtualization vs containerization

overhead. Containers, on the other hand, isolate processes at the OS level (Morabito & Beijar 2016). Containers allow independent applications to be isolated with their own virtual network interfaces, process spaces, and file systems because they share the same host machine's operating system kernel. Containers allow for a higher number of virtualized instances with lower image volumes, all executing on a single machine, thanks to the shared kernel feature (Chen & Zhou 2021).

1.7 Edge Computing in IoT-Based Intelligent Systems: Case Studies

With the rise of intelligent systems, edge computing offers the most efficient computing and storage solutions for devices with limited computational capabilities. This section delves into the applications of edge computing in IoT-based intelligent systems, including healthcare, manufacturing, agriculture, and transportation. We chose these four case studies because they have a substantial impact on improving human life.

1.7.1 Edge Computing in IoT-Based Healthcare

The term "geriatric care" refers to a branch of healthcare that emphasizes meeting the special mental, physical, and social needs of the aged. Geriatric care, which is specifically designed to meet the unique demands of elderly individuals, seeks to enhance their general well-being and health while successfully treating age-related illnesses and diseases. Its ultimate goal is to give them the means to maintain their independence, preserve their well-being, and enjoy the greatest degree of comfort as they age (Paulauskaite-Taraseviciene et al. 2023). In the area of "geriatric care," the danger of falling is regarded as a crucial concern. Unfortunately, many older people fall and unfortunately pass away because they lack good balance. While the fall itself may be the primary cause, the severity of the outcome stems from the inability to recover, leading to deteriorating physical and cognitive health. Numerous studies back up the idea that elderly people can potentially avoid physical consequences like brain injuries if immediate aid is given within 7 minutes of a fall. As a result, the death and disease rates among the aging population would both dramatically decline.

In (Naveen et al. 2021) the authors suggest an intelligent edge-monitoring system that utilizes cameras to detect instances of falling. Real-time video analysis is essential for continuously capturing photos, classifying them as either normal (sleeping) or abnormal (falling) circumstances, and instantly sounding an alarm in emergency scenarios. Due to the significant amount of data involved, relying solely on cloud processing would be impractical because of the resulting delays.

As a result, cameras serve as IoT devices, gathering data and sending it to nearby edge computing equipment for processing locally. In this approach, the edge computing server is equipped with deep learning models that have already undergone pretraining and are specifically created to detect falls with high accuracy and low latency. They successfully reduce transmission delays by maintaining the computing process' independence from the cloud. This approach not only allows them to generate valuable insights on-site, reducing response time and latency, but also addresses privacy concerns by handling sensitive and personal data at the edge.

1.7.2 Edge Computing in IoT-Based Manufacturing

Regarding intelligent manufacturing, the growing number of terminal network devices has presented new issues in terms of data center operation and maintenance, scalability, and dependability. To tackle these challenges, edge computing has advanced, which moves computation from centralized data centers to the network's edge. This approach enables intelligent services to be deployed near the manufacturing units, meeting essential demands such as highly responsive cloud services, data analytics via edge nodes, and a privacy-policy plan (Chen et al. 2018). In (Wang et al. 2020), the authors introduced a cutting-edge visual sorting mcthod created especially for adaptable manufacturing systems. The system utilizes a visual sorting approach based on CNNs. To support this system, they developed a cloud-edge computing platform, which makes it easier to quickly compute and continuously maintain and improve services.

1.7.3 Edge Computing in IoT-Based Agricultural

The IoT is typically applied in agricultural development through a monitoring network comprised of a considerable number of sensor nodes. As a result, agriculture increasingly moves away from a production model that is centered on humans and toward one that is centered on information and software (Zhang et al. 2020). Concerning the notion of edge computing in the context of agricultural IoT, numerous researchers have made contributions from diverse perspectives. (Zamora-Izquierdo et al. 2019) developed a system that meets the demanding needs of precision agriculture (PA) by integrating automation, IoT technologies, edge and cloud computing through virtualization. Three essential layers make up a multitier platform that has been developed: (1) a local layer of cyber-physical systems (CPS) that is connected to agricultural greenhouses. (2) The authors suggest a new edge computing architecture in which control modules are placed on virtualized nodes near the access network. (3) A cloud section outfitted with powerful computing and data analytics tools to help farmers make smart crop management decisions. The entire system was successfully tested in a real greenhouse located in southeast

Spain. Through specialized software that was accessible to the end farmers via the platform, this innovation made it possible to control a closed hydroponic system in real time. To validate the effectiveness of the architecture, two tomato crop cycles were conducted. The results showed remarkable benefits compared to a traditional open crop approach. Significant water savings of over 30% were achieved, which is particularly crucial in their semiarid region. Additionally, certain nutrients saw improvements of up to 80%, thanks to the system's efficient management.

1.7.4 Edge Computing in IoT-Based Transportation

The Internet of Vehicles (IoV), a new paradigm introduced by the IoT, employs edge computing to offer groundbreaking applications for transportation systems. Using sensors and geofencing technologies, IoV connects various cars with Roadside Units (RSUs) and other vehicles in an Intelligent Transportation System (ITS). Edge cloudlets are used by IoV for service provisioning and orchestration. Currently, substantial research on smart vehicles is being undertaken in both academic and industrial domains (Rafique et al. 2020). Traffic flow detection plays a vital role in ITS. By obtaining real-time urban road traffic flow data, ITS can intelligently guide measures to alleviate traffic congestion and reduce environmental pollution. In (Chen et al. 2021), the YOLOv3 (You Only Look Once) model was used by the authors to create a vehicle-detecting method. The YOLOv3 model was trained on an extensive dataset of traffic data and subsequently pruned to achieve optimal performance on edge devices. Additionally, by retraining the feature extractor, they improved the DeepSORT (Deep Simple Online and Realtime Tracking) algorithm, enabling multi-object vehicle tracking. Through the integration of vehicle detection and tracking algorithms, they developed a counter for real-time vehicle tracking capable of accurately detecting traffic flow. Finally, the Jetson TX2 edge device platform received and implemented the vehicle detection network and multi-object tracking network.

1.8 Challenges and Future Research Directions

In the preceding section, we outlined four possible uses of edge computing in IoT-based systems. To achieve the full potential of IoT, we emphasize the need for seamless collaboration between IoT devices and edge computing. Now, in this section, we will summarize some of the challenges faced in implementing edge computing in IoT-based systems and propose potential solutions and research opportunities. These include resource allocation, heterogeneity, privacy and security, and microservices:

- **Resource Allocation**: Edge devices play an essential role in enabling latency-critical services. The majority of end IoT devices often experience resource limitations; for example, local CPU computation capabilities and battery capacity are frequently constrained. Some workloads can be offloaded to more powerful edge devices to bypass these constraints and meet the performance requirements of applications. Edge computing improves IoT device capabilities, allowing them to handle more resource-intensive applications. However, practically speaking, edge computing devices have a finite amount of processing power. As a result, it is unable to handle the massive computing tasks generated by all of the end devices in its service region. As a result, the allocation of resources becomes highly crucial in such environments. Traditional optimization approaches like convex optimization and Lyapunov optimization have been used to tackle the computation offloading problem and find the best scheme. However, these methods are limited when it comes to making optimal decisions in dynamic environments. In contrast, modern resource allocation algorithms, powered by artificial intelligence and deep learning, such as deep reinforcement learning, offer more effective solutions for achieving optimal allocation.
- **Heterogeneity**: The existence of numerous computing technologies in edge computing, such as distinct hardware architectures and operating systems, has made it difficult to develop a viable approach that can be employed with ease in diverse scenarios. This problem can be solved by developing a programming model for edge nodes using software-based techniques, which enables workloads to be executed effectively on numerous hardware configurations simultaneously.
- **Privacy and Security**: At the network's edge, the primary services that require assurance are the protection of usage privacy and data security. When IoT is used in a home, the usage data collected can be used to infer important private information. For instance, examining the patterns of energy or water use can show whether the home is vacant or occupied. This presents a significant challenge in providing services while safeguarding privacy. Having a reliable architecture is crucial before users can feel confident in embracing new technologies. "Privacy by design" can be considered as a reliable approach to enhance security in edge computing. It involves incorporating privacy features directly into the design, taking preventive measures instead of just reacting after privacy breaches, and ensuring data privacy throughout its entire lifecycle.
- **Microservices**: Recently, both edge and cloud services have been changing from monolithic, stand-alone systems to loosely coupled, independent microservices. When running complex computations such as deep learning, there are several software requirements, and it's important to find a way to separate different deep learning services when using shared resources. Currently, the microservice framework, which can be used to host complex services on the edge, is still developing and in its early stages. However, it shows great potential for efficiently introducing services in the future.

1.9 Conclusion

As IoT continues to grow, edge computing is increasingly regarded as a promising and viable solution to address the complexities of managing numerous sensors and devices, along with the demands for resources they require. Edge computing, in contrast to standard cloud computing, involves placing data processing and storage to the edge of the network, bringing them closer to the end users. Thus, by dispersing compute nodes across the network, it is possible to reduce message exchange latency and relieve the computational load on the centralized data center. In conclusion, our chapter has explored the computing paradigms for IoT, edge computing paradigms like cloudlet and MEC, the architecture of edge computing-based IoT, the benefits it offers, and the enabling technologies such as artificial intelligence and lightweight virtualization. Additionally, it presented case studies showcasing how edge computing is applied in intelligent systems based on IoT and highlighted the issues with current research and suggested future directions for further exploration in this field.

References

Abbas, Nasir, et al. 2018. Mobile edge computing: A survey. *IEEE Internet of Things Journal* 5 (1): 450–465. https://doi.org/10.1109/JIOT.2017.2750180.

Alessio, Botta, et al. 2014. On the integration of cloud computing and internet of things. In *Proceedings of the Future Internet of Things and Cloud (FiCloud)*, 23–30.

Alhussaini, Rafal, et al. 2018. Data transmission protocol for reducing the energy consumption in wireless sensor networks. In *New Trends in Information and Communications Technology Applications*. Ed. by Safaa O. Al-mamory, et al., 35–49. Cham: Springer International Publishing.

Alwakeel, Ahmed M., 2021. An overview of fog computing and edge computing security and privacy issues. *Sensors* 21 (24). https://doi.org/10.3390/s21248226.

Atlam, Hany F., et al. 2018. Fog computing and the internet of things: A review. *Big Data and Cognitive Computing* 2 (2). https://doi.org/10.3390/bdcc2020010.

Chen, Baotong, et al. 2018. Edge computing in IoT-based manufacturing. *IEEE Communications Magazine* 56 (9): 103–109. https://doi.org/10.1109/MCOM.2018.1701231.

Chen, Chen, et al. 2021. An edge traffic flow detection scheme based on deep learning in an intelligent transportation system. *IEEE Transactions on Intelligent Transportation Systems* 22 (3): 1840–1852. https://doi.org/10.1109/TITS.2020.3025687.

Chen, Jiasi, and Xukan Ran. 2019. Deep learning with edge computing: A Review. *Proceedings of the IEEE* 107 (8): 1655–1674. https://doi.org/10.1109/JPROC.2019.2921977.

Chen, Shichao, and Mengchu Zhou. 2021. Evolving container to unikernel for edge computing and applications in process industry. *Processes* 9 (2). https://doi.org/10.3390/pr9020351.

Delfin, S., et al. 2019. Fog computing: A new era of cloud computing. In *2019 3rd International Conference on Computing Methodologies and Communication (ICCMC)*, 1106–1111. https://doi.org/10.1109/ICCMC.2019.8819633.

Donta, P.K., Monteiro, E., Dehury, C.K., and Murturi, I. 2023. Learning-driven ubiquitous mobile edge computing: Network management challenges for future generation Internet of Things. *International Journal of Network Management* 33 (5): e2250.

Edge Computing Market Size, Share & Trends Analysis Report By Component (Hardware, Software, Services, Edge-managed Platforms). 2023. By Application, By Industry Vertical, By Region, And Segment Forecasts, 2023–2030. Accessed: July 15, 2023. https://www.grandviewresearch.com/industry-analysis/edge-computing-market.

Elazhary, Hanan. 2019. Internet of Things (IoT), mobile cloud, cloudlet, mobile IoT, IoT cloud, fog, mobile edge, and edge emerging computing paradigms: Disambiguation and research directions. *Journal of Network and Computer Applications* 128: 105–140.

Fazeldehkordi, Elahe, and Tor-Morten Grønli. 2022. A survey of security architectures for edge computing-based IoT. *IoT* 3 (3): 332–365. https://doi.org/10.3390/iot3030019.

Haibeh, Lina A., et al. 2022. A survey on mobile edge computing infrastructure: Design, resource management, and optimization approaches. *IEEE Access* 10: 27591–27610. https://doi.org/10.1109/ACCESS.2022.3152787.

Hassan, Najmul, et al. 2019. Edge computing in 5G: A review. *IEEE Access* 7: 127276–127289. https://doi.org/10.1109/ACCESS.2019.2938534.

Idrees, Ali Kadhum, Alhussaini Rafal, et al. 2020. Energy-efficient two-layer data transmission reduction protocol in periodic sensor networks of IoTs. *Personal and Ubiquitous Computing* 27 (2): 139–158.

Idrees, Ali Kadhum, Chady Abou Jaoude, et al. 2020. Data reduction and cleaning approach for energy-saving in wireless sensors networks of IoT. In *2020 16th International Conference on Wireless and Mobile Computing, Networking and Communications (WiMob)*, 1–6. https://doi.org/10.1109/WiMob50308.2020.9253429.

Idrees, Ali Kadhum, and Lina Waleed Jawad. 2023. Energy-efficient data processing protocol in edge-based IoT networks. *Annals of Telecommunications*, 1–16. https://doi.org/10.1007/s12243-023-00957-8.

Idrees, Ali Kadhum, and Marwa Saiccd Khlief. 2023a. Efficient compression technique for reducing transmitted EEG data without loss in IoMT networks based on fog computing. *The Journal of Supercomputing* 79 (8): 9047–9072.

Idrees, Ali Kadhum, and Marwa Saieed Khlief. 2023b. Lossless EEG data compression using clustering and encoding for fog computing based IoMT networks. *International Journal of Computer Applications in Technology* 72 (1): 77–78.

Idrees, Ali Kadhum, Sara Kadhum Idrees, et al. 2022. An edge-fog computing-enabled lossless EEG data compression with epileptic seizure detection in IoMT networks. *IEEE Internet of Things Journal* 9 (15): 13327–13337.

Idrees, Ali Kadhum, Tara Ali–Yahiya, et al. 2022. DaTOS: Data transmission optimization scheme in tactile internet-based fog computing applications. In *2022 IEEE 33rd Annual International Symposium on Personal, Indoor and Mobile Radio Communications (PIMRC)*, 01–06. Piscataway: IEEE.

Kadhum Idrees, A., and Saieed Khlief, M. 2023. A new lossless electroencephalogram compression technique for fog computing-based IoHT networks. *International Journal of Communication Systems* 36 (15): e5572.

Khan, Wazir Zada et al. 2019. Edge computing: A survey. *Future Generation Computer Systems* 97: 219–235.

Lorenzo, Beatriz et al. 2018. A robust dynamic edge network architecture for the internet of things. *IEEE network* 32 (1): 8–15.

Mach, Pavel, and Zdenek Becvar. 2017. Mobile edge computing: A survey on architecture and computation offloading. *IEEE Communications Surveys & Tutorials* 19 (3): 1628–1656. https://doi.org/10.1109/COMST.2017.2682318.

Mao, Yuyi et al. 2017. Mobile edge computing: Survey and research outlook. CoRR abs/1701.01090. arXiv: 1701.01090. http://arxiv.org/abs/1701.01090.

Marchisio, Alberto et al. 2019. Deep learning for edge computing: Current trends, cross-layer optimizations, and open research challenges. *2019 IEEE Computer Society Annual Symposium on VLSI (ISVLSI)* 553–559. https://doi.org/10.1109/ISVLSI.2019.00105.

Morabito, Roberto, and Nicklas Beijar. 2016. Enabling data processing at the network edge through lightweight virtualization technologies. In *2016 IEEE International Conference on Sensing, Communication and Networking (SECON Workshops)*, 1–6. https://doi.org/10.1109/SECONW.2016.7746807.

Naveen, Soumyalatha, et al. 2021. Low latency deep learning inference model for distributed intelligent IoT edge clusters. *IEEE Access* 9: 160607–160621.

Paulauskaite-Taraseviciene, Agne et al. 2023. Geriatric care management system powered by the IoT and computer vision techniques. *Healthcare* 11 (8): 1152. MDPI

Qiu, Tie et al. 2020. Edge computing in industrial internet of things: Architecture, advances and challenges. *IEEE Communications Surveys & Tutorials* 22 (4): 2462–2488. https://doi.org/10.1109/COMST.2020.3009103.

Rafique, Wajid, et al. 2020. Complementing IoT services through software defined networking and edge computing: A comprehensive survey. *IEEE Communications Surveys & Tutorials* 22 (3): 1761–1804. https://doi.org/10.1109/COMST.2020.2997475.

Ravi, Banoth, et al. 2023. Stochastic modeling for intelligent software-defined vehicular networks: A survey. *Computers* 12 (8): 162.

Satyanarayanan, Mahadev, et al. (2009). The case for vm-based cloudlets in mobile computing. *IEEE Pervasive Computing* 8 (4): 14–23.

Shawqi Jaber, Alaa, and Ali Kadhum Idrees. 2020. Adaptive rate energy-saving data collecting technique for health monitoring in wireless body sensor networks. *International Journal of Communication Systems* 33 (17): e4589. https://doi.org/10.1002/dac.4589.

Srirama, Satish Narayana. n.d. A decade of research in fog computing: Relevance, challenges, and future directions. *Software: Practice and Experience*, 1–23. https://doi.org/10.1002/spe.3243. eprint: https://onlinelibrary.wiley.com/doi/pdf/10.1002/spe.3243. https://onlinelibrary.wiley.com/doi/abs/10.1002/spe.3243.

Wang, Fangxin, et al. 2020. Deep learning for edge computing applications: A state-of-the-art survey. *IEEE Access* 8: 58322–58336. https://doi.org/10.1109/ACCESS.2020.2982411.

Wang, Yuanbin, et al. 2020. A CNN-based visual sorting system with cloud-edge computing for flexible manufacturing systems. *IEEE Transactions on Industrial Informatics* 16 (7): 4726–4735. https://doi.org/10.1109/TII.2019.2947539.

Yousefpour, Ashkan, et al. 2019. All one needs to know about fog computing and related edge computing paradigms: A complete survey. *Journal of Systems Architecture* 98: 289–330.

Yu, Wei, et al. 2017. A survey on the edge computing for the Internet of Things. *IEEE Access* 6: 6900–6919.

Zamora-Izquierdo, Miguel A. et al. 2019. Smart farming IoT platform based on edge and cloud computing. *Biosystems Engineering* 177: 4–17. https://doi.org/10.1016/j.biosystemseng.2018.10.014.

Zhang, Xihai, et al. 2020. Overview of edge computing in the agricultural Internet of Things: Key technologies, applications, challenges. *IEEE Access* 8: 141748–141761. https://doi.org/10.1109/ACCESS.2020.3013005.

Chapter 2
Federated Learning Systems: Mathematical Modeling and Internet of Things

Quentin De La Cruz and Gautam Srivastava

2.1 Introduction

Federated learning is a revolutionary and an alternative approach to artificial intelligence that promises to transform the way we process and analyze data in our current society. In the era of digital society and in a world where data has become a precious resource, but where confidentiality and the protection of personal data are crucial subjects, federated learning offers a unique solution to reconcile these two issues.

Federated learning is based on the principle of data decentralization and collaboration between several entities. Instead of centralizing all data in a single location, federated learning allows data to be kept on local devices, such as smartphones, computers, or connected objects (watches, thermostats, etc.). Learning models are then developed and improved by leveraging local information without individual data being exposed or shared (Collins et al. 2022).

One of the most important benefits of federated learning is preserving user privacy. By avoiding massive transfer of data to central servers, individuals retain control of their sensitive personal information. This responds to a growing privacy concern in today's society, especially with regulations such as the Personal Information Protection and Electronic Documents Act (PIPEDA) in Canada and the General Data Protection Regulation (GDPR) in Europe (Khan et al. 2021).

Q. De La Cruz
Department of Mathematics & Computer Science, Brandon University, Brandon, MB, Canada
e-mail: DelacruzQ@brandonu.ca

G. Srivastava (✉)
Department of Mathematics & Computer Science, Brandon University, Brandon, MB, Canada

Department of Computer Science and Math, Lebanese American University, Beirut, Lebanon
e-mail: SRIVASTAVAG@brandonu.ca

P. K. Donta et al. (eds.), *Learning Techniques for the Internet of Things*,
https://doi.org/10.1007/978-3-031-50514-0_2

In addition, federated learning also solves problems related to data fragmentation. In many domains, data is distributed among different organizations or entities, which makes it difficult to train an overall model from this dispersed data. Using federated learning, models can be trained locally on each entity and then merged to create a better-performing global model (Chen et al. 2021).

The future of federated learning is very bright. It paves the way for many potential applications in various fields, such as health, autonomous transport, energy, finance, and many others. For example, in healthcare, federated learning would allow the development of predictive models for diseases without the disclosure of sensitive patient medical data. Similarly, in the realm of autonomous transport, federated learning could be used to improve real-time navigation systems by aggregating data from individual vehicles without compromising driver privacy (See Fig. 2.1).

However, even if federated learning appears to be a promising solution, it presents many challenges to overcome. Coordinating model updates, dealing with bias issues, and ensuring data security are all areas that need attention. Overall, federated learning represents a major advancement in the fields of artificial intelligence and machine learning, offering a solution to reconcile the efficiency of data processing.

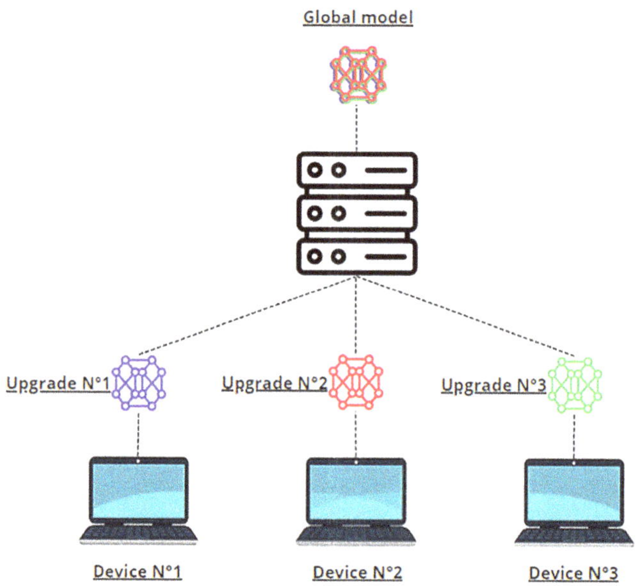

Fig. 2.1 Schema of global work of federated learning

2.2 Federated Learning

2.2.1 *Definition of Federated Learning*

Federated learning is an innovative paradigm in the fields of artificial intelligence and machine learning introduced in 2016 by Google (Mammen 2021). It allows you to create powerful prediction models without having to centralize sensitive data on a single, central server. Instead, data resides on local devices, such as cell phones, personal computers, or wearables. Learning is thus carried out in a decentralized manner.

The fundamental concept of federated learning relies on the collaboration between devices to form an overall model. Rather than sharing raw data, devices send model parameter updates to a central server, which then analyzes these updates to create an improved overall model. This helps preserve the confidentiality of individual data while benefiting from the power of learning on a large set of distributed data (Li et al. 2020).

Federated learning has many advantages. First, it solves the dilemma between centralized data collection and privacy protection. Users retain control of their data because it does not leave their devices. Additionally, federated learning makes it possible to train models on sensitive data that could not otherwise be shared, such as medical or financial data. In addition, federated learning also offers efficiency and speed benefits. Since the calculation is performed locally on each device, there is no need to transfer large amounts of data to a central server, which reduces bandwidth consumption and communication costs. In addition, model updates can be made in real time, allowing rapid response to new data and continuous improvement of overall model performance. However, federated learning also presents unique challenges. Data variability between devices can cause issues with model bias and quality. Additionally, coordinating model updates and managing data privacy are complex aspects to consider (Konečný et al. 2016).

Despite these challenges, federated learning offers enormous potential for solving complex problems while maintaining data privacy. By enabling decentralized collaboration, it paves the way for new applications in areas such as healthcare, the Internet of Things, and distributed data analytics. The future of federated learning is bright, and it will continue to evolve as a powerful and privacy-respecting method of learning.

2.2.2 *The Different Forms of Federated Learning*

Different forms of federated learning vary in how data and models are shared and collaborated between local devices and the central server. Here are some of the common forms of federated learning:

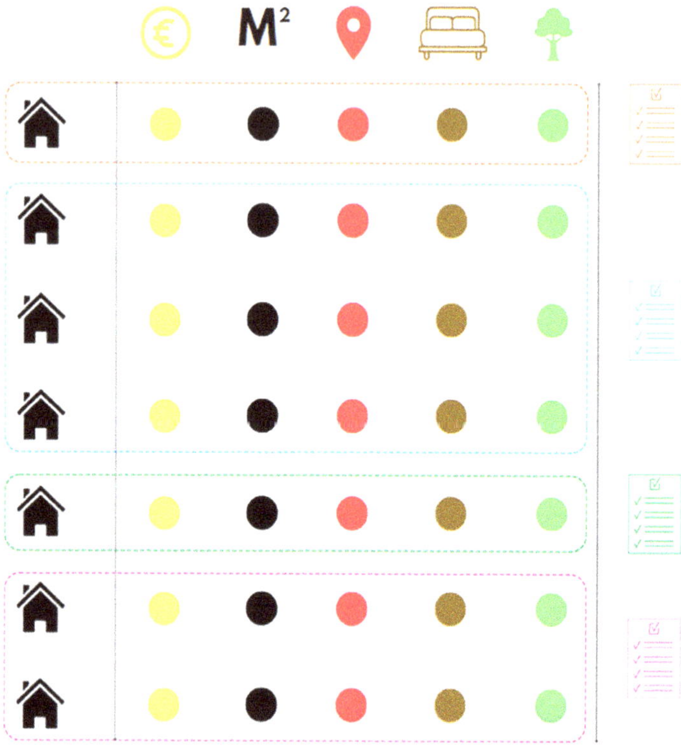

Fig. 2.2 Horizontal federated learning

Horizontal federated learning: In this approach, local devices have similar but different sets of data representing a different view of the problem. Local models are trained on each device's respective data, and parameter updates are aggregated to form a global model. This helps combine local knowledge and take advantage of variations in data between devices. The graphical model of horizontal federated learning is represented in Fig. 2.2.

Vertical federated learning: This form of federated learning is used when different entities have complementary data, usually characterized by different attributes but related to the same problem. Local models are trained on the entities' respective attributes, and then the information is exchanged to create a complete and more accurate global model. The graphical model of vertical federated learning is represented in Fig. 2.3.

Multiparty Federated Learning: This approach to federated learning is for scenarios where multiple parties collaborate to train an overall model without directly sharing their data. Each party trains a local model on their respective data, and then the models are combined using secure fusion techniques to form a global model without compromising data privacy.

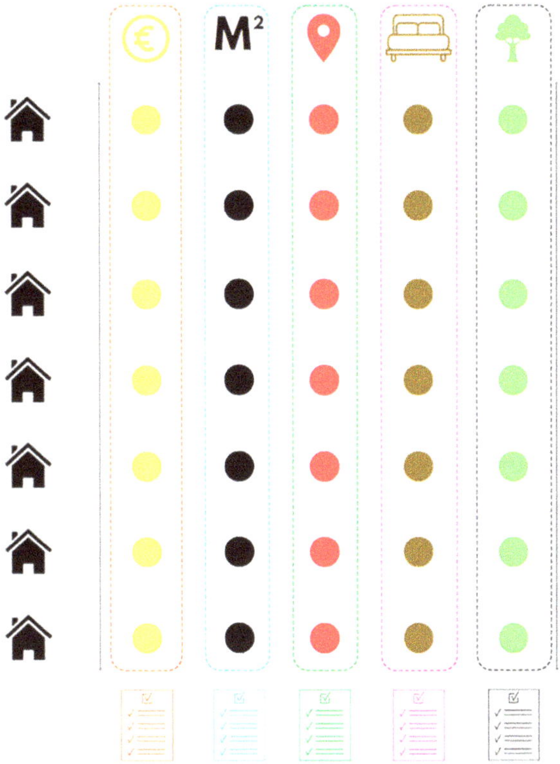

Fig. 2.3 Vertical federated learning

Knowledge transfer-based federated learning: In this type of federated learning, a pre-existing model, mostly pre-trained on a large amount of data, is used as a starting point for learning on local devices. Local models then adapt this pre-existing model to local data, to improve performance on specific tasks. This saves time and resources by leveraging prior knowledge of the overall model.

It is important to note that these different forms of federated learning can be combined and adapted according to the specific needs of each scenario. The main objective of implementing all these different approaches is to enable collaborative learning while preserving the confidentiality and security of individual data.

2.3 Mathematical Modeling

2.3.1 Architecture

Federated learning is therefore a form of machine learning. From this we can deduce on the one hand the mathematical model that each device follows, but also the global

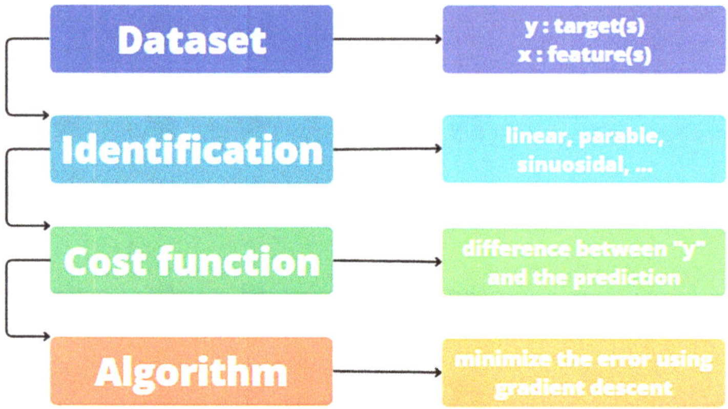

Fig. 2.4 Steps of the process

mathematical model. The objective is to formalize the collaboration between local devices and central server. Next, we present an example of a mathematical model on a local device.

In this example we will take as data the price of a house according to its surface in m^2. These are purely imaginary data. The mathematical model will be broken down into four steps as shown in Fig. 2.4.

Step 1

The first step is to collect data on the subject we want to study. The larger the collection of data, the more accurate the preferred model will be. Indeed, in the same way as a survey, the larger the population studied, the more the result will be representative of reality. In this example, the price of the house is going to be the target (y), and the surface is going to be the feature (x).

We present a list of house price data according to its surface as shown in Table 2.1.

Step 2

The second step is to identify the model of the function that best represents this data. For this, we can use a graphical method. We are therefore going to represent this data in a two-axis graph, with surface on the abscissa and the price on the ordinate. This gives us the graph as shown in Fig. 2.5.

This graphical representation therefore allows us to easily identify which style of function we can associate this data with. In our case, we can associate this representation with a linear function. Indeed, data seems to follow a straight line with the equation: $y = ax + b$

We can represent this function (given in red) as Fig. 2.6.

The final goal of this mathematical model is that the machine determines as accurately as possible the parameters a and b of this function.

Table 2.1 A list of house price data according to its surface

House N^o	Price ($CA)	Surface (m^2)
1	$ 150,000,00	95 m^2
2	$ 225,000,00	150 m^2
3	$ 450,000,00	250 m^2
4	$ 325,000,00	185 m^2
5	$ 175,000,00	100 m^2
6	$ 210,000,00	123 m^2
7	$ 250,000,00	143 m^2
8	$ 175,000,00	140 m^2
9	$ 156,000,00	135 m^2
10	$ 160,000,00	137 m^2
11	$ 186,000,00	145 m^2
12	$ 152,000,00	128 m^2
13	$ 125,000,00	75 m^2
14	$ 115,000,00	45 m^2
15	$ 135,000,00	95 m^2
16	$ 143,000,00	68 m^2
17	$ 385,000,00	225 m^2
18	$ 122,000,00	82 m^2
19	$ 75,000,00	46 m^2
20	$ 115,000,00	86 m^2
21	$ 97,000,00	72 m^2
22	$ 182,000,00	110 m^2
23	$ 86,000,00	69 m^2
24	$ 345,000,00	191 m^2
25	$ 173,000,00	102 m^2
26	$ 145,000,00	88 m^2
27	$ 186,000,00	132 m^2
28	$ 114,000,00	117 m^2
29	$ 175,000,00	96 m^2
30	$ 268,000,00	175 m^2

Step 3

In the third step, we need to define the cost function. The purpose of this function is to measure the errors between real ordinates of the points and those of the theoretical line.

This corresponds to the black lines on the graph shown in Fig. 2.7: For this, we will use the mean squared error function, which corresponds to Eq. (2.1):

$$J_{(a,b)} = \frac{1}{2m} \sum_{i=1}^{m} \left(f(x^{(i)}) - y^{(i)} \right)^2 \tag{2.1}$$

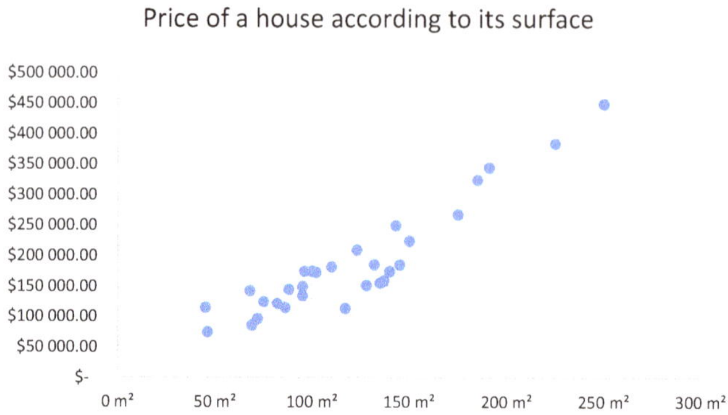

Fig. 2.5 Data representation

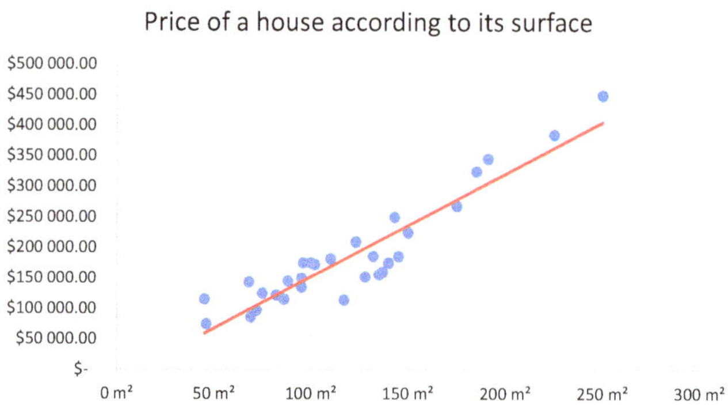

Fig. 2.6 Data representation using $y = ax + b$

This function is therefore a square function of parabolic form. This is a very important element for our fourth step.

Step 4

The purpose of this fourth step is to minimize the cost function. Since this function is of parabolic form, it consists in finding the minima of the function.

Figure 2.8 is a possible representation of a line of the type $y = ax + bx + c$; it is not the correct representation of our cost function. The minima of this function is represented by the value "a" on the graph shown in Fig. 2.8.

To determine point "a", we will use gradient descent. To use this method, we must define learning step "α". Using too large a pitch will make the model very inaccurate. However, choosing too small a pitch will make the model very slow. It is therefore necessary to correctly estimate the step to take, to find the right compromise between precision and time. So using gradient descent, we get:

$$a_{i+1} = a_i - \alpha \frac{\partial J_{(a,b)}}{\partial a} \quad \text{and} \quad b_{i+1} = b_i - \alpha \frac{\partial J_{(a,b)}}{\partial b}$$

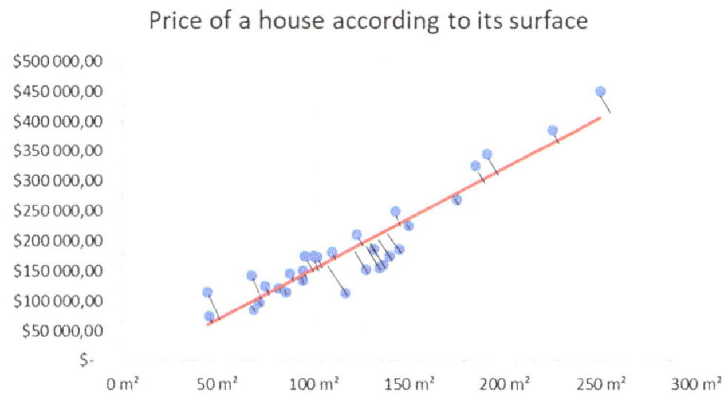

Fig. 2.7 Representation of errors between the theoretical and the real

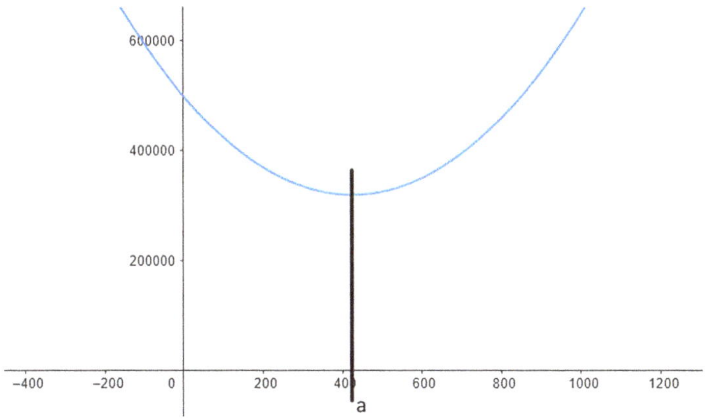

Fig. 2.8 Representation of the minima of the square function

Knowing that:

$$\frac{\partial J_{(a,b)}}{\partial a} = \frac{1}{m}\sum x(ax + b - y) \quad \text{and} \quad \frac{\partial J_{(a,b)}}{\partial b} = \frac{1}{m}\sum x(ax + b - y)$$

Graphically, this is represented in Fig. 2.9.

On this graph, we can see the convergence to the desired objective, as well as the importance of choosing the right learning step.

The next step is to automate this process by creating a schedule. This will make it possible to optimize all the parameters while adapting the entry of new data. Thus, we would have a precise predictive model, which adapts in real time to the evolution of the market.

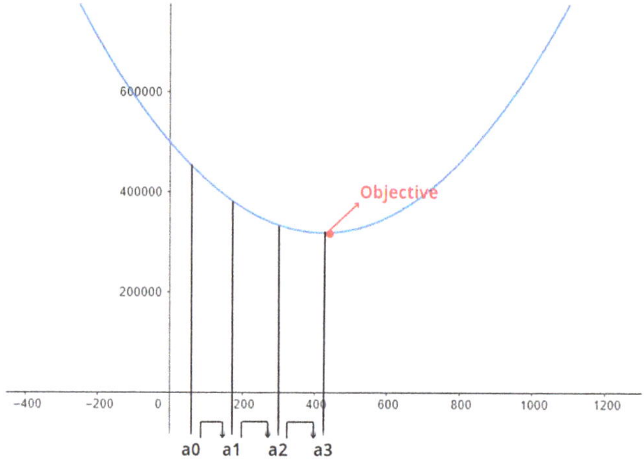

Fig. 2.9 Representation of the minima of the square function

2.4 Internet of Things

2.4.1 Introduction

The Internet of Things (IoT) represents a major advancement in the world of technology, transforming the way everyday objects interact, communicate, and influence our environment. IoT will ultimately allow us to evolve into a world where household appliances can not only perform their usual tasks but also exchange information to make intelligent decisions. Vehicles will also be able to connect to improve road safety and energy efficiency. Or we can use geo-climatic sensors scattered in the environment to extract valuable data to monitor and preserve our planet. Currently, there are nearly seven billion connected IoT devices and three billion smartphones around the world (Lim et al. 2020).

The essence of IoT lies in the transformation of inanimate objects into actors of a global network. Thanks to wireless communication technologies such as Wi-Fi, Bluetooth, and 5G, these objects can now connect to the Internet, thus sharing data and information in real time. Built-in sensors and devices give them the ability to collect data about their surroundings, such as temperature, humidity, geolocation, and more. This information can then be used for a variety of purposes, from improving performance and efficiency to creating more personalized experiences for users.

The impact of IoT is immense and affects multiple sectors. In health and wellness, connected medical devices can monitor patient health in real time and enable doctors to make informed decisions. In the manufacturing industry, IoT is driving automation and predictive maintenance, which increases operational efficiency and reduces costs. Smart cities use IoT to manage urban resources more sustainably, monitoring traffic, air quality, and other parameters to improve city life.

However, IoT also raises major challenges. Data security and privacy are central concerns as the proliferation of connected devices increases potential entry points for cyberattacks. Moreover, managing and analyzing huge amounts of data generated by these objects requires robust infrastructures and advanced analysis capabilities.

In summary, the Internet of Things promises to redefine how we interact with the world around us, creating an interconnected ecosystem where things, data, and people converge to shape a smarter, more responsive future.

2.4.2 *Link Between IoT and Federated Learning*

Federated learning and the Internet of Things are two areas of technology that are closely related and complement each other, especially in the context of managing and analyzing data generated by connected objects. Federated learning is a decentralized machine learning approach in which learning models are trained locally on edge devices (like smartphones, IoT sensors, etc.) rather than centralizing all data on a server. The local models are then securely aggregated to form an improved global model. This has several advantages, including preserving user privacy by avoiding the transmission of sensitive data to a central server.

IoT involves the connectivity of many physical objects to the Internet network to collect and share data. However, this immense amount of data generated by connected objects can be difficult to manage and analyze centrally. This is where federated learning comes in (Li et al. 2023). Connected objects in IoT can be thought of as distributed "nodes" that generate local data. Using federated learning, these nodes can collaborate to train improved learning models while keeping the data in place, reducing the need to transfer large amounts of data to a central server. Not only does this improve model efficiency, but it can also help solve bandwidth, latency, and privacy issues associated with centralizing data.

In summary, federated learning and the Internet of Things combine to enable efficient processing of data generated by connected objects while ensuring privacy and reducing the load on networks. This synergy is particularly relevant in the context of IoT, where distributed collaboration can lead to more robust and better learning models.

2.5 Conclusion

Federated learning and machine learning are areas that will evolve and grow together in the future. It is therefore important to fully understand the issues and challenges. By promoting the use of federated learning rather than classic learning, we help ensure greater security. Although it is always possible to steal sensitive data on a device, the data no longer circulates in its entirety, which removes some of the risk. Indeed, as only the improvements are shared with the global server, the data remains in the same place. What reassures the population at a time when the confidentiality of data and the protection of privacy are major social issues? There are as many mathematical models as there are different functions; however, the methodology remains the same. Indeed, in our example, we used a linear function to make it as simple as possible. But it is quite possible to use the same process with a much more complex function. The goal is each time to be able to develop an algorithm aimed at reducing the cost function as much as possible. Regarding the Internet of Things, it is obvious that it will be increasingly present in our daily lives with the development, for example, of connected watches or even smart thermometers, which will confront us with issues of preserving the privacy of more and more important. It is for this purpose that federated learning is an essential tool since it will guarantee confidentiality by preventing data sharing in its basic principle while allowing less bandwidth to be used to transmit data models derived from these data. The use of federated learning is therefore quite in its infancy and should normally be a bright future in data processing and privacy.

References

Chen, Xiangcong, et al. 2021. IoT cloud platform for information processing in smart city. *Computational Intelligence* 37 (3): 1428–1444.

Collins, Liam, et al. 2022. Fedavg with fine tuning: Local updates lead to representation learning. *Advances in Neural Information Processing Systems* 35: 10572–10586.

Khan, Latif U, et al. 2021. Federated learning for internet of things: Recent advances, taxonomy, and open challenges. *IEEE Communications Surveys & Tutorials* 23 (3): 1759–1799.

Konečný, Jakub, et al. 2016. Federated learning: Strategies for improving communication efficiency. arXiv preprint. arXiv:1610.05492.

Li, Tian, et al. 2020. Federated learning: Challenges, methods, and future directions. *IEEE Signal Processing Magazine* 37 (3): 50–60.

Li, Ying, et al. 2023. Federated domain generalization: A survey. arXiv preprint. arXiv:2306.01334.

Lim, Wei, Yang Bryan, et al. 2020. Federated learning in mobile edge networks: A comprehensive survey. *IEEE Communications Surveys & Tutorials* 22 (3): 2031–2063.

Mammen, Priyanka Mary. 2021. Federated learning: Opportunities and challenges. arXiv preprint. arXiv:2101.05428.

Chapter 3
Federated Learning for Internet of Things

Ying Li, Qiyang Zhang, Xingwei Wang, Rongfei Zeng, Haodong Li, Ilir Murturi, Schahram Dustdar, and Min Huang

3.1 Introduction

The Internet of Things (IoT) possesses the immense potential to revolutionize numerous industries and aspects of daily life by facilitating the seamless integration of the physical world with digital systems (Tataria et al. 2021). It allows for the creation of smart homes, smart cities, industrial automation, precision agriculture, healthcare monitoring, and an array of other innovative applications. To effectively

Y. Li
College of Computer Science and Engineering, Northeastern University, Shenyang, China

Distributed Systems Group, TU Wien, Vienna, Austria
e-mail: liying1771@163.com

Q. Zhang
State Key Laboratory of Network and Switching, Beijing University of Posts and Telecommunications, Beijing, China

Distributed Systems Group, TU Wien, Vienna, Austria
e-mail: qyzhang@bupt.edu.cn

X. Wang (✉) · H. Li
College of Computer Science and Engineering, Northeastern University, Shenyang, China
e-mail: wangxw@mail.neu.edu.cn; 1ihaodong0811@163.com

R. Zeng
College of Software, Northeastern University, Shenyang, China
e-mail: zengrf@swc.neu.edu.cn

I. Murturi · S. Dustdar
Distributed Systems Group, TU Wien, Vienna, Austria
e-mail: imurturi@dsg.tuwien.ac.at; dustdar@dsg.tuwien.ac.at

M. Huang
College of Information Science and Engineering, Northeastern University, Shenyang, China
e-mail: mhuang@mail.neu.edu.cn

P. K. Donta et al. (eds.), *Learning Techniques for the Internet of Things*,
https://doi.org/10.1007/978-3-031-50514-0_3

implement these intelligent applications, a substantial quantity of IoT devices is indispensable (Saad et al. 2019; Al-Fuqaha et al. 2015). According to recent statistics, the rapid growth of the IoT is expected to result in an astonishing number of 125 billion IoT devices by 2030 (SEMICONDUCTORDIGEST n.d.). Alongside this massive proliferation of devices, the amount of data generated by these IoT devices is predicted to be monumental. It is estimated that by 2025, the total data volume generated by connected IoT devices worldwide will reach an astounding 79.4 zettabytes (ZBs) (Statista n.d.). The exponential expansion of network size and data volume within the IoT systems presents an exceptional opportunity to harness the power of artificial intelligence (AI) algorithms. These algorithms have the capability to efficiently process and analyze immense data quantities, thereby extracting valuable insights and facilitating decision-making processes with remarkable efficacy.

In the traditional approach, data gathered by IoT devices is transmitted to cloud servers or data centers, where it is uploaded and processed in a centralized manner. However, this approach is no longer sustainable due to several reasons (Ying et al. 2023): Firstly, data owners are becoming increasingly concerned about privacy issues associated with transmitting their data to centralized servers. Secondly, the traditional approach introduces significant propagation delays, which are unacceptable for applications requiring real-time decision-making. Lastly, transferring large volumes of data to the centralized server for processing puts a strain on the backbone network, impacting its performance and capacity. To address the privacy and latency issues associated with traditional IoT, mobile edge computing (MEC) (Abbas et al. 2017; Cao et al. 2019; Donta et al. 2023) emerged as a paradigm where data processing and analysis occur closer to the data source, reducing data transmission, latency, and reliance on centralized infrastructure. However, it may still involve transmitting raw data to centralized locations for model training, raising privacy concerns.

Against the backdrop of increasingly stringent data privacy regulations, federated learning (FL) (McMahan et al. 2017a; Kairouz et al. 2021) has emerged as a promising solution to tackle privacy concerns in IoT environments. FL, as a privacy-preserving distributed machine learning paradigm, facilitates collaborative and decentralized ML while ensuring that raw data remains within the client's domain, thereby not being transmitted to a central server (Zeng et al. 2021). In FL, the learning process takes place locally on each client within the network, where each client trains its own local models utilizing its own data, while the central server exclusively aggregates and shares the new global model updates. This approach guarantees the preservation of data privacy since sensitive information remains on the clients and is not exposed to the central server or other clients in the FL network. Moreover, FL maintains data utility by aggregating model updates from each client, enabling the central server to create an updated global model that captures knowledge from diverse distributed data, resulting in improved accuracy and generalization capabilities. Specifically, the several benefits that FL offers for IoT as outlined below:

- **Enhanced Data Privacy:** FL ensures data privacy and reduces the risk of data breaches or unauthorized access by keeping raw data on the clients and eliminating the need to transmit sensitive information to a central server, thereby preserving data privacy and enhancing security measures.
- **Reduced Latency and Bandwidth Requirements:** FL minimizes the need for frequent data transmission between clients and the central server by performing local model training on each client, resulting in reduced latency and bandwidth requirements. This makes FL highly suitable for real-time or latency-sensitive IoT applications, ensuring efficient and responsive data processing.
- **Efficient Resource Utilization:** FL optimizes resource utilization by leveraging the computational power of edge devices within the IoT network, distributing the learning process. This reduces the burden on the central server and makes FL well-suited for resource-constrained IoT devices, ensuring efficient utilization of limited resources.
- **Robustness to Device Heterogeneity:** FL is designed to handle the heterogeneity present in IoT networks, accommodating devices with diverse characteristics such as varying hardware configurations or data distributions. FL achieves this by allowing local model training on individual devices, enabling each device to contribute to the global model irrespective of its specific capabilities or data characteristics. This ensures effective utilization of the collective knowledge within the IoT network while accommodating device heterogeneity.
- **Improved Scalability:** FL facilitates large-scale collaboration across numerous IoT devices, enabling each device to actively participate in the training process and contribute its local model update to enhance the global model. The scalable approach efficiently utilizes the vast amount of distributed data available in IoT environments, resulting in improved model performance and leveraging the collective intelligence of the entire IoT network.

Overall, FL provides significant benefits for IoT, including preserving data privacy, reducing latency, optimizing resource efficiency, handling device heterogeneity, and enabling scalability. These advantages make FL a valuable approach for effectively leveraging distributed IoT data while ensuring privacy and maximizing learning performance. In this work, we present state-of-the-art advancements in FL for IoT. The rest of this work is organized as follows. Section 3.2 provides an introduction to preliminary work on FL for IoT. Section 3.3 explores various applications of FL for IoT. Section 3.4 provides the current research challenges and future directions in the field of FL for IoT. Finally, Sect. 3.5 concludes the paper.

3.2 Federated Learning and Internet of Things: Preliminaries

In this section, we first present the fundamental knowledge of FL and IoT. Next, we briefly introduce the overview of FL for IoT.

3.2.1 *Federated Learning*

Recent advancements in AI and the proliferation of IoT devices have led to exponential growth in data. In addition, concerns over data privacy and security have also risen. In response to these concerns, FL provides a viable solution to address these challenges by facilitating collaborative ML without compromising individual privacy. FL leverages the distributed nature of data and allows local learning on IoT devices, promoting data privacy while facilitating collaborative intelligence. Here, we introduce the fundamental concept of FL and subsequently present several significant categories of FL specifically for IoT networks. Specifically, the architectural overview of FL for IoT is provided as shown in Fig. 3.1.

3.2.1.1 Fundamental FL Concept

The FL system for the IoT network consists of five distinct entities that collectively contribute to its operation and effectiveness:

1. **Admin**: The administrator serves as the overseer of the FL system's overall operation, including managing the coordination among the various entities involved, ensuring system stability and security, and addressing any technical issues or updates that may arise.
2. **Model Engineer**: The model engineer is responsible for developing the ML model, defining the training protocol for the FL system, and executing model evaluation.
3. **Aggregation Server/Blockchain**: The aggregation server or blockchain coordinates the FL training process by collecting and aggregating the model updates from the participating clients.
4. **Clients**: Clients represent the devices or organizations that contribute their local data and computational resources to the FL training process (Zeng et al. 2020).
5. **End users**: End users refer to individuals or organizations that utilize the trained ML model to make predictions or decisions.

3.2.1.2 The Typical Process of FL Training for IoT

Let $\mathcal{K} = \{1, 2, \ldots, K\}$ represent the set of clients actively participating in the collaborative training of FL models, leveraging their IoT devices to perform IoT tasks. Each client $k \in \mathcal{K}$ possesses a local dataset $\mathcal{D}^k$ that may undergo changes over time. The size of the local dataset is denoted by $|\mathcal{D}^k|$. For local model training, each client can selectively choose a subset $\varkappa^k \subseteq \mathcal{D}^k$ from its local dataset, and the size of the chosen subset is indicated by $|\varkappa^k|$. Next, we present the typical process of FL training for IoT.

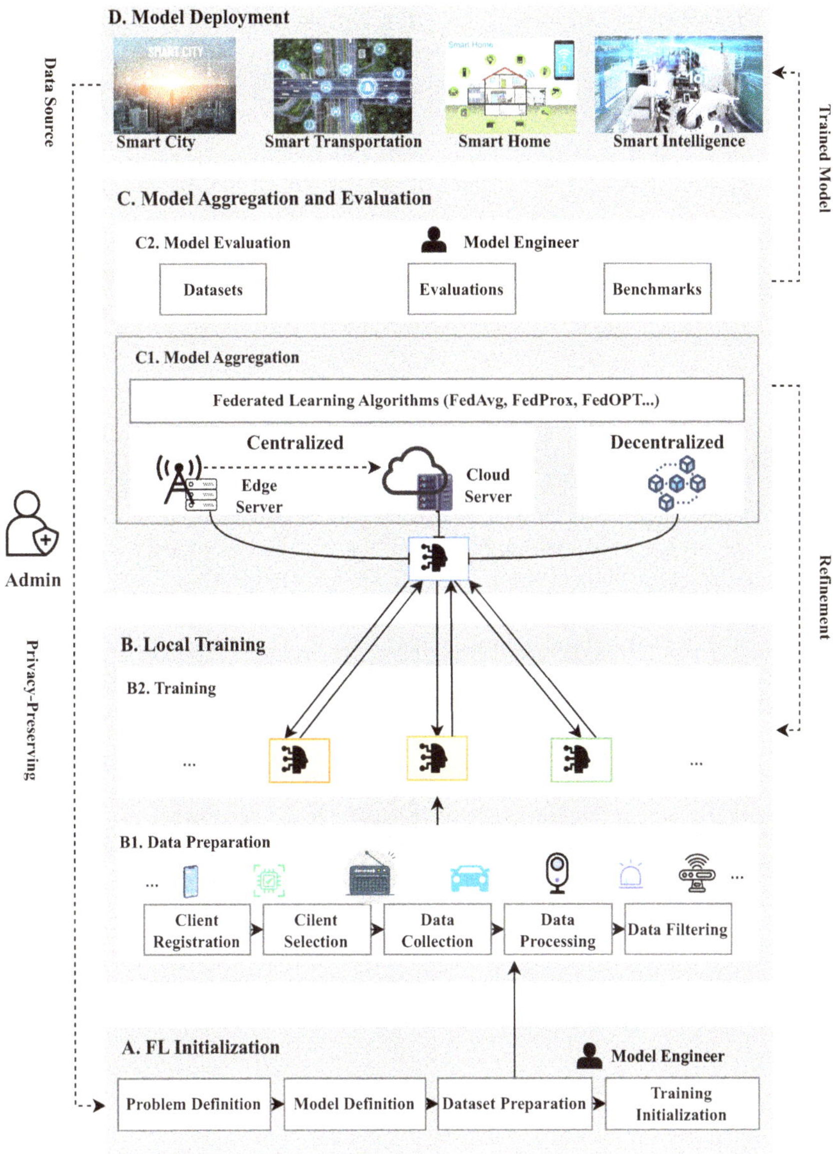

Fig. 3.1 The architecture of federated learning for IoT

Step 1: Client Selection. Client selection plays a crucial role in determining the participating clients in the training process, which influences the performance of the trained model. Let $\mathcal{K}^s$ denote the set of selected clients, and $|\mathcal{K}^s|$ represents the number of clients chosen for participation.

Step 2: Download Global Model. During this step, the clients initiate the process by downloading the global model that was aggregated by the central server in the previous round t. (In the first round, the global model is randomly initialized.)

$$w_t^k = w_t \tag{3.1}$$

where w_t represents the downloaded global model in round t.

Step 3: Local Training. After downloading the global model, the clients undertake local training based on their local datasets, utilizing the downloaded model as a new starting point:

$$w_{t+1}^k = w_t^k - \eta \nabla L^k \left(w_t^k ; \varkappa^k \right). \tag{3.2}$$

where η is the step size, $L^k(w_t^k; \varkappa^k)$ is the local loss function of client k in the round t, and w_{t+1}^k denotes the trained local model of client k in the round t.

Step 4: Upload Local Model Updates. The trained local models are then sent back to the aggregation server or blockchain for aggregation.

Step 5: Global Aggregation. The aggregation server or blockchain combines the model updates from participating clients using an appropriate algorithm, thereby creating a unified global model that represents the collective knowledge of all clients:

$$w_{t+1} = \frac{\sum_{k \in \mathcal{K}} |\varkappa^k| w_{t+1}^k}{\sum_{k \in \mathcal{K}} |\varkappa^k|} \tag{3.3}$$

where w_{t+1} denotes the aggregated global model in the round $t + 1$.

The training process in FL typically consists of multiple rounds, each consisting of T iterations, to achieve convergence, and the termination of the training process depends on the specific objectives and requirements related to accuracy and training time. The objective of FL is to obtain the optimal weights for the global model w^* by minimizing the global loss function $L(w)$ (McMahan et al. 2017b):

$$L(w) = \frac{\sum_{k \in \mathcal{N}^s} |\varkappa^k| L^k(w; \varkappa_k)}{\sum_{k \in \mathcal{N}^s} |\varkappa^k|}. \tag{3.4}$$

where $L^k(w; \varkappa_k)$ represents the loss function for a subset $\varkappa_k$ of client k when the global model's weight is given to w.

$$w^* = \arg\min_w L(w). \tag{3.5}$$

3.2.1.3 The Architecture of Federated Learning for IoT Networks

Figure 3.1 portrays a comprehensive and well-structured depiction of the architecture of FL for IoT networks. This architecture facilitates the integration of FL techniques into IoT systems, enabling collaborative and privacy-preserving ML across distributed IoT devices. The architecture comprises several key components, each playing a vital role in the FL process.

FL Initialization: FL initialization refers to the process of setting up the initial conditions and parameters before commencing the FL process. This stage is crucial as it establishes the foundation for subsequent iterations of model training and aggregation in a FL system. The initialization process in FL typically involves the following key steps: Define the problem by identifying data sources, and target tasks, and specifying performance metrics for model evaluation (step 1). After problem definition, the model engineer designs a model architecture for FL, which includes selecting optimization algorithms, defining model parameters, and determining data partitioning among participating clients (step 2). Then, the dataset is prepared by the data owners, who are responsible for the collection or generation of the data specifically intended for training the model (step 3). Afterwards, the training process is initiated by the central server, which provides the participating clients with the initial model parameters, either through random initialization or by leveraging pre-training on a large dataset (step 4).

Local Training: Local learning in FL refers to the process by which clients perform model training using their locally available data. Prior to local training, a crucial step is to perform data preparation, which encompasses client registration, client selection, data collection, data processing, and data filtering, ensuring the availability of diverse and relevant data that is appropriately formatted for subsequent local training in the FL process. Client registration refers to the initial step in FL where eligible clients or IoT devices voluntarily enroll themselves in the FL system, typically by registering with the central server or a designated entity (step 1). After that, client selection in FL is executed as a strategic process that involves carefully choosing a subset of clients from the registered pool for each iteration, considering criteria such as device capabilities, data quality, and diversity, to ensure their representative and effective participation (step 2). Next, the process of data collection gathers data from the selected clients, where each client contributes its locally stored or generated data (step 3). Subsequently, data processing involves the necessary preprocessing and transformation of collected data to prepare it for model training, aiming to enhance data quality and facilitate efficient learning (step 4). Last but not least, data filtering plays a critical role in data preparation by selectively removing or filtering out data samples or features based on predefined criteria, effectively eliminating outliers, noise, or irrelevant information that could potentially disrupt the training process or compromise privacy (step 5). After the completion of data preparation, each selected client independently trains its local model based on the data available locally.

Model Aggregation and Evaluation: Following the completion of local training, the subsequent step in FL entails aggregating the local model updates to create a new global model. This aggregation process can be carried out using various approaches, including the use of a cloud server, an edge server, a combination of cloud server and edge server, and even leveraging blockchain technology (the details as introduced in Sect. 3.2.2). As a subsequent step, the aggregated global model is evaluated to assess its performance and generalization ability (Ying et al. 2023a). Evaluation metrics, such as accuracy, precision, and F1 score, are commonly used to measure the model's effectiveness in achieving the desired task objectives. Additionally, the evaluation phase also involves comparing the performance of the FL model with other benchmark models or existing approaches to validate its efficacy and identify areas for improvement. If the specific objectives and requirements concerning the performance (such as accuracy) are achieved, the training process could be terminated.

Model Deployment: Upon completion of the training phase, the trained model can be deployed for making predictions on some IoT applications that perform FL model training or previously unseen IoT applications. However, in certain scenarios, it may be necessary to fine-tune the model using new data in order to adapt to evolving conditions or enhance its performance.

3.2.2 *Types of Federated Learning for IoT*

In this subsection, we present the classification of FL approaches based on their networking structure, centralization levels, and participating clients. By comprehending these categories, informed decisions can be made when implementing FL in IoT applications.

3.2.2.1 Types of FL for IoT Based on Networking Structure

From a networking structure perspective, FL can be categorized into two main classes, including centralized FL and decentralized FL, as illustrated in Fig. 3.2.

Centralized FL refers to the FL setting where a central server acts as the main coordinator during the learning process. In this approach, the training data remains distributed across multiple clients, but the coordination and aggregation of model updates are performed by the central server. The FL framework entails the central server distributing the global model to the clients, who subsequently perform local training using their own local datasets. After training, the clients transmit their locally updated model to the central server. The central server then aggregates these model updates, resulting in an improved global model. This iterative process of model distribution, local training, and aggregation is repeated across multiple rounds to enhance the performance of the global model.

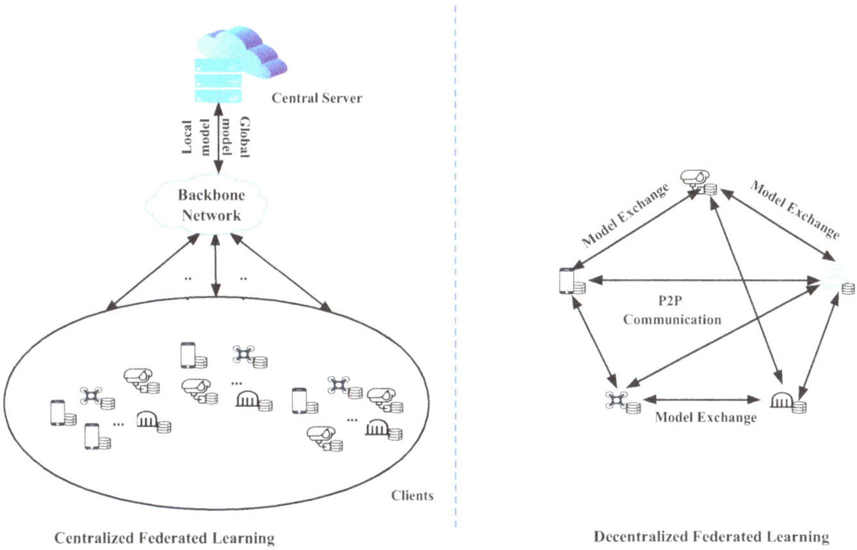

Fig. 3.2 Types of federated learning models for IoT networks

Decentralized FL, on the other hand, involves a more distributed and peer-to-peer approach. In this class of FL, there is no central server that coordinates the learning process. Instead, the participating clients form a network and collaborate directly with each other to train a shared global model. The clients exchange model updates with their neighboring devices and use those updates to refine their own local models. The collaboration and communication between the clients can occur in various ways, such as through P2P communication (blockchain) and direct device-to-device communication (Bluetooth, Wi-Fi Direct) in the network. The decentralized nature of this approach provides benefits such as improved privacy, reduced reliance on a single point of failure, and potential scalability advantages.

Both centralized FL and decentralized FL offer distinct advantages and considerations. The selection between these two classes hinges upon several factors, including the nature of the data, privacy requirements, communication capabilities, computational resources, and specific use case requirements. These factors play a pivotal role in determining the most suitable approach for a given scenario.

3.2.2.2 Types of Centralized Federated Learning

Centralized federated learning is a widely adopted architecture in IoT systems, encompassing various implementations such as cloud-based FL, edge-based FL, and cloud-edge-based FL. These architectures leverage the centralized coordination and management provided by a central server while incorporating different computing and communication trade-offs to suit specific IoT scenarios.

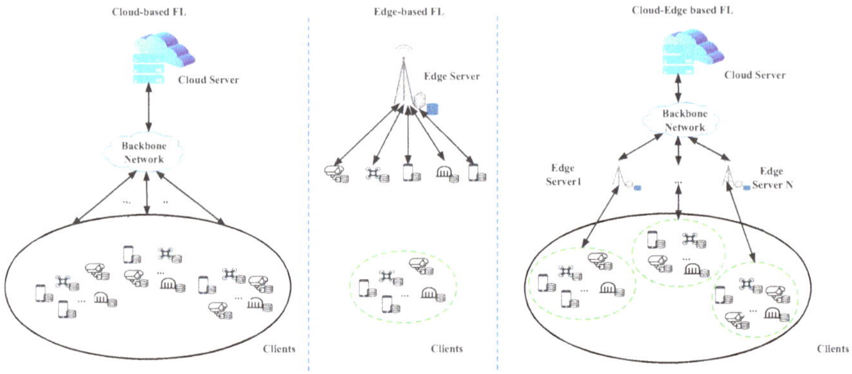

Fig. 3.3 The overview of federated learning for centralized IoT networks

In cloud-based FL, a large number of clients, potentially reaching millions (Bonawitz et al. 2019), contribute large datasets required for DL, as depicted on the left side in Fig. 3.3. However, communication with cloud servers is slow and unpredictable, resulting in inefficient training processes due to network congestion. The communication efficiency and convergence rate in Federated Averaging (FedAVG) involve a trade-off where more local computation is performed at the cost of reduced communication. Despite this, cloud-based FL benefits from the ability to access vast training samples on cloud servers.

On the other hand, edge-based FL has emerged as a response to the increasing demand for decentralized and real-time ML capabilities in IoT and edge computing environments, as depicted in the middle of Fig. 3.3. In edge-based FL, the server is placed closer to the edge, such as base stations. This architecture reduces computation latency as it aligns with the communication latency to the edge parameter server. While edge-based FL offers the advantage of faster local model updates, it has limitations in terms of the number of clients' access to each server, resulting in performance losses.

To address these challenges, a hierarchical FL system, called cloud-edge-based hierarchical FL (Liu et al. 2020; Wu et al. 2020), has been proposed, as depicted on the right side in Fig. 3.3. The architecture integrates the strengths of both cloud-based and edge-based FL approaches. It effectively harnesses the extensive training data available on cloud servers while enabling rapid model updates through local clients deployed on edge servers. Compared to cloud-based FL, the cloud-edge-based hierarchical FL significantly reduces expensive communication with the cloud servers. This reduction is accomplished through the integration of efficient client updates by edge servers, resulting in noteworthy decreases in both runtime and the number of local iterations required. Conversely, the cloud-edge-based hierarchical FL framework surpasses edge-based FL in terms of model training efficacy due to the cloud servers' access to more extensive data.

Centralized FL architectures, including cloud-based FL, edge-based FL, and cloud-edge-based FL, offer distinct advantages and trade-offs in IoT systems.

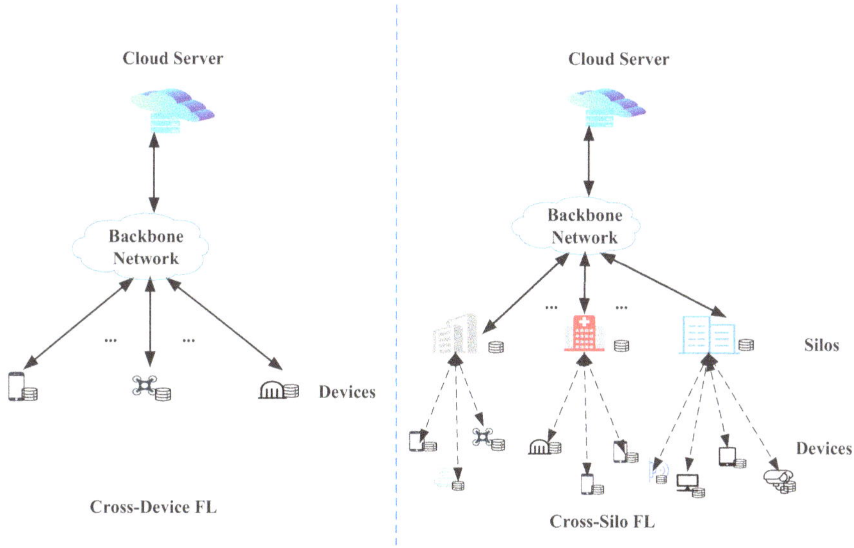

Fig. 3.4 Types of federated learning for IoT networks based on participating clients

Cloud-based FL excels in scalability and model performance but raises concerns regarding privacy, latency, and communication. In contrast, edge-based FL prioritizes privacy preservation, low latency, and efficient bandwidth utilization, but faces challenges related to resource constraints and hardware heterogeneity. Cloud-edge-based FL strikes a balance between privacy, latency, communication, and resource utilization, yet necessitates careful orchestration and deployment considerations. By comprehending the unique characteristics of each architecture and considering specific requirements, it helps to make well-informed decisions to select the most suitable FL approach for their IoT systems.

3.2.2.3 Types of Federated Learning for IoT Based on Participating Clients

According to the setting based on participating clients, FL for IoT can be classified into two types, cross-device FL and cross-silo FL, as illustrated in Fig. 3.4.

Cross-device FL refers to the scenario where the distributed devices participating in the FL process belong to different individuals or organizations, where the number of clients is big and the data size provided by each client is small (Rehman et al. 2021; Yang et al. 2022). These devices can be personal smartphones, tablets, or other IoT devices owned by different users. Each device holds its own local data and contributes to the FL process by performing local model training using its own data. The model updates are then transferred and aggregated across the devices to obtain a global model. Cross-device FL enables collaborative learning while preserving data privacy since the data remains on the devices and is not centralized.

Unlike cross-device FL, cross-silo FL involves the collaboration of multiple organizations that possess separate data silos, typically with a smaller number of organizations but with larger data volumes and many IoT devices within each organization. Each data silo represents a distinct dataset owned by a different organization (Li et al. 2023), such as different hospitals, cities, or industries. In this scenario, the organizations collaborate to train a shared global model by exchanging model updates while keeping their data locally. The data from each silo is not shared with other organizations, maintaining data privacy and security. Cross-silo FL enables the collaborative training of a more comprehensive model by leveraging diverse datasets from multiple organizations without directly sharing the raw data.

The classification of FL into cross-device and cross-silo types provides a clear distinction between scenarios involving individual devices owned by different users and scenarios involving separate organizations with their own data silos. The selection between these types depends on specific contextual factors such as data ownership, collaboration requirements, privacy considerations, and the nature of the FL for IoT applications.

3.2.3 FL Framework for IoT

This subsection provides a comprehensive overview of different frameworks that have been developed specifically for the implementation of FL for IoT networks:

(1) **FedML:** FedML is an open-source research framework that helps in developing and implementing FL algorithms (He et al. 2020). It consists of two main components: FedML-core and FedML-API. FedML-core is the low-level API component responsible for distributed communication and model training. FedML-API, built upon FedML-core, is the high-level API component that simplifies the implementation of distributed algorithms in FL. FedML is distinguished by its ability to facilitate FL on real-world hardware platforms. Notably, FedML incorporates two on-device FL testbeds, namely, FedML-Mobile and FedML-IoT, both of which are constructed using actual hardware platforms. This feature strengthens FedML's practicality and enables researchers to conduct FL experiments in authentic mobile and IoT environments.

(2) **Flower:** Flower is an open-source Python library developed by IBM Research that simplifies the implementation of FL systems by providing a high-level interface and abstraction layer (Beutel et al. 2020). It supports popular ML frameworks like PyTorch and TensorFlow, handling model update aggregation, client sampling, communication protocols, and FL system evaluation. Flower's architecture allows for experiments at global and local levels, separating client selection, parameter aggregation, and evaluation through strategy abstraction. It accommodates heterogeneous client platforms and implementations, manages

complexities like connection handling, and offers a simplified environment for researchers.

(3) **TensorFlow-Federated (TFF):** TFF is an open-source framework developed by Google that extends TensorFlow for FL (Blog n.d.). It offers tools and libraries for building and deploying ML models in federated settings, incorporating federated computations and aggregations. TFF consists of two layers: FL, providing high-level interfaces for seamless integration of existing ML models, and FC, offering lower-level interfaces for custom-federated algorithms using TensorFlow and distributed communication operators. This modular approach promotes flexibility and adaptability, empowering users to leverage FL according to their specific needs and research goals.

(4) **PySyft:** PySyft is an open-source Python library that aims to provide privacy-preserving ML and secure multiparty computation techniques (Ryffel et al. 2018). It integrates with popular DL frameworks such as PyTorch and TensorFlow, allowing users to perform privacy-enhancing tasks such as FL, encrypted computation, and differential privacy. PySyft leverages secure multiparty computation protocols, homomorphic encryption, and other cryptographic techniques to ensure the confidentiality and privacy of data in distributed learning scenarios. It offers an essential toolkit for building privacy-enhancing applications and fostering trust in collaborative ML environments.

(5) **LEAF:** LEAF, which stands for Low-resource Environments for Aggregation and FL, is a research framework and benchmark suite designed for FL under resource-constrained environments (Caldas et al. 2018). LEAF offers a curated collection of datasets, benchmarks, and evaluation metrics explicitly designed to evaluate the efficacy of FL algorithms in scenarios characterized by limited computational resources, constrained bandwidth, or energy constraints. By providing a standardized platform, LEAF facilitates benchmarking and comparative analysis of diverse algorithms and methodologies, thereby fostering advancements in FL techniques for resource-constrained settings. This framework plays a pivotal role in promoting research and innovation in privacy-preserving machine learning within challenging resource limitations.

(6) **FATE:** FATE, short for Federated AI Technology Enabler, is an open-source FL platform developed by WeBank's AI department (FedAI n.d.). FATE is a framework that aims to address the challenges of privacy, security, and trust in FL, providing a secure and reliable environment for FL system development. By offering a comprehensive suite of tools and components, including FL algorithms, distributed computing protocols, secure computation mechanisms, and privacy protection techniques, FATE enables the development and deployment of large-scale FL systems across diverse domains such as finance, healthcare, and smart cities. FATE plays a significant role in advancing FL research and innovation, contributing to the establishment of robust and privacy-preserving FL practices in various academic and industrial contexts.

(7) **Paddle FL:** Paddle FL is a federated learning framework developed by PaddlePaddle, an open-source deep learning platform (Ma et al. 2019). It is a comprehensive framework that facilitates FL using PaddlePaddle. It supports

various FL scenarios and integrates with PaddlePaddle's distributed computing capabilities to provide efficient strategies for model update aggregation, communication, and synchronization. With flexible options for model architectures, optimization algorithms, and customization, Paddle FL enables developers to create tailored FL systems. It focuses on scalability, efficiency, and privacy preservation, allowing for the training of large-scale models on distributed data sources while ensuring data security.

3.3 Federated Learning for IoT Applications

This section offers a comprehensive discussion on the integration of FL into various essential IoT applications. These applications encompass smart healthcare, vehicular IoT, smart cities, smart industries, and cybersecurity.

3.3.1 FL for Smart Healthcare

The IoT revolution has demonstrated significant potential for numerous healthcare applications, leveraging the vast amount of medical data collected through IoT devices. However, the increasing demands for privacy and security of healthcare data have resulted in each IoT device becoming an isolated data island. To tackle this challenge, the emergence of FL has introduced new possibilities for healthcare applications (Yuan et al. 2020; Chen et al. 2020; He et al. 2023b). FL enables collaborative and privacy-preserving machine learning, with immense potential to transform the landscape of smart healthcare. It empowers healthcare service providers to collectively leverage their data and knowledge, thereby enhancing the performance of diagnoses (Elayan et al. 2021) while adhering to stringent data privacy regulations and ethical considerations (Singh et al. 2022).

3.3.2 FL for Vehicular IoT

Vehicular IoT systems, encompassing cooperative autonomous driving and intelligent transport systems (ITS), are particularly susceptible to privacy breaches due to the abundance of devices and privacy-sensitive data. FL holds significant promise as an effective approach to address privacy concerns and optimize resource utilization in future vehicular IoT systems (Du et al. 2020). By preserving data privacy, fostering collaboration, and leveraging localized computing capabilities, FL can enable the realization of efficient and privacy-preserving cooperative autonomous driving (Li et al. 2021; Nguyen et al. 2022) and intelligent transport networks (Manias & Shami 2021; Zhao et al. 2022). However, further research and development efforts are necessary to tailor FL algorithms to the specific requirements of vehicular IoT systems and overcome challenges related to scalability, heterogeneity, and trustworthiness. By addressing these challenges, FL can pave the way for the widespread deployment of secure and privacy-preserving vehicular IoT systems, contributing to safer and more efficient transportation networks.

3.3.3 FL for Smart City

Smart cities are rapidly evolving ecosystems that leverage various IoT technologies to enhance urban services and infrastructure. However, the massive amount of data collected by IoT devices raises significant concerns regarding privacy and resource efficiency. FL has emerged as a promising approach to address privacy concerns and optimize resource utilization in smart city environments, offering significant potential for enhancing the efficiency and privacy of smart city applications (Jiang et al. 2020). By enabling distributed model training and preserving data privacy, FL can facilitate the development of more efficient and privacy-preserving smart city systems. The adoption of FL in smart city deployments requires further research and development to address challenges related to heterogeneity, model consistency, network dynamics, and trustworthiness. By overcoming these challenges, FL can contribute to the realization of intelligent and privacy-conscious smart city ecosystems, promoting sustainable urban development and improving the quality of life for citizens (Imteaj & Amini 2019; Qolomany et al. 2020; He et al. 2023a).

3.3.4 FL for Smart Industry

Smart industry, powered by industrial Internet of Things (IIoT) technologies, poses unique challenges concerning privacy and resource efficiency. FL presents a promising approach to address these challenges by offering privacy preservation and resource optimization in smart industry applications (Pham et al. 2021), enhancing privacy preservation while improving resource efficiency in industrial IoT deployments. However, additional research and development efforts are necessary to overcome challenges related to network heterogeneity, model synchronization, and security. Addressing these challenges would enable FL to unlock the full potential of smart industry, fostering efficient and privacy-conscious industrial processes, and facilitating data-driven decision-making for enhanced productivity and competitiveness (Li et al. 2022; Yang et al. 2021; Qolomany et al. 2020; Ma et al. 2021).

3.3.5 *FL for Cybersecurity*

Cybersecurity has become a critical issue in the digital age, requiring effective solutions to detect threats and protect privacy. With the continuous expansion of IoT services and applications, the decentralization paradigm has attracted a lot of attention from government, academia, and industry in cybersecurity and ML for IoT. FL has gained prominence as a promising approach for addressing cybersecurity challenges, offering innovative solutions to enhance the security and efficiency of IoT systems. The concept of federated cybersecurity (FC) (Ghimire & Rawat 2022) is considered revolutionary, as it paves the way for a more secure and efficient future in IoT environments by effectively detecting security threats, improving accuracy, and enabling real-time response in network systems (Belenguer et al. 2022; Attota et al. 2021; Issa et al. 2023; Liu et al. 2020). Future advancements in FL algorithms and privacy-enhancing techniques will further strengthen the effectiveness and scalability of FL for cybersecurity applications, contributing to a more secure digital landscape.

3.4 Research Challenges and Directions

Despite the aforementioned benefits, the implementation of FL for IoT still faces numerous challenges, as outlined below.

3.4.1 *Heterogeneity of IoT Devices*

The heterogeneity observed among IoT devices poses significant challenges to the implementation of FL in IoT applications. This heterogeneity encompasses both data heterogeneity and device heterogeneity. To address these challenges, it is essential to develop adaptive FL algorithms capable of accommodating the diverse capabilities of IoT devices. Additionally, FL algorithms need to consider the limitations of resource-constrained environments and limited power sources. To mitigate these constraints, it is crucial to incorporate energy-efficient strategies and optimization techniques that minimize computational and communication overhead. Moreover, the dynamic nature of IoT networks introduces further challenges related to device mobility and connectivity fluctuations. FL algorithms should account for device mobility, enabling seamless model synchronization and training continuity even during device joins or departures, as well as intermittent connectivity. To overcome the heterogeneity of IoT devices, future research should prioritize the development of adaptive and robust FL algorithms (Sun et al. 2020) capable of effectively handling varying capabilities (Li et al. 2022; Wang et al. 2021; Pang et al. 2020; Chen et al. 2021), resource constraints (Imteaj et al. 2022; Savazzi et al. 2020), and dynamic network conditions (Wang et al. 2021).

3.4.2 Limited Computational Resources

The implementation of FL for IoT applications encounters challenges arising from the limited computational resources available on IoT devices. These devices possess constraints in processing power, memory, and energy, which impede the execution of complex ML algorithms necessary for FL. To address this issue, it is crucial to develop resource-efficient FL algorithms that employ techniques such as model compression, lightweight architectures, and efficient communication protocols. These techniques aim to minimize the computational overhead associated with FL operations. The heterogeneity of computational resources among devices further complicates the design of FL algorithms, necessitating the adoption of adaptive approaches capable of adjusting computational requirements based on device capabilities and availability. Moreover, ensuring energy efficiency is of paramount importance, and FL algorithms should incorporate strategies such as reducing device participation frequency and duration, employing compressed model updates, and leveraging local computation to minimize energy consumption. Therefore, future research should focus on the development of resource-constrained algorithms (Imteaj et al. 2021) that achieve a balance between computational efficiency, model accuracy, and energy consumption, while also exploring techniques for adaptive resource allocation (Nguyen et al. 2020), and energy optimization (Yu et al. 2021) to facilitate the effective deployment of FL in resource-constrained IoT environments.

3.4.3 Communication and Bandwidth Limitations

The successful implementation of FL for IoT applications faces significant challenges attributed to limitations in communication and bandwidth (Brown et al. 2020). IoT devices operate within resource-constrained environments characterized by restricted network bandwidth, intermittent connectivity, and diverse communication protocols. To address these challenges, communication-efficient FL algorithms can minimize data transmission and reduce reliance on continuous connectivity through techniques such as model compression (Itahara et al. 2020; Bernstein et al. 2018), client selection (McMahan et al. 2017c; Li et al. 2021), and sparse updates (Thonglek et al. 2022). Additionally, adaptive strategies for communication scheduling can optimize bandwidth utilization (Hönig et al. 2022; Diao et al. 2020). These approaches enable communication-efficient FL algorithms that minimize data transmission, reduce reliance on continuous connectivity, and optimize bandwidth utilization. By leveraging these techniques, FL can be effectively applied to

IoT environments, unlocking the potential for distributed machine learning while accommodating the unique constraints of resource-constrained IoT devices.

3.4.4 *Privacy and Security Concerns*

Privacy and security concerns pose significant challenges to the implementation of FL for IoT applications (Briggs et al. 2021). FL involves the sharing and aggregation of sensitive data from multiple devices, raising concerns about data privacy and potential security breaches. To address these concerns, robust privacy-preserving techniques should be developed, such as differential privacy (Zhao et al. 2020; Zhou et al. 2020) and blinding technique (Fu et al. 2020; Zhou et al. 2020). These techniques ensure that individual device data remains private and secure during the FL process. In addition, securing FL against potential attacks and malicious participants is crucial. FL algorithms should incorporate mechanisms for detecting and mitigating adversarial behavior, such as anomaly detection (Liu et al. 2020; Cui et al. 2021). Furthermore, implementing robust authentication and access control mechanisms prevents unauthorized devices from participating in the FL process (Li et al. 2022). Compliance with data privacy regulations and ethical considerations is essential in FL for IoT. Adhering to regulatory frameworks like the General Data Protection Regulation (GDPR) and integrating privacy-by-design principles ensures transparent and privacy-preserving FL processes.

3.4.5 *Scalability and Management*

The successful implementation of FL for IoT applications is impeded by scalability and management concerns. Scalability encompasses the FL system's ability to effectively handle a large number of participating clients and increasing data volumes. As the IoT system expands with a growing number of clients and data sources, FL algorithms need to efficiently manage the aggregation of model updates and ensure timely convergence. Thus, the development of scalable FL architectures and distributed optimization techniques become crucial to accommodate the growing scale of IoT deployments. Furthermore, effective management of FL systems is paramount for their seamless operation. This entails various tasks such as device registration, model synchronization, performance monitoring, and fault tolerance. The development of comprehensive management frameworks and protocols is necessary to ensure the reliability, availability, and performance of FL systems within dynamic IoT environments. To address the challenges associated with scalability and management in FL for IoT, future research should prioritize the development of scalable and efficient algorithms capable of handling large-scale deployments and increasing data volumes (Imteaj et al. 2022; Savazzi et al.

2020; Rahman et al. 2020). Additionally, robust management frameworks need to be designed to facilitate seamless client management, model synchronization, and system monitoring, thus contributing to the successful deployment and operation of FL in IoT environments (Li et al. 2022; Rey et al. 2022; Khan et al. 2020; Cui et al. 2021).

3.4.6 Federated Domain Generalization

Federated domain generalization (FDG) is a critical consideration in implementing FL for IoT applications as it pertains to the ability of FL models to effectively generalize across diverse data domains collected from various clients or locations within IoT environments (Ying et al. 2023b). Domain shift can lead to performance degradation when models are deployed in new or unseen domains. Addressing FDG necessitates the development of robust techniques like domain adaptation (Wu & Gong 2021; Zhang et al. 2023), transfer learning (Shenaj et al. 2023; Zhang & Li 2021), and meta-learning (Chen et al. 2021; Lin et al. 2020), which aim to enhance the generalization capabilities of FL models across diverse domains by leveraging knowledge from multiple domains and incorporating domain-awareness mechanisms. Addressing data distribution heterogeneity in FL is essential to prevent biased models that excel on certain devices but underperform on others, stemming from variations in data distributions. Techniques like data augmentation (Duan et al. 2019; Yang & Soatto 2020) and adaptive aggregation (Yang et al. 2022; Shenaj et al. 2023) can be employed to mitigate distributional differences and improve the generalization performance of FL models across devices. Future research should prioritize the development of techniques and algorithms that effectively address domain shifts and data distribution heterogeneity in order to enhance the generalization capabilities of FL models, ensuring robust performance across diverse domains and IoT devices.

3.5 Conclusion

FL is a significant research area within the IoT environment. This work provides a comprehensive introduction to the field of FL for IoT, serving as a valuable resource for researchers seeking in-depth insights into FL in the IoT environment. By covering the theoretical foundations of FL, the architecture of FL for IoT, the different types of FL for IoT, FL frameworks tailored for IoT, diverse FL for IoT applications, and future research challenges and directions pertaining to FL for IoT, it provides a comprehensive view of the field. This work offered herein aims to offer valuable insights to researchers and inspire further research for novel advancements in privacy-preserving FL techniques for IoT.

References

Abbas, Nasir, et al. 2017. Mobile edge computing: A survey. *IEEE Internet of Things Journal* 5 (1): 450–465.

Al-Fuqaha, Ala et al. 2015. Internet of things: A survey on enabling technologies, protocols, and applications. *IEEE Communications Surveys & Tutorials* 17 (4): 2347–2376.

Attota, Dinesh Chowdary, et al. 2021. An ensemble multi-view federated learning intrusion detection for IoT. *IEEE Access* 9: 117734–117745.

Belenguer, Aitor, et al. 2022. A review of federated learning in intrusion detection systems for IoT. arXiv preprint. arXiv:2204.12443.

Bernstein, Jeremy, et al. 2018. signSGD: Compressed optimisation for non-convex problems. In *International Conference on Machine Learning*, 560–569. PMLR.

Beutel, Daniel J., et al. 2020. Flower: A friendly federated learning research framework. arXiv preprint. arXiv:2007.14390.

Blog, Tensorflow (n.d.). Introducing TensorFlow Federated. https://blog.tensorflow.org/2019/03/introducing-tensorflow-federated.html. Accessed Jun 15, 2023.

Bonawitz, Keith, et al. 2019. Towards federated learning at scale: System design. *Proceedings of Machine Learning and Systems* 1: 374–388.

Briggs, Christopher, et al. 2021. A review of privacy-preserving federated learning for the Internet-of-Things. In: *Federated Learning Systems: Towards Next-Generation AI*, 21–50.

Brown, Tom, et al. 2020. Language models are few-shot learners. *Advances in Neural Information Processing Systems* 33: 1877–1901.

Caldas, Sebastian, et al. 2018. Leaf: A benchmark for federated settings. arXiv preprint. arXiv:1812.01097.

Cao, Bin, et al. 2019. Intelligent offloading in multi-access edge computing: A state-of-the-art review and framework. *IEEE Communications Magazine* 57 (3): 56–62.

Chen, Yiqiang, et al. 2020. Fedhealth: A federated transfer learning framework for wearable healthcare. *IEEE Intelligent Systems* 35 (4): 83–93.

Chen, Zheyi, et al. 2021. Towards asynchronous federated learning for heterogeneous edge-powered internet of things. *Digital Communications and Networks* 7 (3): 317–326.

Cui, Lei, et al. 2021. Security and privacy-enhanced federated learning for anomaly detection in IoT infrastructures. *IEEE Transactions on Industrial Informatics* 18 (5): 3492–3500.

Diao, Enmao, et al. 2020. HeteroFL: Computation and communication efficient federated learning for heterogeneous clients. arXiv preprint. arXiv:2010.01264.

Donta, Praveen Kumar, et al. 2023. Learning-driven ubiquitous mobile edge computing: Network management challenges for future generation Internet of Things. *International Journal of Network Management* 33 (5): e2250.

Du, Zhaoyang, et al. 2020. Federated learning for vehicular internet of things: Recent advances and open issues. *IEEE Open Journal of the Computer Society* 1: 45–61.

Duan, Moming, et al. 2019. Astraea: Self-balancing federated learning for improving classification accuracy of mobile deep learning applications. In *2019 IEEE 37th International Conference on Computer Design (ICCD)*, 246–254. Piscataway: IEEE.

Elayan, Haya, et al. 2021. Deep federated learning for IoT-based decentralized healthcare systems. In *2021 International Wireless Communications and Mobile Computing (IWCMC)*, 105–109. Piscataway: IEEE.

FedAI. n.d. *An Industrial Grade Federated Learning Framework*. https://fate.fedai.org. Accessed Jun 15, 2023.

Fu, Anmin, et al. 2020. VFL: A verifiable federated learning with privacy-preserving for big data in industrial IoT. *IEEE Transactions on Industrial Informatics* 18 (5): 3316–3326.

Ghimire, Bimal, and Danda B. Rawat. 2022. Recent advances on federated learning for cybersecurity and cybersecurity for federated learning for internet of things. *IEEE Internet of Things Journal* 9 (11): 8229–8249.

He, Chaoyang, et al. 2020. FedML: A research library and benchmark for federated machine learning. arXiv preprint. arXiv:2007.13518.

He, Qiang, Yu Wang, et al. 2023a. is in Early Access.

He, Qiang, Zheng Feng, et al. 2023b. is in Early Access.

Hönig, Robert et al. 2022. DAdaQuant: Doubly-adaptive quantization for communication-efficient Federated Learning. In *International Conference on Machine Learning*, 8852–8866. PMLR.

Imteaj, Ahmed, and M. Hadi Amini. 2019. Distributed sensing using smart end-user devices: Pathway to federated learning for autonomous IoT. In *2019 International Conference on Computational Science and Computational Intelligence (CSCI)*, 1156–1161. Piscataway: IEEE.

Imteaj, Ahmed, et al. 2022. Federated learning for resource-constrained IoT devices: Panoramas and state of the art. In *Federated and transfer learning. Adaptation, learning, and optimization*, ed. Razavi-Far, R., Wang, B., Taylor, M.E., Yang, Q. vol. 27, 7–27. Cham: Springer.

Imteaj, Ahmed, Urmish Thakker, et al. 2021. A survey on federated learning for resource-constrained IoT devices. *IEEE Internet of Things Journal* 9 (1): 1–24.

Issa, Wael, et al. 2023. Blockchain-based federated learning for securing internet of things: A comprehensive survey. *ACM Computing Surveys* 55 (9): 1–43.

Itahara, Sohei, et al. 2020. Lottery hypothesis based unsupervised pre-training for model compression in federated learning. In *2020 IEEE 92nd Vehicular Technology Conference (VTC2020-Fall)*, 1–5. Piscataway: IEEE.

Jiang, Ji Chu, et al. 2020. Federated learning in smart city sensing: Challenges and opportunities. *Sensors* 20 (21): 6230.

Kairouz, Peter, et al. 2021. Advances and open problems in federated learning. *Foundations and Trends® in Machine Learning* 14 (1–2): 1–210.

Khan, Latif U., et al. 2020. Resource optimized federated learning-enabled cognitive internet of things for smart industries. *IEEE Access* 8: 168854–168864.

Li, Ang, et al. 2021. Hermes: an efficient federated learning framework for heterogeneous mobile clients. In *Proceedings of the 27th Annual International Conference on Mobile Computing and Networking*, 420–437.

Li, Huilin, et al. 2023. Privacy-preserving cross-silo federated learning atop blockchain for IoT. *IEEE Internet of Things Journal* 10 (24), 21176–21186. https://doi.org/10.1109/JIOT.2023.3279926.

Li, Yijing, et al. 2021. Privacy-preserved federated learning for autonomous driving. *IEEE Transactions on Intelligent Transportation Systems* 23 (7): 8423–8434.

Li, Ying, et al. 2023. VARF: An incentive mechanism of cross-silo federated learning in MEC. *IEEE Internet of Things Journal* 10 (17), 15115–15132. https://doi.org/10.1109/JIOT.2023.3264611.

Li, Ying, Xingwei Wang, Rongfei Zeng, Praveen Kumar Donta, et al. 2023a. Federated domain generalization: A survey. arXiv preprint. arXiv:2306.01334.

Li, Ying, Xingwei Wang, Rongfei Zeng, Praveen Kumar Donta, et al. 2023b. Federated domain generalization: A survey. arXiv preprint. arXiv:2306.01334.

Li, Ying, Yaxin Yu, and Xingwei, Wang. 2023. Three-tier storage framework based on TBchain and IPFS for protecting IoT security and privacy. *ACM Transactions on Internet Technology* 23 (3): 1–28.

Li, Yuanjiang, et al. 2022. An effective federated learning verification strategy and its applications for fault diagnosis in industrial IOT systems. *IEEE Internet of Things Journal* 9 (18): 16835–16849.

Li, Zonghang, et al. 2022. Data heterogeneity-robust federated learning via group client selection in industrial IoT. *IEEE Internet of Things Journal* 9 (18): 17844–17857.

Lin, Sen, et al. 2020. A collaborative learning framework via federated meta-learning. In *2020 IEEE 40th International Conference on Distributed Computing Systems (ICDCS)*, 289–299. Piscataway: IEEE.

Liu, Lumin, et al. 2020. Client-edge-cloud hierarchical federated learning. In *ICC 2020-2020 IEEE International Conference on Communications (ICC)*, 1–6. Piscataway: IEEE.

Liu, Yi, et al. 2020. Deep anomaly detection for time-series data in industrial IoT: A communication-efficient on-device federated learning approach. *IEEE Internet of Things Journal* 8 (8): 6348–6358.
Ma, Lianbo, et al. 2021. TCDA: Truthful combinatorial double auctions for mobile edge computing in industrial Internet of Things. *IEEE Transactions on Mobile Computing* 21 (11): 4125–4138.
Ma, Yanjun, et al. 2019. PaddlePaddle: An open-source deep learning platform from industrial practice. *Frontiers of Data and Domputing* 1 (1): 105–115.
Manias, Dimitrios Michael, and Abdallah Shami. 2021. Making a case for federated learning in the internet of vehicles and intelligent transportation systems. *IEEE Network* 35 (3): 88–94.
McMahan, Brendan, et al. 2017a. Communication-efficient learning of deep networks from decentralized data. In *Artificial Intelligence and Statistics*, 1273–1282. PMLR.
McMahan, Brendan, et al. 2017b. Communication-efficient learning of deep networks from decentralized data. In *Artificial Intelligence and Statistics*, 1273–1282. PMLR.
McMahan, Brendan, et al. 2017c. Communication-efficient learning of deep networks from decentralized data. In *Artificial Intelligence and Statistics*, 1273–1282. PMLR.
Nguyen, Anh, et al. 2022. Deep federated learning for autonomous driving. In *2022 IEEE Intelligent Vehicles Symposium (IV)*, 1824–1830. Piscataway. IEEE.
Nguyen, Van-Dinh, et al. 2020. Efficient federated learning algorithm for resource allocation in wireless IoT networks. *IEEE Internet of Things Journal* 8 (5): 3394–3409.
Pang, Junjie, et al. 2020. Realizing the heterogeneity: A self-organized federated learning framework for IoT. In *IEEE Internet of Things Journal* 8 (5): 3088–3098.
Pham, Quoc-Viet, et al. 2021. Fusion of federated learning and industrial internet of things: a survey. arXiv preprint. arXiv:2101.00798.
Qolomany, Basheer, et al. 2020. Particle swarm optimized federated learning for industrial IoT and smart city services. In *GLOBECOM 2020-2020 IEEE Global Communications Conference*, 1–6. Piscataway: IEEE.
Rahman, Sawsan Abdul, et al. 2020. Internet of things intrusion detection: Centralized, on-device, or federated learning? *IEEE Network* 34 (6): 310–317.
Rehman, Muhammad Habib ur, et al. 2021. TrustFed: A framework for fair and trustworthy cross-device federated learning in IIoT. *IEEE Transactions on Industrial Informatics* 17 (12): 8485–8494.
Rey, Valerian, et al. 2022. Federated learning for malware detection in iot devices. *Computer Networks* 204: 108693.
Ryffel, Theo, et al. 2018. A generic framework for privacy preserving deep learning. arXiv preprint. arXiv:1811.04017.
Saad, Walid, et al. 2019. A vision of 6G wireless systems: Applications, trends, technologies, and open research problems. *IEEE Network* 34 (3): 134–142.
Savazzi, Stefano, et al. 2020. Federated learning with cooperating devices: A consensus approach for massive IoT networks. *IEEE Internet of Things Journal* 7 (5): 4641–4654.
SEMICONDUCTORDIGEST. n.d. Number of connected IoT devices will surge to 125 billion by 2030. https://sst.semiconductor-digest.com/2017/10/number-of-connected-iot-devices-will-surge-to-125-billion-by-2030. Accessed Jun 11, 2023
Shenaj, Donald, et al. 2023. Learning across domains and devices: Style-driven source-free domain adaptation in clustered federated learning. In *Proceedings of the IEEE/CVF Winter Conference on Applications of Computer Vision*, 444–454.
Singh, Saurabh, et al. 2022. A framework for privacy-preservation of IoT healthcare data using Federated Learning and blockchain technology. *Future Generation Computer Systems* 129: 380–388.
Statista. n.d. Data volume of internet of things (IoT) connections worldwide in 2019 and 2025. https://www.statista.com/statistics/1017863/worldwide-iot-connected-devices-data-size. Accessed Jun 11, 2023
Sun, Wen, et al. 2020. Adaptive federated learning and digital twin for industrial internet of things. *IEEE Transactions on Industrial Informatics* 17 (8): 5605–5614.

Tataria, Harsh, et al. 2021. 6G wireless systems: Vision, requirements, challenges, insights, and opportunities. *Proceedings of the IEEE* 109 (7): 1166–1199.

Thonglek, Kundjanasith, et al. 2022. Sparse communication for federated learning. In *2022 IEEE 6th International Conference on Fog and Edge Computing (ICFEC)*, 1–8. Piscataway: IEEE.

Wang, Han, et al. 2021. Non-IID data re-balancing at IoT edge with peer-to-peer federated learning for anomaly detection. In *Proceedings of the 14th ACM Conference on Security and Privacy in Wireless and Mobile Networks*, 153–163.

Wang, Zhiyuan, et al. 2021. Resource-efficient federated learning with hierarchical aggregation in edge computing. In *IEEE INFOCOM 2021-IEEE Conference on Computer Communications*, 1–10. Piscataway: IEEE.

Wu, Guile, and Shaogang Gong. 2021. Collaborative optimization and aggregation for decentralized domain generalization and adaptation. In *Proceedings of the IEEE/CVF International Conference on Computer Vision*, 6484–6493.

Wu, Qiong, et al. 2020. Personalized federated learning for intelligent IoT applications: A cloud-edge based framework. *IEEE Open Journal of the Computer Society* 1: 35–44.

Yang, Hui, et al. 2021. BrainIoT: Brain-like productive services provisioning with federated learning in industrial IoT. *IEEE Internet of Things Journal* 9 (3): 2014–2024.

Yang, Seunghan, et al. 2022. *Client-agnostic Learning and Zero-shot Adaptation for Federated Domain Generalization*. Submitted to ICLR 2023.

Yang, Wenti, et al. 2022. A practical cross-device federated learning framework over 5g networks. *IEEE Wireless Communications* 29 (6): 128–134.

Yang, Yanchao, and Stefano Soatto. 2020. FDA: Fourier domain adaptation for semantic segmentation. In *Proceedings of the IEEE/CVF Conference on Computer Vision and Pattern Recognition*, 4085–4095.

Yu, Liangkun, et al. 2021. Jointly optimizing client selection and resource management in wireless federated learning for internet of things. *IEEE Internet of Things Journal* 9 (6): 4385–4395.

Yuan, Binhang, et al. 2020. A federated learning framework for healthcare IoT devices. arXiv preprint. arXiv:2005.05083.

Zeng, Rongfei, Chao Zeng, et al. 2021. A comprehensive survey of incentive mechanism for federated learning. arXiv preprint. arXiv:2106.15406.

Zeng, Rongfei, Shixun Zhang, et al. 2020. FMore: An incentive scheme of multi-dimensional auction for federated learning in MEC. In *2020 IEEE 40th International Conference on Distributed Computing Systems (ICDCS)*, 278–288. Piscataway: IEEE.

Zhang, Liling, et al. 2023. Federated learning for IoT devices with domain generalization. *IEEE Internet of Things Journal* 10 (11) 9622–9633. https://doi.org/10.1109/JIOT.2023.3234977.

Zhang, Wei, and Xiang Li. 2021. Federated transfer learning for intelligent fault diagnostics using deep adversarial networks with data privacy. *IEEE/ASME Transactions on Mechatronics* 27 (1): 430–439.

Zhao, Jianxin, et al. 2022. Participant selection for federated learning with heterogeneous data in intelligent transport system. *IEEE transactions on intelligent transportation systems* 24 (1): 1106–1115.

Zhao, Yang, et al. 2020. Local differential privacy-based federated learning for internet of things. *IEEE Internet of Things Journal* 8 (11): 8836–8853.

Zhou, Chunyi, et al. 2020. Privacy-preserving federated learning in fog computing. *IEEE Internet of Things Journal* 7.11, pp. 10782–10793.

Chapter 4
Machine Learning Techniques for Industrial Internet of Things

Megha Sharma, Abhishek Hazra, and Abhinav Tomar

4.1 Introduction

The Internet of Things (IoT) refers to a network of interconnected nodes, such as smart sensors, actuators, and digital devices, which can exchange data over a network without requiring human-to-human or human-to-computer interaction. Every IoT device must maintain a special identification number to connect to the Internet (Hazra & Amgoth 2022). IoT interventions have led to significant developments in several unrelated technology areas. These interconnections have improved living conditions and enhanced our understanding of surroundings and health, driving humankind toward a smarter life. IIoT, or Industrial IoT, is an enormous network of smart devices that holds great promise for revolutionizing industries by enhancing productivity, efficiency, and predictive abilities (Hazra et al. 2022,b). One of the key facilitators of realizing IIoT potential is the use of ML techniques. As a subset of artificial intelligence (AI), ML enables devices to learn from data, adapt, and make decisions without explicit programming. Current ML techniques allow manufacturers to optimize specific assets and entire manufacturing processes using insights gained from ML. Using intelligent sensors, machines, and gadgets, smart factories can gather production-related data continuously. Combining ML and IIoT has transformed manufacturing, enabling us to deal with enormous amounts of data in new ways (Boyes et al. 2018). The architectural depiction of IIoT architecture is illustrated in Fig. 4.1.

M. Sharma · A. Tomar
Netaji Subhas University of Technology, Delhi, India
e-mail: megha.sharma.phd22@nsut.ac.in; abhinav.tomar@nsut.ac.in

A. Hazra (✉)
Indian Institute of Information Technology, Sri City, India

P. K. Donta et al. (eds.), *Learning Techniques for the Internet of Things*,
https://doi.org/10.1007/978-3-031-50514-0_4

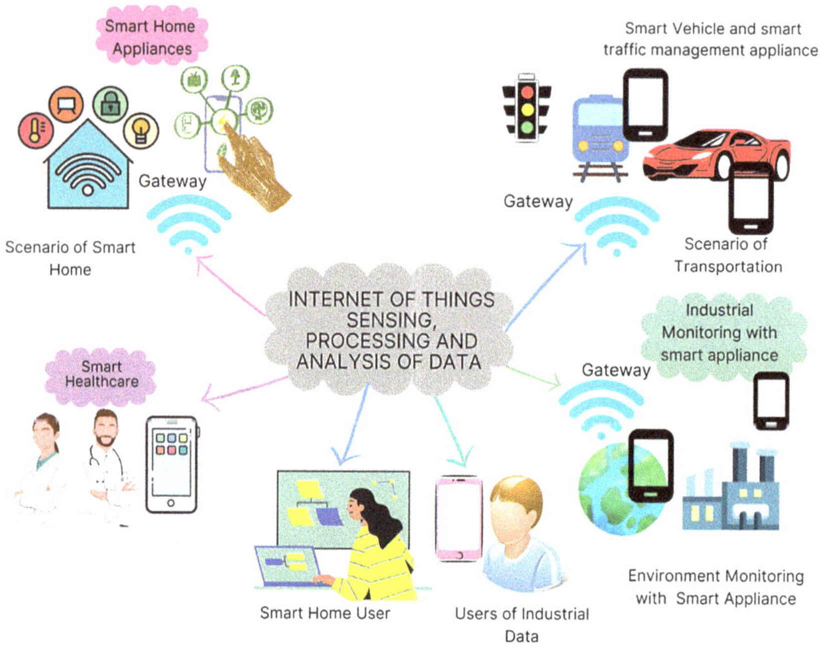

Fig. 4.1 IIoT architecture

4.1.1 Evolution of IoT to IIoT

The evolution of IoT to IIoT began with the emergence of connecting commonplace gadgets to the Internet for convenience. The emergence of IIoT was fueled by industries realizing the potential of IoT to improve workflows and cut costs. IIoT strongly emphasizes addressing industrial needs such as real-time data processing and harsh circumstances (Jaidka et al. 2020). Advanced sensors, dependable networking choices, and data security were all prioritized. The emphasis on IIoT enabled seamless modernization by promoting interoperability and integration with legacy systems. Data analytics and AI have taken center stage in IIoT for predictive maintenance and process optimization (Lin et al. 2017). To safeguard vital industrial assets, safety and security were of the utmost importance in the IIoT. In today's world, IIoT is widely used across many sectors, revolutionizing the industrial, energy, transportation, and healthcare industries. The continual development of IIoT is fueled by improvements in sensors, connectivity, AI, and data analytics, which also boosts industrial productivity and efficiency (Carbonell et al. 1983). Table 4.1 summarizes the difference between IoT and IIoT.

Table 4.1 Comparison between IoT and IIoT t echnologies

Parameter	IoT	IIoT
Key focus	Home automation and general consumer use	Applications for business and industry in numerous sectors
Objectives	Enhancing connectivity, automation, and comfort	Enhancing industrial productivity, safety, and efficiency
Data collection	Gathers and examines data from a range of consumer devices	Focuses on data collected in real time by industrial machinery and sensors
Use Cases	Linked products, wearable technology, smart houses	Condition monitoring, process optimization, and predictive maintenance
Device deployment	Involves a lot of internet-connected consumer products	Significant use of sensors and gadgets in industrial settings is necessary
Communication protocols	Bluetooth, Wi-Fi, Zigbee	MQTT, OPC UA

4.1.2 Significance of ML in IIoT

With IIoT, ML algorithms play an essential role in analyzing massive data streams from connected equipment, resulting in actionable insights that improve decision-making. ML enables IIoT systems to adapt quickly to changing circumstances through real-time data analysis (Xue & Zhu 2009). Furthermore, pattern recognition in large industrial datasets improves resource allocation and process optimization. ML-driven strategies increase quality control by allowing producers to spot flaws and ensure precise product quality. Through the use of ML in IIoT, difficult tasks can be executed by intelligent machines and robots without human intervention, increasing productivity and reducing errors (Javaid et al. 2022). ML algorithms continuously improve performance by learning from data and adapting to changing circumstances. This capacity for self-improvement is particularly useful in dynamic industrial situations where conditions and requirements change. The importance of ML will only continue to grow as IIoT applications develop further. This will spur innovation, boost productivity, and influence industrial automation and decision-making in the future (Muttil & Chau 2007).

4.1.3 Computational Offloading in ML for IIoT Application

A key idea in ML for IIoT applications is computational offloading. It entails transferring computationally demanding tasks from edge devices with limited resources to more powerful cloud or edge servers. Because of computational offloading, IIoT applications can scale and effectively manage greater data volumes (Kumar et al. 2013). Offloading work to servers with additional computational power will ensure smooth operation and the analysis of large datasets as data generation increases with the deployment of more IoT devices (Akherfi et al. 2018). This method assists in overcoming the constraints of edge devices, such as constrained memory and processing power, allowing them to effectively carry out ML algorithms that would otherwise be difficult to perform locally. Edge devices may analyze data faster and respond quickly by outsourcing complicated ML computations to more powerful servers. This enables real-time data analysis and decision-making. The battery life of resource-constrained devices can be extended by offloading computationally expensive operations to energy-efficient cloud or edge servers, optimizing energy usage in IIoT deployments (Schneider 2017). Moreover, it improves resource utilization by enabling the execution of complex computations on shared cloud or edge servers rather than by installing costly hardware on each IIoT device. The type of computational offloading used in IIoT applications is significant since different sorts have different benefits and are better suited for different situations (Jaidka et al. 2020):

- **Binary Offloading:** With binary offloading, the complete computation task is sent from the edge device to the edge node or cloud for processing. This strategy is advantageous when the edge device has little computational power and cannot conduct any computation locally. Binary offloading enables the edge device to utilize the more powerful capabilities at the destination for estimation while concentrating only on data gathering and transmission. This also decreases the amount of local storage needed, which helps to reduce the cost of the edge device while increasing the speed and accuracy of the computation. Moreover, this strategy reduces the amount of power consumption of the edge device, meaning that fewer maintenance activities are needed Bi and Zhang (2018).
- **Partial Offloading:** Sending only a piece of the computing task from the edge device to the cloud or edge node is called partial offloading. Some computations are carried out locally on the edge device, but the more resource-intensive portions are offloaded for processing at the destination. In situations when the edge device can handle some computing but needs assistance for complicated tasks, partial offloading finds a balance between edge intelligence and resource utilization (Kuang et al. 2019).
- **Fog Federation:** In a fog computing environment, fog federation involves distributed processing among several fog nodes or edge devices. Collaborative computing, where edge devices cooperate to share computation chores, is made possible by this kind of offloading. Fog federation is advantageous when a network of edge devices can handle the computational load collectively, ensuring efficient resource utilization and decreasing reliance on a single powerful server (Mukherjee et al. 2020; Srirama n.d.).

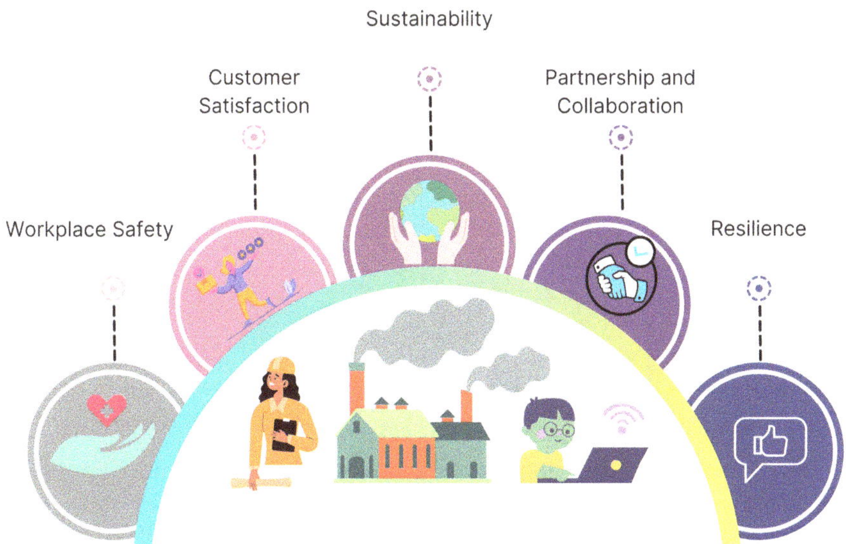

Fig. 4.2 Industrial objectives

4.1.4 Objective of This Chapter

Industrial organizations strive to achieve diverse objectives to ensure their long-term success and overall influence. The primary aim is prioritizing workplace safety by establishing surroundings that promote employee well-being through secure and healthy conditions. At the same time, it is crucial to prioritize customer satisfaction, which requires ongoing endeavours to meet or beyond consumer expectations by improving products/services and providing prompt customer support. Industries are encouraged to reduce their environmental impact and adopt environmentally friendly practices as part of their core goal of sustainability. Another strategic objective is establishing partnerships, which fosters mutually advantageous relationships to improve corporate skills and broaden market penetration. Moreover, resilience is a primary goal that highlights the significance of adaptability and swiftly recovering from disturbances or crises, guaranteeing the organization's uninterrupted functioning. The core objective of industry, as pictorially explained in Fig. 4.2.

4.1.5 Contributions

With the emergence of IoT and the importance of ML techniques in the field of industrial automation, we aim to present a brief discussion of the combined benefits of IIoT and ML in today's society by examining the application of IIoT. The major contributions are as follows:

- We present a holistic evaluation, starting from IoT and transitioning toward IIoT. We discuss the significance of ML techniques and their applicability in IIoT automation.
- We provide a fundamental discussion regarding ML, covering various ML techniques and how to analyze their performance. This part also delves into the set of ML hyperparameters and how to tune them.
- We briefly discuss several application areas of IIoT, exploring how to incorporate the advantages of ML techniques and their capabilities.
- Furthermore, we address the potential challenges faced by organizations in achieving a completely autonomous industrial process. We also discuss related challenges for industrial operations and potential solutions to address them.

The remaining sections are organized as follows: Sect. 4.2 discusses the fundamental concepts of ML. In Sect. 4.3, we briefly discuss several IIoT applications and how we can further extend their functionalities with the help of ML techniques. Potential challenges and future research opportunities of ML-based IIoT networks are presented in Sect. 4.4. Finally, Sect. 4.5 concludes the discussion with a short summary.

4.2 Fundamental Concept of Machine Learning

In the modern world, ML is undoubtedly the best way to tackle industrial challenges that cannot be solved manually. Unlike conventional algorithms, ML algorithms rely heavily on data produced by humans, nature, and other algorithms, rather than following predefined rules (Hazra et al. 2022, 2023). This method has three major components: (1) input data, which represents input features and output labels, (2) a model that deduces patterns from the data, and (3) a learning algorithm that enhances the model's efficiency. The ML process includes data collection, preprocessing, model choice, training, and evaluation (Kollmannsberger et al. 2021).

ML algorithms require data to learn, which can be gathered from various sources, including sensors, databases, and social media. Additionally, they need a training model or algorithm to learn from the data. Models are selected based on the type of problem being solved, such as estimation or downtime estimation (Khattab & Youssry 2020). It is necessary to evaluate ML algorithms to determine their effectiveness and precision. This entails evaluating the model's predictive power against existing data. To maximize their performance, machine learning (ML) algorithms require adjustments to hyperparameters, which involve adjusting the model's parameters (Hussain et al. 2020). The pictorial representation of the benefits of ML in industrial operations is shown in Fig. 4.3.

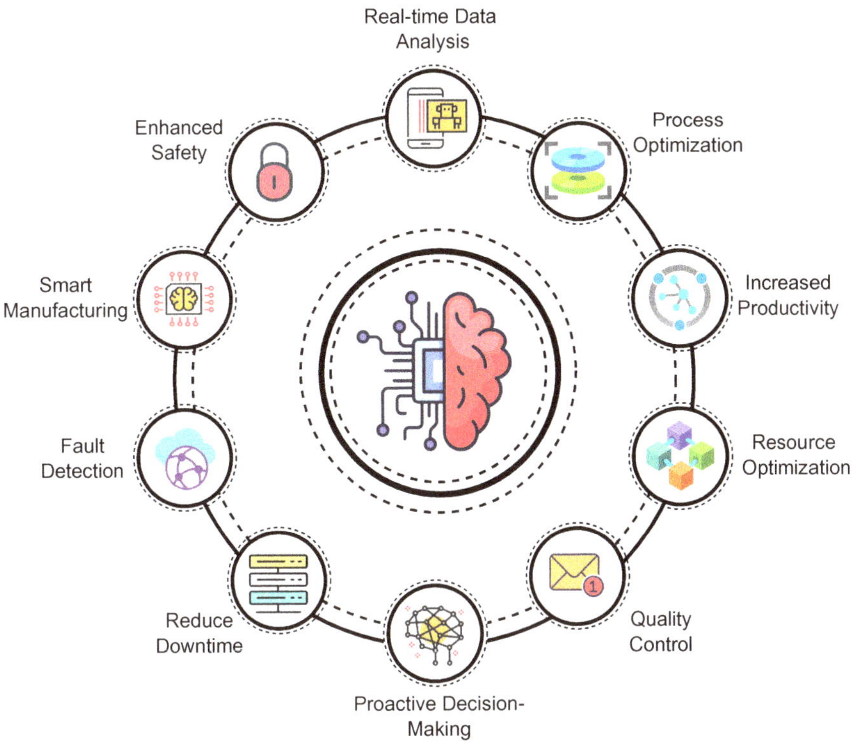

Fig. 4.3 Benefits of ML in industrial operations

4.2.1 Key Machine Learning Technique for IIoT

Various ML algorithms (such as supervised, unsupervised, and reinforcement learning) can be applied to a range of problems. Similarly, ML algorithms for IIoT include supervised learning for fault detection and predictive maintenance, unsupervised learning for anomaly detection and clustering of similar devices, reinforcement learning for autonomous control, and time-series analysis for sensor reading prediction and forecasting (Novo et al. n.d.). On the other hand, transfer learning enables knowledge transfer between related domains, while deep learning excels at complicated pattern recognition tasks like image and speech analysis. Online learning offers real-time adaptation to continuous data streams, while federated learning ensures privacy and security in collaborative IIoT contexts (Lin et al. 2017). The correct ML algorithm can greatly improve productivity and enable data-driven decision-making in industrial settings, depending on the unique IIoT use case, data characteristics, and intended outcomes (Hou et al. 2019). The mentioned ML techniques are briefly presented in the following subsections:

- **Supervised Learning:** Supervised learning involves forecasting a target variable based on labeled data. The machine is trained using information already assigned to the right response (Tran et al. 2022). To analyze the training data (a set of training examples) and generate an accurate result from labeled data, the machine is then given another set of samples (data). Regression and classification are the two categories of algorithms employed in supervised learning. Regression techniques forecast continuous values, whereas classification algorithms predict definite values. In IIoT systems, supervised learning can anticipate equipment breakdowns, spot anomalies, and adjust production settings. It helps businesses and industries run more efficiently and reliably by utilizing real-time data from sensors and other sources (Yang et al. 2017).
- **Unsupervised Learning:** Unlike supervised learning, unsupervised learning does not need labeled data. It offers a strong tool for data exploration, allowing users to find undiscovered patterns and associations without human intervention (Hore & Wakde 2022). To operate on unlabeled data, algorithms are used that are excellent at spotting patterns and abnormalities. Moreover, unlabeled data can be processed instantly and is time- and cost-saving. Unsupervised learning performs better than supervised learning at spotting unexpected data trends or novel occurrences. However, it is more difficult, is pricier, and demands higher expertise levels than supervised learning. It is used in IIoT systems to group data, find anomalies, and recognize patterns (Hassan et al. 2021).
- **Reinforcement Learning:** To maximize a reward signal, reinforcement learning is an ML technique that involves an agent learning to make decisions through trial and error. It has drawn interest in IIoT because it can potentially improve resource allocation, scheduling, and decision-making in IIoT networks (Zhang et al. 2021). The agent is trained to identify the best response for a given situation, which then yields the highest reward. This makes reinforcement learning suitable for IIoT applications where the environment is dynamic and unpredictable. Reinforcement learning can be applied to various IIoT tasks, such as predictive maintenance and anomaly detection. It can also be used to optimize routing and energy consumption in IIoT networks. In addition, RL is helpful in optimizing the performance of robotic systems in contexts connected to the IIoT (Liu et al. 2019).

4.2.2 Experiment Analysis of Machine Learning Methods

It is necessary to conduct an experimental study of currently available ML methods to better understand their capabilities and limitations. Through simulations, testing of actual systems, and comparing alternatives, this can be accomplished in a number of ways. An effective ML system should be able to adapt to varying contexts and recognize patterns from incomplete data (Churcher et al. 2021). Developing an effective ML system relies on continuously evaluating and refining underlying algorithms:

- **Evaluation on Benchmark Datasets:** For ML techniques to be effective, they must be compared to a benchmark activity or scenario. These standards are widely used in academic research and meet strict criteria for definition and performance measurement. By comparing the results of these studies, various ML approaches can be evaluated and contrasted, highlighting their respective advantages and disadvantages. Researchers and industry practitioners can benefit from this insight into the ML field, enabling them to more effectively apply different techniques to their projects. Additionally, it stimulates further research and development in the field by identifying areas for improvement in existing ML algorithms (Fumera & Roli 2005).
- **Testing on Real-World Environment:** ML techniques can also be tested on operational systems such as robots or self-driving cars. Results from these trials can shed light on the effectiveness of ML approaches for scalability, noise management, and outcome prediction. In order to refine future strategies and approaches, this knowledge can be used to inform and refine current ones. Additionally, ML technologies can be used to identify areas for improvement and optimization in existing operational systems. As a result, safety and efficiency can be improved (Short et al. 2018).
- **Comparison among ML Techniques:** By comparing ML methods with traditional teaching methods like supervised learning and reinforcement learning, its effectiveness can be evaluated. A comparison of an ML approach to another method can give an insight into how it compares in terms of speed, scalability, and reliability. Furthermore, ML can uncover more efficient ways of using data and can be used to solve more complex problems than traditional methods. This is why ML is becoming increasingly popular and is seen as a potential game-changer in many industrial applications (Obaid et al. 2018).

Experimental studies of current ML approaches can shed light on their performance and limits, pointing the way to potential research and development directions. These studies can inform researchers about which areas of ML are most promising for further research and development and what approaches are most likely to succeed. Considering these factors can assist in determining which research directions should be pursued. In addition, ML algorithms can be fine-tuned by adjusting a few critical parameters (Handelman et al. 2018). These criteria include:

1. **Learning Rate:** The training rate of a neural network affects the size of the training steps used to update the network's weights. While a high learning rate speeds up convergence, it can also overestimate the ideal solution. On the other hand, reducing the network's learning rate can postpone convergence but improve its likelihood of locating the most appropriate answer. Careful selection of the learning rate is essential for successful ML model training. It is critical to find a balance between speed and accuracy. Too high a learning rate can lead to instability, while too low a learning rate can lead to slow convergence (Ananya et al. 2020).
2. **Discount Factor:** The discount factor determines how much weight is given to future benefits. The agent will prioritize the immediate benefits if the discount

factor is high and the long-term benefits if the discount factor is low. When the discount factor is high, future rewards are valued less than current rewards, whereas when it is low, future rewards are valued more. Thus, the discount factor acts as a tool to control ML agents' behavior. We can observe the use of reinforcement learning in reinforcement learning-based techniques (Pitis 2019).

3. **Batch Size:** The batch size determines how many samples are used in each iteration of gradient descent optimization. As a result of the small batch size, there may be greater variation in the updating rate. Larger batch sizes can reduce the amount of variation in the update rate since the parameters of the model are updated more consistently. However, a larger batch size can also slow down the training process and may not lead to better results. Therefore, finding the optimal batch size for the given task is necessary.
4. **Dropout Rate:** Dropout is a regularization method used in neural networks in which a small percentage of neurons are removed randomly. The dropout rate is one hyperparameter that can be adjusted to alter dropout probability. The dropout rate is the probability of removing a neuron from the network. A lower dropout rate results in a smaller portion of neurons being removed, while a higher rate will result in more neurons being removed. This helps to reduce overfitting and prevents the network from memorizing the data.
5. **Number of Iterations/Epochs:** The learning algorithm is supposed to iterate over the entire dataset a number of times when it is training, usually indicated by the number of iterations or epochs used. The more epochs used, the more accurate the model will be. However, increasing the number of epochs can also increase the computational costs, making the training process slower and more expensive. Therefore, finding the right balance between accuracy and cost is important.
6. **Number of Neurons and Hidden Layers:** The effectiveness of the ML algorithm is directly related to the values of these parameters, which govern the neural network's structure. It is possible to increase performance significantly by adjusting the appropriate hyperparameters. However, it is not always easy to determine the optimal values for these parameters. It is often necessary to experiment to find the parameters that can maximize the accuracy of the model. Additionally, it is important to use appropriate regularization techniques to ensure that the model is not overfitting the data.

4.2.3 State-of-the-Art Research Initiatives

ML is a fast-developing discipline, and researchers have made significant contributions to its development as shown in Table 4.2. It entails using labeled data for training a model, where input data is matched with output labels (Hazra et al. 2021). After that, the model is trained to look for patterns and connections between the input and output labels, allowing it to correctly forecast or categorize data when provided with fresh, unobserved data. Using supervised learning models,

Table 4.2 Analysis of existing works based on performance

Existing works	Summary	T1	T2	T3	T4	T5	Limitation
Boyes et al. (2018)	Service deployment strategy	×	✓	✓	×	×	Data quality and availability
Hazra et al. (2023)	Task offloading strategy	×	✓	×	×	×	Partial offloading
Lin et al. (2017)	Integration of Fog/edge and IoT	×	✓	×	×	✓	Security and privacy
Xue and Zhu (2009)	Improving performance of ML algorithm	×	×	✓	×	×	Overfitting
Hazra et al. (2022a)	Energy optimized offloading	✓	✓	✓	×	×	IoT controller
Javaid et al. (2022)	Fog-level execution	✓	×	×	×	×	Energy consumption
Muttil and Chau (2007)	Selection of input variables	×	✓	✓	×	×	Interoperability
Kumar et al. (2013)	Computational offloading	×	✓	✓	×	×	Resource constrained
Bi and Zhang (2018)	Edge intelligence and offloading	×	×	✓	×	✓	Load balancing
Schneider (2017)	QoE in fog computing	✓	×	×	×	×	Partial offloading
Kuang et al. (2019)	Reliability and security	✓	✓	✓	✓	✓	Resource optimization
Mukherjee et al. (2020)	Offloading in fog networks	✓	✓	✓	✓	✓	Edge intelligence
Hussain et al. (2020)	IoT security and privacy	✓	✓	✓	✓	✓	Computational complexity

(continued)

Table 4.2 (continued)

Boyes et al. (2018)	Service deployment strategy	×	✓	✓	×	×	Data quality and availability
Hou et al. (2019)	Typical IoT architecture	✓	✓	✓	✓	✓	IoT security
Tran et al. (2022)	Anamoly detection in IIoT	✓	✓	✓	✓	✓	Energy consumption
Yang et al. (2017)	Safety of commercial IoT devices	✓	✓	✓	✓	✓	Battery life extension
Hassan et al. (2021)	Multilevel protocol in IIoT network	✓	✓	✓	✓	✓	Resource optimization
Babbar et al. (2022)	Smart healthcare system	✓	✓	✓	✓	✓	Partial offloading
Abuhasel and Khan (2020)	Secure task scheduling	✓	✓	✓	✓	✓	Load balancing
Kozma et al. (2019)	Supply chain management	✓	✓	✓	✓	✓	Real-time visibility
Lu et al. (2023)	Communication, resource allocation	✓	✓	✓	✓	✓	Reliability
Chen and Wan (2019)	Integrate ML into IIoT	✓	✓	✓	✓	×	Interoperability
Amjad et al. (2021)	IoT application protocol	✓	✓	✓	✓	✓	Reliability
Costa et al. (2020)	Improve quality of process in IIoT	✓	✓	✓	✓	✓	Resource optimization
Huang et al. (2020)	Fault detection in IIoT environment	✓	✓	✓	✓	✓	Resource constrained
Huang et al. (2022)	Energy efficient framework for IIoT	✓	✓	✓	✓	✓	Reliability
Sun et al. (2019)	ML meet IIoT	✓	✓	✓	✓	✓	Partial offloading

"✓" incorporates the features into the research
"×" didn't incorporate the features into the research
"T1" signifies supervised learning
"T2" signifies unsupervised learning
"T3" signifies reinforcement learning
"T4" signifies computational efficiency
"T5" signifies interoperability

future predictions based on labeled data can be made without the need for manual categorization labor.

4.2.3.1 Supervised Learning

Recently many efforts in supervised learning have been put forward for industrial environments to utilize the power of data produced by linked devices fully. For instance, (Tran et al. 2022) have discussed the identification of time-series anomalies in an IIoT system. Similarly, (Sun et al. 2019) have proposed an IIoT intelligent computing architecture where edge servers and remote clouds work together. An AI-driven offloading system automatically distributes traffic to edge servers or distant clouds while taking service accuracy into account as a new metric. On the other hand, (Aouedi et al. 2023) have proposed a supervised model. This model extensively used unlabeled data without privacy issues and little labeled data. Additionally, unlike the traditional federated learning model, the supervised learning task with our model utilized the server in addition to the model aggregation task. The proposed model's capacity to recognize network traffic and various threats was assessed. Additionally, it examined how well the model performed under different conditions. The experimental findings using two real datasets showed that using unlabeled data during training can improve the performance of the learned model and reduce communication overhead.

4.2.3.2 Unsupervised Learning

In reference (Huang et al. 2022) have proposed the Energy-efficient And Trustworthy Unsupervised Anomaly Detection Framework (EATU), and it not only boasts low energy consumption but also enhances the reliability and accuracy of anomaly detection in the IIoT. Similarly, (Amruthnath & Gupta 2018) have presented, developed, and implement predictive maintenance methodology using unsupervised learning. Moreover, (Yang et al. 2020) have suggested a cutting-edge compute offloading framework for distributing hierarchical ML tasks for the IIoT. A piecewise convex optimization problem is created to reduce the overall processing time while considering the ML model complexity, data quality, computing power at the device level and MES, and communications bandwidth. This is necessary because the processing time for ML tasks is affected by both communications and computing.

4.2.3.3 Reinforcement Learning

Recently, (Zhang et al. 2021) have explained how to manage and train a huge amount of data generated by IIoT. This article suggests a federated learning method with deep reinforcement learning (DRL) assistance for wireless network situations. The primary method for selecting IIoT equipment nodes is DRL based on Deep Deterministic Policy Gradient (DDPG). Additionally, (Liu et al. 2019) have proposed a revolutionary DRL-based performance optimization framework that increased the blockchain's scalability while maintaining the system's decentralization, security, and latency. This framework was designed for blockchain-enabled IIoT systems. Firstly, establish a numerical measurement for the effectiveness of blockchain technologies in our suggested framework. Following that, the DRL technique was used to choose the block producers, the consensus process, and the block size and interval to maximize the on-chain transactional throughput of the blockchain system.

4.3 Machine Learning in IIoT Applications

IIoT applications can benefit from ML techniques to increase manufacturing quality, security, and sustainability, as shown in Fig. 4.4. To use ML in the IIoT, selecting the appropriate ML technique for the application is critical. A solid grasp of the data flowing from IT systems is vital. Additionally, it is crucial to select the appropriate evaluation metrics for the particular IIoT application and consider the data's properties, such as class imbalance, temporal correlation, and high dimensionality (Sharma et al. 2021).

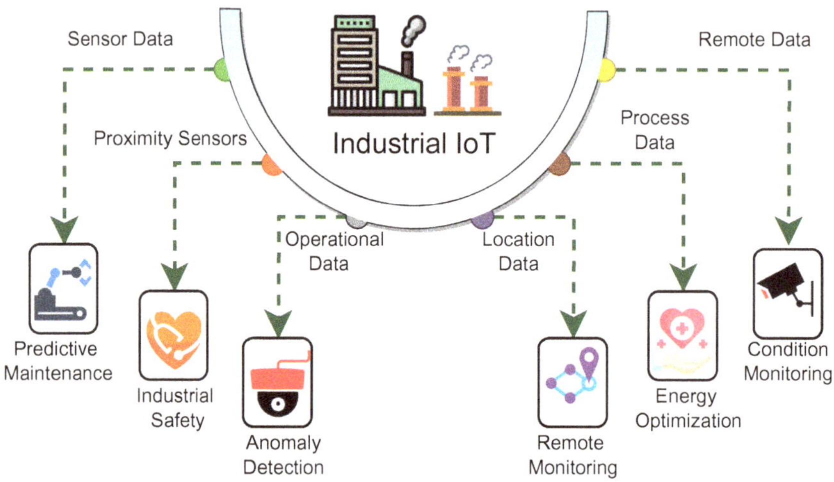

Fig. 4.4 Industrial IoT applications

4.3.1 Predictive Maintenance

To prevent unplanned outages of machines and plants, predictive maintenance aims to predict breakdowns before they may occur. The goal is to avoid breakdowns by their timely prediction and, simultaneously, to maximize service life. The projections are based on data, including present situations and accumulated knowledge. Massive volumes of data are produced in IIoT setups. Normally, sensors have constant data streams. Data continuously generated, often at a high rate, is called a data stream. Real-time information gathering and quick reaction are essential in fully automated industrial environments (Chehri & Jeon 2019). In these IIoT environments, the requirement for real-time communication and the availability of continuously flowing data streams contrast with taking a snapshot of the full dataset and executing calculations with unpredictable response times. To cope with such demands, self-adaptive algorithms constantly learning and refining their models are necessary. Such algorithms should also exhibit real-time behavior and excellent performance.

4.3.2 Smart Healthcare

It improves the detection of diseases, the creation of custom treatment regimens, and remote patient monitoring. It forecasts the need for equipment repair and streamlines healthcare supply networks. Drug discovery is aided by ML, which uses NLP to extract information from medical records. Predicting patient outcomes and lowering readmissions are two benefits of predictive analytics. It enhances cybersecurity and finds fraud to protect patient data (Babbar et al. 2022). Personalized health interventions are made possible through behavior analysis powered by ML. Real-time data analysis improves patient care and operational effectiveness in smart healthcare systems.

4.3.3 Smart Manufacturing

Smart manufacturing, called Industry 4.0, optimizes the entire manufacturing process by combining cutting-edge technologies like IoT, AI, data analytics, and cyber-physical systems (Hazra et al. 2023). It uses IoT devices and sensors to collect real-time data, enabling data-driven decision-making and process monitoring. The data is analyzed using advanced analytics and AI, which offers insightful information for process improvement. Virtual representations are created using digital twin technologies for simulation and optimization. The supply chain is optimized via smart manufacturing, allowing for quick customization and flexible production. Collaboration between people and machines is emphasized for effective production.

Real-time monitoring and control increase energy efficiency (Abuhasel and Khan 2020). Real-time defect detection helps with quality control. Strong cybersecurity safeguards protect sensitive manufacturing data. Smart manufacturing increases efficiency, lowers waste, and improves sustainability, increasing the competitiveness of businesses in the global market.

4.3.4 Supply Chain Optimization

Supply chain optimization entails streamlining several supply chain elements to increase productivity and save costs. It all starts with precise demand forecasting to match inventory levels and production schedules with customer demand. Effective inventory control lowers carrying costs and stockouts by ensuring the appropriate amount of stock is available when needed. Supplier management goals are to find trustworthy suppliers, negotiate good terms, and guarantee on-time deliveries. Process simplification in the manufacturing industry improves output and reduces lead times. The best possible transportation options must be chosen to maximize logistics and transportation efficiency (Kozma et al. 2019). Warehouse optimization reduces handling expenses and order processing time while enhancing storage and order fulfilment procedures. With supply chain partners, cooperation and information sharing improve coordination and responsiveness. Real-time analytics and data allow for prompt decision-making and proactive responses to shifts in supply or demand.

4.3.5 Ultralow Latency Data Transmission

IIoT applications can support time-sensitive procedures like high-frequency trading and 5G-enabled use cases with ultralow latency connectivity. It improves product quality, decreases downtime, and optimizes industrial operations using quick feedback loops and real-time analysis (Lu et al. 2023). With the help of this technology, remote monitoring and control, urgent situations can be handled quickly. It also helps with asset tracking, supply chain optimization, and energy management. Greater network density, lower latency, and quicker data transfer speeds are promised by 5G networks, which can help IIoT applications like ML and AI.

4.4 Challenges and Future Research Opportunities

ML confronts unique challenges when used in the context of IIoT applications. Numerous IIoT devices, particularly those installed at the edge of networks, have

Table 4.3 Challenges and role of ML in IIoT

IIoT challenges	How ML solves IIoT challenges
Data overload	Massive volumes of data may be processed in real-time by ML algorithms, allowing for effective data analysis and useful insights
Interoperability	Data from many sources and protocols can be combined in ML models, enabling communication and interoperability between various IIoT devices
Real-time decision-making	ML makes real-time data processing and analysis possible, enabling crucial operations and systems to make decisions instantly
Energy efficiency	In IIoT devices and systems, ML algorithms can optimize energy usage, resulting in more energy-efficient operations and lower operational expenses
Human-machine collaboration	With the help of ML technologies, humans and machines may work together to execute complicated tasks, increasing worker productivity and safety
Data quality assurance	ML help with data validation and cleansing, ensuring high-quality data for precise and trustworthy insights

constrained memory, processing, and energy capacities. ML models for resource-intensive cloud environments won't work perfectly on these limited devices (Chen & Wan 2019). To enable effective and practical deployment of ML in IIoT applications, it is essential to develop ML algorithms that are compact and optimized for low power consumption while preserving accuracy. For effective ML-based solutions in IIoT, balancing the trade-off between model complexity and performance in resource-constrained contexts is a critical problem. Techniques like model compression, quantization, and edge computing paradigms are essential to overcome these hardware constraints and guarantee ML's successful integration in IIoT devices and edge environments. Table 4.3 explains the potential challenges in IIoT and how ML solve these challenges.

4.4.1 Data Collection and Quality

Traditional data storage and processing techniques might not handle the sheer amount and variety of data produced by several sources, including IoT devices, social media platforms, and enterprise systems. The heterogeneity of data formats and structures necessitates extensive preprocessing and integration to assure compatibility and utility. Data biases can produce biased and unfair ML models, influencing decision-making processes and reinforcing preconceptions already held. Addressing data bias entails carefully attempting to spot and lessen these biases, encouraging justice and moral use of ML models. Additionally, data imbalance is frequently seen during classification tasks when some classes have noticeably fewer samples than others. The model's overall performance may suffer due to biased models that perform well for the majority class but poorly for the minority classes. Noisy data, which could contain errors or outliers, is another data quality difficulty, necessitating data cleaning and outlier detection approaches. In dynamic situations, data drift, a phenomenon where data distribution varies over time, is widespread, demanding ongoing monitoring and ML model adaption. The future of ML offers promising prospects for development and innovation (Hazra et al. 2022a). New opportunities will become possible as ML research and technology advance, including explainable AI, where efforts are focused on creating ML models that offer clear and understandable justifications for their choices. This will increase trust, confidence, and understanding of AI-driven systems, particularly in crucial industries like finance and healthcare.

4.4.2 Interoperability

Interoperability is essential for fostering cooperation among various businesses and stakeholders. Interoperability promotes partnerships and encourages sharing of data and expertise, resulting in creative solutions and breakthroughs in multiple disciplines. Various industrial equipment and devices that use different communication protocols and data formats are frequently used in IIoT contexts. Integrating these numerous systems can be challenging and time-consuming, so they operate as a single unit. Timing and synchronization of data across many systems and devices are essential in IIoT applications. A difficult task is ensuring precise and synchronized data communication. The development of 5G and beyond will offer faster and more reliable connectivity, enabling seamless communication between IIoT devices and permitting real-time data exchange, among another important future potentials for IIoT. Intelligent automation will result from integrating AI and ML in IIoT, enabling systems to make decisions independently, streamline operations, and boost productivity. Edge computing will spread more widely, allowing data processing and analysis to occur nearer to the data source, lowering latency and bandwidth needs and increasing system efficiency (Amjad et al. 2021).

4.4.3 Real-Time Processing

In modern computing and data analytics, real-time processing refers to the immediate or almost immediate management and analysis of data as it is generated or received. Real-time processing allows for prompt answers and decision-making in time-sensitive applications because data is analyzed and acted upon instantly, without any noticeable delay (Costa et al. 2020). Real-time processing is essential for the IIoT because it makes proactive decision-making possible and optimizes industrial operations. IIoT devices and sensors produce massive volumes of data in real time, giving important information on ambient conditions, equipment health, and industrial processes (Huang et al. 2020). Industries can get practical insights and react quickly to shifting situations by analyzing this data in real time, increasing operational effectiveness and decreasing downtime. Ultrafast and dependable data transmission will be made possible by the widespread use of 5G networks, enabling real-time processing in various applications, including augmented reality, virtual reality, and autonomous systems. The combination of real-time processing, AI, and ML will produce enhanced AI-driven real-time insights, enabling systems to make independent judgments and predictions immediately (Lin et al. 2023).

4.5 Conclusion

In this work, we presented a survey on ML techniques in the IIoT environment. We started this work by defining IoT and IIoT from the perspective of ML. We later discussed the evolution of IoT to IIoT, the significance of ML in IIoT, and computational offloading in ML for IIoT applications such as binary offloading, partial offloading, etc., followed by the discussion on key-enabling techniques of ML. Moreover, we discuss the experimental analysis of ML methods and state of the art of research initiatives. Also, we discuss several potential applications and obstacles, such as data quality and availability, interoperability, real-time processing, and future opportunities to succeed in the IIoT environment.

Acknowledgments The author would like to thank NSUT and IIIT Sri City for providing the necessary support to conduct this research work.

References

Abuhasel, Khaled Ali, and Mohammad Ayoub Khan. 2020. A secure Industrial Internet of Things (IIoT) framework for resource management in smart manufacturing. *IEEE Access* 8: 117354–117364. https://doi.org/10.1109/ACCESS.2020.3004711.

Akherfi, Khadija, et al. 2018. Mobile cloud computing for computation offloading: Issues and challenges. *Applied Computing and Informatics* 14 (1): 1–16.

Amjad, Anam, et al. 2021. A systematic review on the data interoperability of application layer protocols in industrial IoT. *IEEE Access* 9: 96528–96545. https://doi.org/10.1109/ACCESS.2021.3094763.

Amruthnath, Nagdev, and Tarun Gupta. 2018. Fault class prediction in unsupervised learning using model-based clustering approach. In *2018 International Conference on Information and Computer Technologies (ICICT)*, 5–12. https://doi.org/10.1109/INFOCT.2018.8356831.

Ananya, A., et al. 2020. SysDroid: A dynamic ML-based android malware analyzer using system call traces. *Cluster Computing* 23 (4): 2789–2808.

Aouedi, Ons, et al. 2023. Federated semisupervised learning for attack detection in Industrial Internet of Things. *IEEE Transactions on Industrial Informatics* 19 (1): 286–295. https://doi.org/10.1109/TII.2022.3156642.

Babbar, Himanshi, et al. 2022. Intelligent edge load migration in SDN-IIoT for smart healthcare. *IEEE Transactions on Industrial Informatics* 18 (11): 8058–8064. https://doi.org/10.1109/TII.2022.3172489.

Bi, Suzhi, and Ying Jun Zhang. 2018. Computation rate maximization for wireless powered mobile-edge computing with binary computation offloading. *IEEE Transactions on Wireless Communications* 17 (6): 4177–4190. https://doi.org/10.1109/TWC.2018.2821664.

Boyes, Hugh, et al. 2018. The industrial internet of things (IIoT): An analysis framework. *Computers in Industry* 101: 1–12.

Carbonell, Jaime G., et al. 1983. An overview of machine learning. In *Machine Learning*, 3–23.

Chehri, Abdellah, and Gwanggil Jeon. 2019. The industrial internet of things: examining how the IIoT will improve the predictive maintenance. In *Innovation in Medicine and Healthcare Systems, and Multimedia: Proceedings of KES-InMed-19 and KES-IIMSS-19 Conferences*, 517–527. Berlin: Springer.

Chen, Baotong, and Jiafu Wan. 2019. Emerging trends of ML-based intelligent services for Industrial Internet of Things (IIoT). In *2019 Computing, Communications and IoT Applications (ComComAp)*, 135–139. https://doi.org/10.1109/ComComAp46287.2019.9018815.

Churcher, Andrew, et al. 2021. An experimental analysis of attack classification using machine learning in IoT networks. *Sensors* 21 (2): 446.

Costa, Felipe S., et al. 2020. Fasten IIoT: An open real-time platform for vertical, horizontal and end-to-end integration. *Sensors* 20 (19): 5499.

Fumera, G., and F. Roli. 2005. A theoretical and experimental analysis of linear combiners for multiple classifier systems. *IEEE Transactions on Pattern Analysis and Machine Intelligence* 27 (6): 942–956. https://doi.org/10.1109/TPAMI.2005.109.

Handelman, Guy S., et al. 2018. Peering into the black box of artificial intelligence: Evaluation metrics of machine learning methods. *AJR. American Journal of Roentgenology* 212 (1): 38–43.

Hassan, Mohammad Mehedi, et al. 2021. An adaptive trust boundary protection for IIoT networks using deep-learning feature-extraction-based semisupervised model. *IEEE Transactions on Industrial Informatics* 17 (4): 2860–2870. https://doi.org/10.1109/TII.2020.3015026.

Hazra, Abhishek, Ahmed Alkhayyat, et al. 2022. Blockchain-aided integrated edge framework of cybersecurity for Internet of Things. *IEEE Consumer Electronics Magazine*, 1–1. https://doi.org/10.1109/MCE.2022.3141068.

Hazra, Abhishek, Mainak Adhikari, and Tarachand Amgoth. 2022. Dynamic service deployment strategy using reinforcement learning in edge networks. In *2022 International Conference on Computing, Communication, Security and Intelligent Systems (IC3SIS)*, 1–6. https://doi.org/10.1109/IC3SIS54991.2022.9885498.

Hazra, Abhishek, Mainak Adhikari, Tarachand Amgoth, and Satish Narayana Sri-rama. 2021. A comprehensive survey on interoperability for IIoT: Taxonomy, standards, and future directions. *ACM Computing Surveys* 55 (1). ISSN: 0360-0300. https://doi.org/10.1145/3485130.

Hazra, Abhishek, Mainak Adhikari, Tarachand Amgoth, and Satish Narayana Sri-rama. 2022a. Fog computing for energy-efficient data offloading of IoT applications in industrial sensor networks. *IEEE Sensors Journal* 22 (9): 8663–8671. https://doi.org/10.1109/JSEN.2022.3157863.

Hazra, Abhishek, Mainak Adhikari, Tarachand Amgoth, and Satish Narayana Sri-rama. 2022b. Intelligent service deployment policy for next-generation industrial edge networks. *IEEE Transactions on Network Science and Engineering* 9 (5): 3057–3066. https://doi.org/10.1109/TNSE.2021.3122178.

Hazra, Abhishek, Mainak Adhikari, Tarachand Amgoth, and Satish Narayana Sri-rama. 2023. Collaborative AI-enabled intelligent partial service provisioning in green industrial fog networks. *IEEE Internet of Things Journal* 10 (4): 2913–2921. https://doi.org/10.1109/JIOT.2021.3110910.

Hazra, Abhishek, Praveen Kumar Donta, et al. 2023. Cooperative transmission scheduling and computation offloading with collaboration of fog and cloud for industrial IoT applications. *IEEE Internet of Things Journal* 10 (5): 3944–3953. https://doi.org/10.1109/JIOT.2022.3150070.

Hazra, Abhishek, and Tarachand Amgoth. 2022. CeCO: Cost-efficient computation offloading of IoT applications in green industrial fog networks. *IEEE Transactions on Industrial Informatics* 18 (9): 6255–6263. https://doi.org/10.1109/TII.2021.3130255.

Hore, Umesh W., and DG Wakde. 2022. An effective approach of IIoT for anomaly detection using unsupervised machine learning approach. *Journal of IoT in Social, Mobile, Analytics, and Cloud* 4: 184–197.

Hou, Jianwei, et al. 2019. A survey on internet of things security from data perspectives. *Computer Networks* 148: 295–306.

Huang, Huakun, et al. 2020. Real-time fault detection for IIoT facilities using GBRBM-Based DNN. *IEEE Internet of Things Journal* 7 (7): 5713–5722. https://doi.org/10.1109/JIOT.2019.2948396.

Huang, Zijie, et al. 2022. An energy-efficient and trustworthy unsupervised anomaly detection framework (EATU) for IIoT. *ACM Transactions on Sensor Networks* 18 (4): 1–18.

Hussain, Fatima, et al. 2020. Machine learning in IoT security: Current solutions and future challenges. *IEEE Communications Surveys & Tutorials* 22 (3): 1686–1721. https://doi.org/10.1109/COMST.2020.2986444.

Jaidka, Himanshu, et al. 2020. Evolution of IoT to IIoT: Applications & challenges. In *Proceedings of the International Conference on Innovative Computing & Communications (ICICC).*

Javaid, Mohd, et al. 2022. Significance of machine learning in healthcare: Features, pillars and applications. *International Journal of Intelligent Networks* 3: 58–73.

Khattab, Ahmed, and Nouran Youssry. 2020. Machine learning for IoT systems. In *Internet of Things (IoT) Concepts and Applications*, 105–127.

Kollmannsberger, Stefan, et al. 2021. Fundamental concepts of machine learning. In *Deep Learning in Computational Mechanics: An Introductory Course*, 5–18.

Kozma, Dániel, et al. 2019. Supply chain management and logistics 4.0 - A study on arrowhead framework integration. In *2019 8th International Conference on Industrial Technology and Management (ICITM)*, 12–16. https://doi.org/10.1109/ICITM.2019.8710670.

Kuang, Zhufang, et al. 2019. Partial offloading scheduling and power allocation for mobile edge computing systems. *IEEE Internet of Things Journal* 6 (4): 6774–6785. https://doi.org/10.1109/JIOT.2019.2911455.

Kumar, Karthik, et al. 2013. A survey of computation offloading for mobile systems. *Mobile Networks and Applications* 18: 129–140.

Lin, Jie, et al. 2017. A survey on Internet of Things: Architecture, enabling technologies, security and privacy, and applications. *IEEE Internet of Things Journal* 4 (5): 1125–1142. https://doi.org/10.1109/JIOT.2017.2683200.

Lin, Yijing, et al. 2023. A novel architecture combining oracle with decentralized learning for IIoT. *IEEE Internet of Things Journal* 10 (5): 3774–3785. https://doi.org/10.1109/JIOT.2022.3150789.

Liu, Mengting, et al. 2019. Performance optimization for blockchain-enabled Industrial Internet of Things (IIoT) systems: A deep reinforcement learning approach. *IEEE Transactions on Industrial Informatics* 15 (6): 3559–3570. https://doi.org/10.1109/TII.2019.2897805.

Lu, Yinzhi, et al. 2023. An intelligent deterministic scheduling method for ultralow latency communication in edge enabled Industrial Internet of Things. *IEEE Transactions on Industrial Informatics* 19 (2): 1756–1767. https://doi.org/10.1109/TII.2022.3186891.

Mukherjee, Mithun, et al. 2020. Revenue maximization in delay-aware computation offloading among service providers with fog federation. *IEEE Communications Letters* 24 (8): 1799–1803. https://doi.org/10.1109/LCOMM.2020.2992781.

Muttil, Nitin, and Kwok-Wing Chau. 2007. Machine-learning paradigms for selecting ecologically significant input variables. *Engineering Applications of Artificial Intelligence* 20 (6): 735–744.

Novo, Oscar, et al. n.d. Capillary networks - bridging the cellular and IoT worlds, year=2015. In *2015 IEEE 2nd World Forum on Internet of Things (WF-IoT)*, 571–578. https://doi.org/10.1109/WF-IoT.2015.7389117.

Obaid, O. Ibrahim, et al. 2018. Evaluating the performance of machine learning techniques in the classification of Wisconsin Breast Cancer. *International Journal of Engineering & Technology* 7 (4.36): 160–166.

Pitis, Silviu. 2019. Rethinking the discount factor in reinforcement learning: A decision theoretic approach. In *Proceedings of the AAAI Conference on Artificial Intelligence*. Vol. 33. 01, 7949–7956.

Schneider, Stan. 2017. The industrial internet of things (IIoT) applications and taxonomy. In *Internet of Things and Data Analytics Handbook*, 41–81.

Sharma, Parjanay, et al. 2021. Role of machine learning and deep learning in securing 5G-driven industrial IoT applications. *Ad Hoc Networks* 123: 102685.

Short, Elaine Schaertl, et al. 2018. Detecting contingency for HRI in open-world environments. In *Proceedings of the 2018 ACM/IEEE International Conference on Human-Robot Interaction*, 425–433.

Srirama, Satish Narayana. n.d. A decade of research in fog computing: Relevance, challenges, and future directions. *Software: Practice and Experience* n/a.n/a. https://doi.org/10.1002/spe.3243. eprint: https://onlinelibrary.wiley.com/doi/pdf/10.1002/spe.3243. https://onlinelibrary.wiley.com/doi/abs/10.1002/spe.3243.

Sun, Wen, et al. 2019. AI-enhanced offloading in edge computing: When machine learning meets industrial IoT. *IEEE Network* 33 (5): 68–74. https://doi.org/10.1109/MNET.001.1800510.

Tran, Duc Hoang, et al. 2022. Self-supervised learning for time-series anomaly detection in Industrial Internet of Things. *Electronics* 11 (14): 2146.

Xue, Ming, and Changjun Zhu. 2009. A study and application on machine learning of artificial intellligence. In *2009 International Joint Conference on Artificial Intelligence*, 272–274. https://doi.org/10.1109/JCAI.2009.55.

Yang, Bo, et al. 2020. Mobile-edge-computing-based hierarchical machine learning tasks distribution for IIoT. *IEEE Internet of Things Journal* 7 (3): 2169–2180. https://doi.org/10.1109/JIOT.2019.2959035.

Yang, Yuchen, et al. 2017. A survey on security and privacy issues in Internet-of-Things. *IEEE Internet of Things Journal* 4 (5): 1250–1258. https://doi.org/10.1109/JIOT.2017.2694844.

Zhang, Peiying, et al. 2021. Deep reinforcement learning assisted federated learning algorithm for data management of IIoT. *IEEE Transactions on Industrial Informatics* 17 (12): 8475–8484. https://doi.org/10.1109/TII.2021.3064351.

Chapter 5
Exploring IoT Communication Technologies and Data-Driven Solutions

Poonam Maurya, Abhishek Hazra, and Lalit Kumar Awasthi

5.1 Introduction

The Internet of Things (IoT) is an upscale and transformative technology that connects devices to share information with one another using a dedicated standard communication protocol. By 2030, more than 45 billion IoT devices are expected to be connected worldwide, marking a significant advancement in wireless technology and its widespread application across various industries (IoT in 2023 and beyond 2023). The choice of communication protocol plays a crucial role in the successful deployment and performance of IoT networks. Selecting appropriate communication protocols that meet application requirements is challenging in seamless IoT functionality. The attributes of a communication protocol play a vital role in determining the essential aspects of a radio network, such as the communication range, data rate, power consumption, latency, cost, scalability, and quality of services. However, achieving all of these network characteristics with equal emphasis using single radio technology is impractical. Therefore, understanding various communication protocols and their strengths and limitations is required for effectively designing and implementing IoT systems that meet specific application requirements (Srirama 2023).

P. Maurya
Aalborg University, Aalborg, Denmark
e-mail: poonamm@es.aau.dk

A. Hazra (✉)
Indian Institute of Information Technology, Sri City, India

L. K. Awasthi
National Institute of Technology Uttarakhand, Srinagar, India
e-mail: lalit@nituk.ac.in

P. K. Donta et al. (eds.), *Learning Techniques for the Internet of Things*,
https://doi.org/10.1007/978-3-031-50514-0_5

The multitude of IoT protocols has its own advantages and limitations. Achieving the right balance among existing IoT protocols is challenging because of trade-offs between various characteristics. Data-driven technologies such as machine learning (ML) and deep learning (DL) are used to overcome these challenges. By integrating data-driven technologies, standard IoT protocols can optimize dynamic resource allocation, enhance security, and predict potential issues. Ultimately, these technologies improve protocol efficiency, responsiveness, and integration with IoT devices, ensuring optimal performance in dynamic IoT ecosystems.

5.1.1 Evolution of Communication Protocol

The evolution of IoT communication protocols, as shown in Fig. 5.1, provides an overview of the chronological development and advancements in communication protocols that have played a pivotal role in shaping the IoT landscape. It is important to examine the transition from conventional protocols to contemporary or emerging technologies, emphasizing significant milestones and how every protocol impacts IoT connectivity. By analyzing the attributes and advantages of different communication protocols, this chapter offers valuable insights into the advancement of IoT. It also offers potential implications for the next generation of interconnected systems.

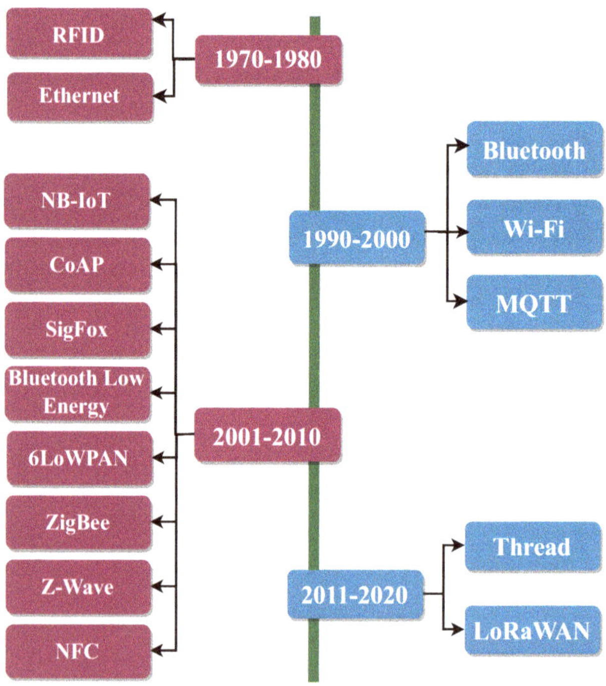

Fig. 5.1 Timeline of the communication protocols

5.1.2 Standard IoT Architecture

The IoT reference architecture is a guideline for building standard IoT infrastructure. The reference architecture facilitates sophisticated IoT applications by providing a scalable and interoperable framework. It facilitates trouble-free information transfer between IoT devices (e.g., sensors and actuators) and cloud servers, faster data processing, and simple incorporation into a wide range of software and services. IoT solutions can have better functionality, security, and dependability if built within a standardized framework. In general, any standard IoT architecture follows a three-layer architecture: the IoT device layer, gateway layer, and cloud server layer. We have also illustrated a standard reference architecture in Fig. 5.2. The smooth integration and communication of IoT devices and data are made possible by the distinct roles played by each layer, discussed in the following paragraphs.

- **IoT Device Layer:** This layer includes physical IoT devices such as sensors and actuators. The devices can be used for a variety of purposes, such as collecting and processing data, storing small amounts of data, and analyzing data as needed. Temperature, humidity, motion, and even the sensor location are just some of the data these devices capture and record. In particular, IoT devices use short- and long-range communication protocols based on type, application, environment, and computation capabilities. Devices with limited processing power and memory establish communications with the gateway layer to transmit their collected data for further processing and analysis.
- **Gateway Layer:** A gateway layer is a layer between the IoT device layer and the cloud service layer. In addition to gathering information, gateway devices can preprocess, forward, translate, and analyze data. It is crucial that data is filtered and compressed at the gateways before being sent to the cloud server so that latency can be minimized and bandwidth can be maximized. In order to make this happen, standard communication protocols such as Bluetooth, radio frequency identification (RFID), WiFi, and Zigbee are widely used in a wide variety of

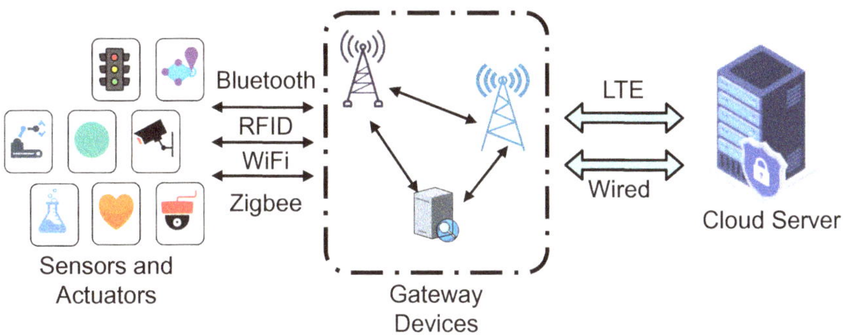

Fig. 5.2 IoT reference architecture

applications. Gateways can also perform edge computing functions to reduce the amount of information sent to the cloud.

- **Cloud Server Layer:** A cloud server layer stores, processes, and analyses IoT data. This level consists of cloud-based platforms and data centers that process and store large amounts of data. After collecting the data, sophisticated algorithms for analytics, ML, and artificial intelligence (AI) techniques are used to draw conclusions from the data. Standard communication technologies such as 4G, 5G, and 6G are widely used to transfer data from IoT gateway devices to cloud data centers. APIs provide remote access to this information to make data-driven decisions and automate processes so users and programs can access cloud services remotely.

5.1.3 Data-Driven Technologies for IoT

Data-driven technologies like AI, ML, and DL significantly influence the IoT ecosystem. Through these innovations, IoT platforms can mine the mountains of information produced by interconnected devices for actionable intelligence that can then be used to automate and improve previously manual processes. They revolutionize the operation and communication of IoT systems by encouraging creativity, enhancing efficiency, and driving intelligent automation.

1. **Artificial Intelligence:** AI is the underlying technology that underpins data-driven IoT applications. AI algorithms enable IoT devices and systems to mimic humanlike cognitive functions, such as problem-solving, pattern recognition, and decision-making. Integrating AI into IoT allows devices to process complex data in real time, adapt to changing conditions, and deliver personalized and context-aware services. Similarly, AI can improve IoT communication protocols' intelligence, flexibility, and efficiency, allowing developers to create more advanced and trustworthy IoT applications.
2. **Machine Learning:** ML is another vital data-driven technology for IoT. It empowers IoT systems to learn from historical data and make predictions or recommendations based on existing data. ML algorithms optimize various IoT tasks, including predictive maintenance, anomaly detection, and resource allocation. ML plays a vital role in improving IoT communication protocols by enhancing their reliability, adaptability, and efficiency in the context of IoT and communication protocols. For example, ML helps with predictive analytics, quality of service (QoS) improvement, dynamic resource allocation, adaptive routing, traffic optimization, security enhancement, fault detection, and self-healing. ML models continuously improve their performance as they gather more data, making them invaluable in dynamic and evolving IoT environments.
3. **Deep Learning:** The DL subset of AI has revolutionized IoT applications by allowing machines to learn directly from raw data without explicit programming. DL models, such as neural networks, excel at image and speech recognition

tasks. In IoT, DL enables sophisticated analysis of multimedia data from cameras and audio sensors. This enables applications like facial recognition, anomaly detection, and natural language processing (Hazra et al. 2021b). In addition, DL enhances IoT communication protocols by addressing complex challenges and improving overall system performance. DL can be used for adaptive beamforming, network optimization, anomaly detection, channel estimation, interference mitigation, etc.

5.1.4 Features of IoT Communication Protocols

IoT communication protocols are equipped with a variety of features (as shown in Fig. 5.3) that facilitate the seamless and dependable operation of interconnected devices and systems. These protocols are intended to address various IoT application scenarios' distinctive challenges and constraints. IoT communication protocols have the following features.

5.1.4.1 Low-Power Consumption

Low-power consumption is one of the crucial features of IoT communication protocols, especially for battery-powered devices (Mao et al. 2021). It extends battery life, ensuring effective and efficient operation over extended periods without frequent replacements or recharging. As a result, IoT applications can be more sustainable and widely adopted in many domains and applications due to this feature, thereby reducing operational costs and enhancing device efficiency.

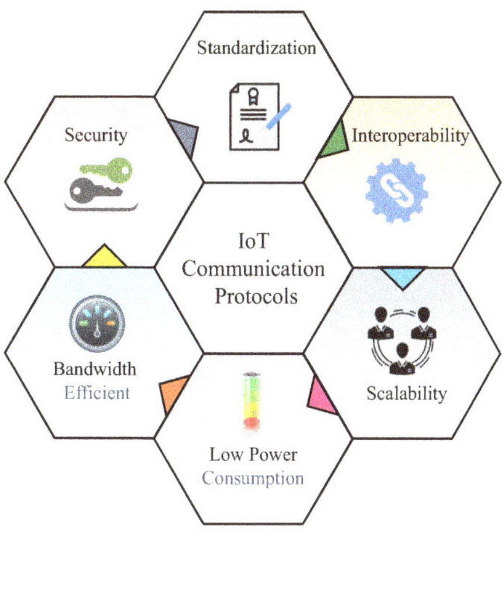

Fig. 5.3 Features of IoT communication protocols

5.1.4.2 Scalability

IoT communication protocols require the capacity to scale up a network. Due to urbanization, the number of connected devices in IoT deployments increases continuously, affecting many communication factors. To maintain the same performance level, the protocols must efficiently handle the increasing data traffic and changes in the channel propagation condition (Maurya et al. 2022b). Scalability allows IoT networks to seamlessly accommodate many devices without compromising performance, ensuring smooth data exchange and communication.

5.1.4.3 Security

Effective IoT communication protocols must possess strong security mechanisms that protect data, guarantee privacy, and maintain IoT devices and network integrity. This entails the incorporation of encryption, authentication, and access control mechanisms to avoid unauthorized access and prevent potential data breaches (Ali et al. 2015). The implementation of such robust security measures enhances communication trustworthiness and reliability.

5.1.4.4 Interoperability and Standardization

The interoperability of IoT communication protocols is a critical component that enables seamless connections between a wide variety of devices produced by a number of different companies (Hazra et al. 2021a). By following standardized communication rules, IoT devices can effectively interact and communicate with each other irrespective of their manufacturer (Al-Qaseemi et al. 2016). This versatility provides compatibility with current technologies and promotes cross-domain integration, which cultivates a coherent and dynamic IoT ecosystem across multiple applications and industries.

Data-driven technologies (e.g., AI, ML, and DL) play a pivotal role in enhancing IoT communication protocols by introducing intelligent and adaptive features that are essential for the successful functionality of IoT applications. ML enables protocols to optimize bandwidth usage, ensure efficient data transmission, and prevent network congestion (Najm et al. 2019). Moreover, the dynamic allocation of resources based on real-time data requirements enhances QoS, guaranteeing an uninterrupted and smooth user experience (Chen et al. 2023). Furthermore, ML minimizes power consumption, making IoT devices more energy-efficient and prolonging battery life. In addition, adaptive routing and load balancing through ML techniques ensure optimal network utilization, which minimizes data transmission delays (Natarajan et al. 2022).

5.1.5 *Contributions*

Considering all the abovementioned challenges, in this work, we summarize the importance of data-driven technologies for IoT communications, with a particular focus on several long-term and short-term communication protocols for a wide range of IoT applications. The major contributions of this chapter are summarized as follows:

- Discuss the standard architecture to understand the foundation of IoT.
- Unleashing the data-driven technology opportunities to overcome IoT communication protocol challenges and limitations.
- Brief overview, classification, and features of communication protocols have been explored.
- Highlight the emerging use case of IoT along with their potential challenges and recent innovative-related research works.

The rest of the chapters are organized as follows. Section 5.2 briefly discusses several IoT communication protocols, including short-range and long-range options. The emerging use cases for IoT are presented in Sect. 5.3. Next, we discuss several potential challenges and possible approaches to solve them. Finally, Sect. 5.5 concludes our work by briefly summarizing our literature.

5.2 Classification of Communication Protocols

There are numerous ways to classify IoT communication protocols, such as network type, range, protocol type, power consumption, and application-specific profiles. In this chapter, we discuss range-based IoT protocol classification as shown in Fig. 5.4. This classification helps you to choose a suitable protocol to meet IoT design and performance requirements.

5.2.1 *Overview of Short-Range IoT Communication Technologies*

Short-range IoT communication technologies encompass diverse wireless protocols for close-proximity data exchanges between IoT devices. They are suitable for applications where devices are in close proximity and require low-power consumption, moderate data rates, and reliable connectivity.

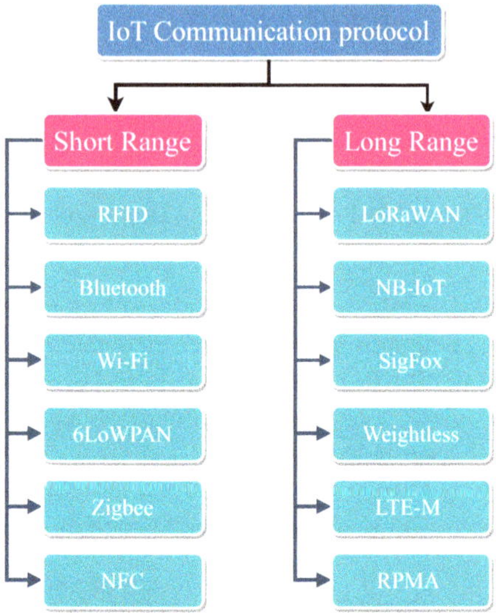

Fig. 5.4 Classification of IoT communication technologies based on communication range

5.2.1.1 Bluetooth

Bluetooth is a wireless communication technology designed for short-distance data transmission between devices, covering distances up to 10 m. Bluetooth establishes connections using radio waves while operating in the 2.4 GHz frequency band. It has advantages such as low transmit power (limited to 2.5 A mW), cost-effectiveness, and simplicity. To minimize interference, Bluetooth uses a frequency-hopping spread spectrum. It supports point-to-point and point-to-multipoint connections, allowing devices to communicate in various modes (A survey on bluetooth multi-hop networks 2019). Data transmission rates range from 1 Mbps to 3 Mbps, depending on the Bluetooth version and the specific use case. Over time, Bluetooth has undergone advancements, such as enhanced data transfer rates, extended coverage range, improved energy efficiency, and support for novel applications. Bluetooth low energy (BLE) is a notable addition that offers a low-power communication option suitable for energy-sensitive applications such as wearables and sensors (Barua et al. 2022; Praveen Kumar et al. 2023).

5.2.1.2 Wi-Fi

Wi-Fi (Wireless Fidelity) is a wireless technology built on the IEEE 802.11 family of standards that describes methods and specifications for wireless communications. The IEEE standard for Wi-Fi, which includes 802.11a, 802.11b, 802.11g, 802.11n, 802.11ac, 802.11ax (Wi-Fi 6), and the soon-to-be-released 802.11be (Wi-Fi 7), specifies data rates, channel bandwidth, modulation methods, and security mechanisms (Omar et al. 2016). Wi-Fi networks use standards like CSMA/CA to ensure data is sent fairly and without collisions. Wi-Fi communication depends on strong security protocols, such as Wired Equivalent Privacy (WEP), Wi-Fi Protected Access (WPA), and its improved version WPA2 (Ramezanpour et al. 2023). These protocols use encryption methods to secure communication and prevent data from being accessed by people who should not access it. There are two ways to set up a Wi-Fi network: infrastructure mode and ad hoc mode. In infrastructure mode, devices join a single access point to talk to each other and get on the Internet. In ad hoc mode, devices can talk directly to each other without navigating through an access point.

5.2.1.3 Zigbee

Zigbee is a communication protocol for LPWANs (low-power wide-area networks) based on the IEEE 802.15.4 standard, which defines the LR-WPAN (low rate wireless personal area network) protocols (Zohourian et al. 2023). It operates in the sub-GHz and ISM bands. Zigbee excels in IoT applications that require low-bandwidth communication over short distances (approximately 10 m). Support for star, tree, and mesh network topologies, together with low-power consumption and long battery life, makes it suitable for a wide range of IoT network deployment scenarios. Zigbee guarantees dependable and extensive coverage by using dynamic data routing over the network.

5.2.1.4 RFID

RFID (radio frequency identification) is a widely used IoT communication protocol for applications requiring short-range communications and automatic identification of objects. RFIS system consists of RFID tags with unique identifiers, RFID readers for capturing data, and a back-end body for information processing and management. RFID operates at a wide range of low, high, and ultrahigh frequencies, depending on applications such as access control, payment systems, and supply chain management. RFID tags are used to store data, including item information, product details, and customer information. The RFID readers then capture the data from the tags, which is then processed and managed by the back-end system. This system is used to track and manage the flow of products and data in a variety of industries. We have made a comparative analysis among the short-range communication technologies in Table 5.1.

5.2.2 Overview of Long-Range IoT Communication Protocols

Long-range IoT technologies provide extended communication capabilities, broader coverage, and energy-efficient operation, making them ideal for applications that demand connectivity over vast and remote regions.

Table 5.1 Comparison between short-range IoT techniques

Feature	Wi-Fi	Bluetooth	Zigbee	RFID
Frequency	2.4 GHz or 5 GHz	2.4 GHz	2.4 GHz	Various frequency ranges
Maximum range	100 meters	100 meters	100 meters	Varies depending on the RFID type
Data rate	Up to several Gbps	Up to 3 Mbps	Up to 250 kbps	Varies
Power	Moderate to high	Low to moderate	Low	Very low
Topology	Point-to-point, point-to-multipoint	Point-to-point, point-to-multipoint	Mesh network	Point-to-point
Security	WPA2, WPA3 encryption, authentication protocols	Encryption, authentication protocols	AES-128 encryption	Limited security features
Mobility	Stationary or mobile devices	Mobile devices	Stationary devices	Stationary devices
Latency	Low latency	Low latency	Low latency	Negligible latency
Network infrastructure	Requires access points and routers	Direct device-to-device communication	Requires a coordinator device and routers	Tags communicate with readers
Range	Local area coverage	Short-range coverage	Local area coverage	Varies (short range to long range)
Application	High-speed Internet access, local area networking	Wireless audio streaming, peripheral device connectivity	Home automation, industrial control systems, smart metering	Asset tracking, inventory management, access control

5.2.2.1 LoRaWAN

LoRa Alliances developed LoRaWAN, an open medium access control layer protocol dedicated to LoRa RF technology owned by Semtech. LoRa is a proprietary physical layer (PHY) modulation technique based on chirp spread spectrum (CSS) (LoRa and LoRaWAN: A Technical Overview 2019). The LoRaWAN protocol, which combines LoRa (PHY) and LoRaWAN (MAC), enables long-range communication between nodes to connect with diverse IoT applications (Maurya et al. 2022a). LoRaWAN supports bidirectional (uplink and downlink) communication and utilizes a star-of-stars (star-of-stars) topology. It operates at unlicensed frequencies (sub-GHz) in accordance with regional specifications (LoRAWAN Regional Parameters 2022). It can connect millions of devices in huge networks and send data from 290 bps to 50 kbps. LoRaWAN has a wide range of applications such as smart meters, smart cities, industries, healthcare, tracking, and others (Maurya and

Kherani 2020). In recent developments, the International Telecommunication Union (ITU) has officially acknowledged LoRaWAN as the global standard for low-power wide-area network (LPWAN) technology (Recommendation ITU-T Y.4480 2021).

5.2.2.2 NB-IoT

Narrowband Internet of Things (NB-IoT) is a low-power wide-area (LPWA) communication technology for IoT applications. It is a standardized, secure, and scalable LPWA solution that uses licensed cellular bands (Misra et al. 2021). NB-IoT's advantages make it suitable for many IoT use cases. NB-IoT offers extensive coverage and penetration, linking deep indoors, underground, and isolated rural locations where other technologies fail. It coexists with other cellular services due to its narrowband nature. Energy-efficient NB-IoT devices can function on battery power for years without battery changes. Smart meters, environmental monitoring, and asset tracking benefit from its low-power consumption and energy efficiency. NB-IoT's increased security makes it suited for data privacy and integrity applications, securing device-to-cloud communication by integrating with cellular infrastructure. Smart cities can use NB-IoT for parking, garbage management, and lighting, while precision farming, crop monitoring, and irrigation control are possible applications. The technology's scalability and flexibility in managing connected devices foster IoT deployment growth. NB-IoT's long-range coverage and excellent security position it as a viable technology for the next wave of IoT applications (El Soussiet al. 2018).

5.2.2.3 Sigfox

Sigfox is an IoT communication protocol that offers low-power wide-area network connectivity for end-to-end IoT applications (Aguilar et al. 2022). It utilizes binary phase-shift keying (BPSK) modulation to enable communication between devices and base stations. Sigfox enhances noise level, scalability, and communication range by employing ultra-narrowband frequencies, making it well-suited to various IoT applications. Sigfox operates in unlicensed ISM bands, such as 915 MHz in North America, 433 MHz in Asia, and 868 MHz in Europe, providing global coverage for IoT deployments (Levchenko et al. 2022). Sigfox is excellent for IoT use cases with devices spread over broad geographical regions or distant places. Sigfox can link sensors across huge farmlands to provide real-time soil moisture, temperature, and crop health data. Low-power usage extends sensor battery life, lowering maintenance expenses.

5.2.2.4 LTE-M

The LTE-M technology, also known as LTE-MTC (machine-type communication), caters to the needs of IoT and M2M applications. Derived from LTE, it focuses

Table 5.2 Comparison of LPWAN communication protocols

Feature	LoRaWAN	NB-IoT	Sigfox	LTE-M
Standard	Open-access	Licensed cellular	Open-access	Licensed cellular
Governing bodies	LoRa alliance	3GPP	Sigfox	3GPP
Frequency	Sub-GHz	Licensed	Unlicensed	Licensed
Modulation	CSS	QPSK	DPSK, GFSK	QPSK, QAM
Bandwidth	125–500 kHz	180 kHz	100 Hz	1.4 to 5 MHz
Data rate	0.3 to 50 Kbps	32 to 250 Kbps	100 bps	62.5 to 1000 Kbps
Range	12–15 km	20–22 km	15–17 km	As per LTE
Power	Low	Low	Very low	Not power efficient
Payload size	51 to 242 bytes	1600 bytes	12 bytes	Up to 1600 bytes
Authentication & encryption	Yes	Yes	Not supported	Yes

on minimizing power consumption to extend IoT device battery life, enabling prolonged operation in remote and harsh environments. LTE-M optimizes spectrum utilization and network efficiency by utilizing narrowband channels and supporting full-duplex communication. It is well-suited to sensors and periodic data updates, designed to accommodate the low data rates common to IoT devices. Emphasizing strong security measures, LTE-M safeguards data and communication integrity, ensuring IoT privacy. Its unified integration with existing LTE networks and compliance with regulatory standards make LTE-M an attractive and versatile option for industries seeking scalable and reliable IoT connectivity solutions. For a better understanding, we have also considered major futures of IoT and compared their performance in Table 5.2.

5.2.3 Literature

LoRaWAN is an emerging LPWAN technology with various challenges that must be overcome for global recognition. In response to this, the works (Li 2022; Farhad and Pyun 2023; Kurniawan and Kyas 2022; Tellache et al. 2022; Kherani and Maurya 2019; Farhad et al. 2022; Carvalho et al. 2021) have highlighted resource allocation, security, and collision issues and provided a data-driven approach to address LoRaWAN challenges and limitations. As illustrated in Fig 5.5, a number of emerging technologies have been developed to address these challenges, including AI, ML, DL, edge computing, and blockchain. Moreover, these technologies will be used to improve the performance of LoRaWAN networks and enable more efficient and secure communication. For example, (Aihara et al. 2019) have presented a Q-learning model designed to efficiently assign orthogonal channels in LoRaWAN networks using CSMA/CA to prevent interference and collisions. This study demonstrated the potential of intelligent algorithms for optimizing resource allocation.

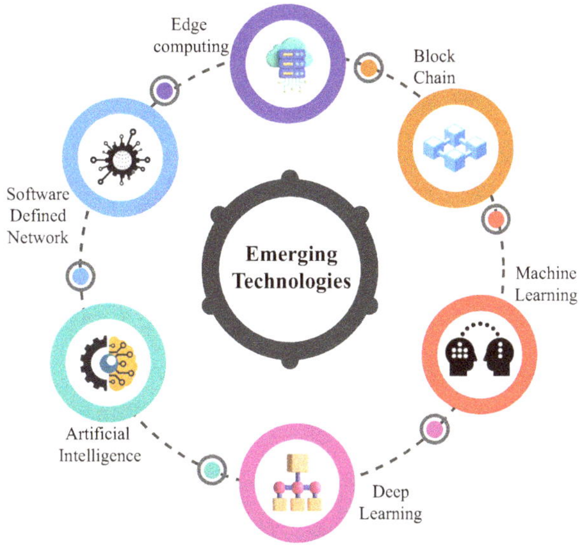

Fig. 5.5 Emerging technologies to overcome IoT protocol challenges

Extensive research has also been done into integrating data-driven technology to mitigate communication issues with competing LoRaWAN technologies such as NB-IoT, Sigfox, etc. In another work, (Caso et al. 2021) optimized the power consumption rate caused by the random-access nature of NB-IoT using an ML approach. Similarly, (Mohammed and Chopra 2023; Ren et al. 2023; Alizadeh and Bidgoly 2023) have also demonstrated the potential of data-driven technology to minimize LPWAN challenges.

Typically, short-range communication technologies operate in the ISM (Industrial, Scientific, and Medical) band, leading to network congestion and other issues. Rigorous investigations are underway to address these concerns. Some research articles (Zhang et al. 2023; Fu et al. 2023; Iannizzotto et al. 2023; Huang and Chin 2023b,a) have demonstrated promising data-driven technologies for overcoming challenges in the domain of short-range communication technologies. (Hasan and Muhammad Khan 2023) have explored a DL approach to detect labeled transmissions from Wi-Fi, wireless sensor networks, and Bluetooth for managing interference in the ISM band. The literature presents several research efforts from different perspectives to address the emerging challenges related to IoT communication protocols. Researchers also employ techniques such as spectrum sensing, cognitive radio, game theory, and evolutionary AI-based solutions. IoT communication challenges remain open to the research community. Additionally, we provide a brief comparative analysis with the most recent state-of-the-art contributions in Table 5.3.

Table 5.3 Summary of the related literature work

Existing works	Summary	T1	T2	T3	T4	T5
Li (2022)	Intelligent resource allocation	LoRaWAN	Deep Q learning	✓	×	×
Hasan and Muhammad Khan (2023)	Detection of labeled transmission	Wi-Fi	DL	×	×	×
Kurniawan and Kyas (2022)	Anomaly detection	LoRaWAN	ML	×	✓	×
Caso et al. (2021)	Power consumption minimizing	NB-IoT	ML	✓	×	×
(Mohammed and Chopra 2023)	Food supply chain security	Nb-IoT	Block-chain	✓	✓	×
(Alizadeh and Bidgoly 2023)	Bit flipping attack detection	LPWAN	DL	✓	✓	×
Rana et al. (2021)	Energy efficiency & interoperability	LPWAN	ML	✓	✓	✓
Hasan and Muhammad Khan (2023)	Detection of labelled transmission	WiFi, BLE	DL	✓	×	×
Fu et al. (2023)	Fingerprint Identification	Radio Frequency	DL	×	✓	×
Reddy et al. (2022)	Interoperability for microgrids	LPWAN	ML, DL, AI	✓	✓	✓
Aihara et al. (2019)	Assign orthogonal channel	LoRaWAN	Q-learning	✓	×	×
Carvalho et al. (2021)	Adaptive data rate	LoRaWAN	Q-learning	✓	×	×
Sudharsan et al. (2022)	Maintain the communication quality	LPWAN	ML	×	×	✓
Tan et al. (2022)	Challenges and Opportunities	Wifi	ML, DL	×	✓	×
Suresh and Chakaravarthi (2022)	Exploring diverse applications	RFID	ML	✓	✓	✓
Li et al. (2023)	Secure Authentication	RFID	ML, DL	×	✓	×
Sivanandam and Ananthan (2022)	Intrusion Detection System	Bluetooth	DL	×	✓	×
Shahjalal et al. (2022)	System security	LoRaWAN	Blockchain	✓	✓	×

"✓" incorporates the features into the research
"×" didn't incorporates the features into the research
"T1" Utilized protocol
"T2" Data driven technology
"T3" Energy optimization
"T4" Security and trust
"T5" Interoperability

5.3 Emerging Use Cases of IoT

IoT applies to a wide range of fields and applications, from smart homes to factories, as illustrated in Fig. 5.6. With the development of IoT, businesses and industries in the field of communication technology are now using it to transmit data across long distances. They also monitor remote patients' health and conduct deep underground oil and mining activities (Hazra et al. 2023). Intelligent communication protocols greatly benefit from ML approaches and are vital for latency-sensitive IoT applications like telemedicine, fraud detection, and analyzing safety and security-related signals.

5.3.1 Industry 5.0

In Industry 5.0, communication protocols and ML methods combine to produce a revolutionary production approach. These protocols allow robots, sensors, and human employees to connect to one another and share information in real time inside a smart factory environment. Incorporating ML into communication protocols opens the door to possibilities like predictive maintenance, smart resource allocation, and flexible production workflows. Industry 5.0 can maximize output, minimize waste, and cut costs through ML algorithms in the production process. Advantages of such integration include the capacity to respond proactively to production difficulties and decrease downtime, thanks to data-driven decisions. Data privacy and security, legacy system integration, and workforce skills gaps are just a few obstacles to overcome before ML technology can be used effectively. Industry 5.0 can only

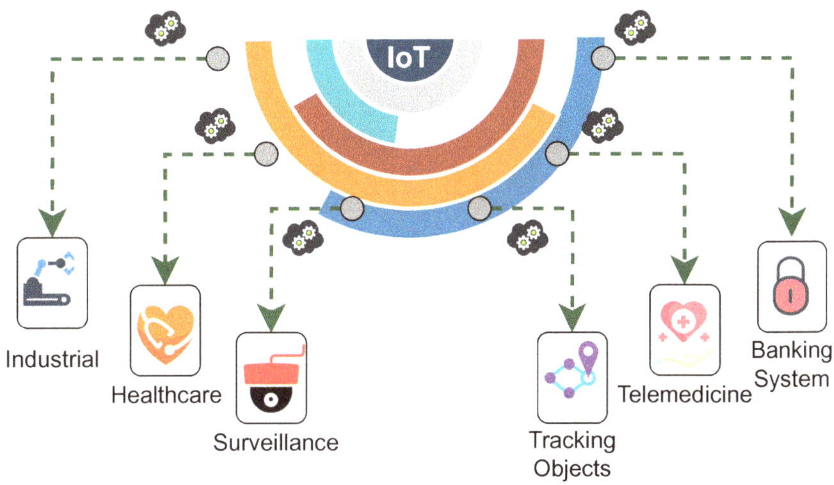

Fig. 5.6 Application areas of IoT

reach its full potential if researchers and businesses work together to address these issues. This is done by encouraging a holistic strategy to optimize communication protocols through machine intelligence. As a result, Industry 5.0 will be realized, and businesses will be able to benefit from ML technology. Additionally, it is imperative to address data privacy and security concerns to protect users from potential misuse of their data.

5.3.2 Smart Healthcare

By integrating ML techniques into communication protocols, smart healthcare systems are reshaping the healthcare sector in a game-changing way (Hazra et al. 2022). These communication protocols ensure that patients, healthcare practitioners, and medical devices all have access to current and accurate information in a timely manner. For smart healthcare, ML algorithms can be used to monitor patients in real time, diagnose diseases, provide individualized treatment suggestions, and analyze health data. With this technology, medical decision-making is optimized, patient outcomes are improved, and healthcare resources are efficiently utilized. ML can improve preventative healthcare through enhanced communication protocols by facilitating early identification and implementing proactive measures. Many obstacles must be overcome, including concerns about privacy and security, interoperability challenges across healthcare systems, and ethical concerns about medical data. As ML is incorporated into communication protocols, smart healthcare will continue to advance, thanks to telemedicine, remote patient monitoring, and individualized treatment that can benefit everyone in the future.

5.3.3 Smart Agriculture

ML techniques in smart agriculture systems have the potential to completely transform the farming environment. Smart agriculture systems improve efficiency, production, and sustainability by incorporating ML algorithms into communication protocols. These protocols are crucial because they enable sensors, actuators, farm equipment, and data management systems to communicate efficiently. Because ML adapts dynamically to changing environmental conditions, these protocols can optimize data transfer and anticipate problems or abnormalities in real time. Farmers and stakeholders can improve agricultural productivity and resource efficiency by making decisions based on accurate data. Agricultural operations can be controlled remotely using modern communication protocols, allowing precision farming, machine-learning-enhanced monitoring, and remote control using ML. However, the full potential of these cutting-edge smart agriculture systems requires overcoming obstacles like data privacy, network security, and integration complexity. Thus, ML in communication protocols has enormous potential for radically altering the farming sector, with numerous positive outcomes for environmentally responsible and productive farming methods.

5.3.4 Intelligent Transportation System

The transportation industry has witnessed a surge in the adoption of ML methods to enhance communication protocols. These protocols play a crucial role in enabling reliable communication among various transportation network components, including infrastructure, vehicles, base stations, and control units (Adhikari et al. 2021). Integrating ML algorithms into communication protocols empowers networks to respond more effectively to changing traffic conditions, optimize data transmission, and improve overall performance. ML-driven communication protocols can reduce latency and enhance reliability by intelligently prioritizing data, predicting network congestion, and dynamically allocating resources. Moreover, ML techniques can be leveraged to analyze past communication patterns, facilitating demand forecasting and protocol adaptation. The combination of ML with communication protocols can usher in an era of improved transportation network safety, efficiency, and connectivity.

5.4 Challenges and Future Opportunities

Designing an autonomous IoT ecosystem with seamless communication services on a large scale poses several new challenges for developers. These challenges include interoperability, energy optimization, scalability, security, etc. We must also address those challenges in order to reach its full potential when it comes to standardizing the IoT ecosystem. It will be difficult to ensure that all IoT devices work together if these challenges aren't addressed. Therefore, it is imperative to develop standards and protocols to ensure reliable communication in an IoT network. Among those, we discussed some basic challenges.

5.4.1 Interoperability

Interoperability is a term that describes a process of information sharing and cooperation between different technologies that is based on cooperation. Interoperability between various devices and systems is essential in IoT communications to facilitate effective and integrated IoT solutions. This allows data to be shared across multiple platforms, enabling a more efficient and connected user experience. With interoperability, devices can communicate with each other, allowing for better automation and integration of services. Interoperability is a key factor for the growth of IoT in the future. Lack of compatibility among IoT communication protocols is one of the biggest obstacles to the widespread adoption of IoT technologies since it makes it challenging to share information between disparate systems and devices. IoT devices and platforms with heterogeneous data streams can integrate and communicate seamlessly with data-driven technologies, which provide smart data processing and analytics. As a result, novel and innovative solutions can be developed for a wide range of applications. In addition, it facilitates the development of novel business models and revenue streams.

5.4.2 Energy-Optimized Data Transmission

Energy-optimized data transfer has been developed to reduce IoT power consumption and increase longevity. It is imperative that data transmission in the IoT is energy-efficient in order for battery-operated IoT devices to continue functioning effectively and sustainably for as long as possible. Data transmission inefficiency is one of the main problems plaguing IoT communication protocols, causing power consumption and battery life issues. In IoT systems, data-driven technologies reduce energy consumption and increase communication efficiency by optimizing data processing, compression, and transmission, thereby addressing the difficulties associated with energy-optimized data transmission. For example, ML in IoT helps improve data accuracy, reduce latency, and increase scalability, thus enhancing the whole system's performance. In addition, it allows the system to learn from its mistakes over time, which leads to continuous improvement.

5.4.3 Zero-Touch IoT Automation

Zero-touch IoT automation can make all the difference when deploying and maintaining IoT devices. With zero-touch IoT automation, device onboarding can be simplified, setup times can be shortened, and human errors can be reduced, making IoT deployments more efficient and scalable. Configuring and provisioning devices manually takes time, introduces errors, and complicates deployment. Without zero-touch IoT automation, IoT communication protocols face several new problems. Zero-touch IoT automation eliminates these issues, making more efficient and streamlined deployments. This allows organizations to focus their time and resources on ongoing device management, ensuring their IoT systems are secure and running optimally. In order to simplify zero-touch IoT automation problems, data-driven technologies use ML algorithms to automate zero-touch IoT automation problems. This allows for more effective and smoother IoT device deployments. This reduced complexity saves companies money, increases operational efficiency, and reduces deployment time. Furthermore, it enables companies to quickly and securely configure and deploy devices, encouraging them to take advantage of the data generated by IoT sensors.

5.4.4 Security and Trust

Protecting sensitive information and establishing encrypted connections between devices is of the utmost importance in many IoT use cases, including healthcare. Strong encryption and authentication mechanisms become crucial in protecting privacy and confidentiality. IoT security challenges can be mitigated using advanced encryption algorithms, multifactor authentication, blockchain technology for data integrity, and data-driven technologies for anomaly detection in real-time threat detection. Continuous monitoring and timely security software and hardware updates can mitigate vulnerabilities and maintain a resilient IoT ecosystem. Regular security audits can help identify and address potential security issues before they become a problem. Additionally, organizations can implement user education and awareness programs to ensure that users understand the importance of security and their role in it.

5.4.5 Scalability

The term *scalability* refers to the ability of the network to provide service to the accommodated number of devices and users without significant degradation in the network's performance and efficiency. Scalability is a fundamental characteristic of any IoT network as it allows the network to expand its capacity and resources to meet growing demands while ensuring efficient communication and reliable services. However, scalability presents challenges like handling large amounts of data and network congestion. These challenges can be overcome using advanced data analytics and data-driven technologies like ML, DL, software-defined networks, etc. In addition, AI-driven provisioning and dynamic resource allocation technology advances lead to higher scalability and responsiveness in IoT systems, allowing them to meet the needs of a growing population. Table 5.4 showcases the capabilities of data-driven technologies for IoT communication, providing a comprehensive insight.

Table 5.4 Potential of data-driven technologies

Challenges	How data-driven technologies solve those challenges	Related works
Interoperability	Technologies like ML and blockchain have the capability to integrate disparate data sources and standard formats and enable real-time processing and anomaly detection to deal with IoT interoperability	(Self-evolving intelligent algorithms for facilitating data interoperability in IoT environments 2018; Rana et al. 2021; Nilsson and Sandin 2018; Bhat et al. 2022; Buhalis and Leung 2018)
Energy optimization	Data-driven technologies analyze sensor and device data, identifying patterns to predict power consumption. Furthermore, data analytics identifies power-hungry components for optimizing energy-intensive tasks	(Raval et al. 2021; Sodhro et al. 2019; Mocanu et al. 2019; Rajab et al. 2021; Tu et al. 2022)
Scalability	Data-driven technologies can effectively manage the transmission and reception of data between the devices by allowing automatic provisioning and smart resource allocation for massive deployment, leading to network performance optimization	(Minhaj et al. 2023; Praveen Kumar et al. 2023; Ajorlou and Abbasfar 2020; Aruna and Pradeep 2020; Lee et al. 2018; Benites and Sapozhnikova 2017)
Security and trust	By adopting real-time threat detection, robust authentication, encryption, and anomaly detection, data-driven techniques address the security issues of various domains of IoT applications like healthcare	(Wheelus and Zhu 2020; Sivaganesan 2021; Chauhan and Ramaiya 2022; Rajawat et al. 2021; Zeadally and Tsikerdekis 2020; Magaia et al. 2020)
Zero-touch IoT automation	Advanced ML or DL techniques can be easily adapted for handling IoT automation issues, eventually reducing human error. Specifically, these techniques can address self-learning, self-fulfilling, and self-assuring issues in IoT	(Sanjoyo and Mambo 2022; Strebel and Magno 2018; Mayer et al. 2019; Yoshino et al. 2020; Ben Saad et al. 2022)

5.5 Conclusion

Over time, the IoT has become one of the emerging topics of interest in both industry and academia. In this context, IoT protocols have gained the utmost attention due to their applicability, demand, and benefits. On the other hand, data-driven technologies have also garnered significant attention for their intelligent decision-making capabilities. These technologies play a vital role in addressing standard challenges in IoT and related communication technologies. This chapter summarizes all these points and presents a comprehensive literature review, specifically focusing on data-driven technologies for IoT communication. First, we provide a brief overview of IoT architecture and the basics of data-driven technologies. Then, we discuss long-range and short-range communication protocols. Additionally, we highlight the importance of data-driven technologies for various IoT applications. Finally, several challenges and opportunities are discussed in the context of IoT and IoT communications. We hope this chapter motivates readers to work toward developing data-driven technologies for IoT communications.

References

A survey on bluetooth multi-hop networks. 2019. In *Ad Hoc Networks*.

Adhikari, Mainak, et al. 2021. A roadmap of next-generation wireless technology for 6G-enabled vehicular networks. *IEEE Internet of Things Magazine* 4 (4): 79–85. https://doi.org/10.1109/IOTM.001.2100075.

Aguilar, Sergio, et al. 2022. Energy consumption model of SCHC packet fragmentation over Sigfox LPWAN. *Sensors* 22 (6). ISSN: 1424–8220. https://www.mdpi.com/1424-8220/22/6/2120.

Aihara, Naoki, et al. 2019. Q-learning aided resource allocation and environment recognition in LoRaWAN with CSMA/CA. *IEEE Access* 7: 152126–152137. https://doi.org/10.1109/ACCESS.2019.2948111.

Ajorlou, Amir, and Aliazam Abbasfar. 2020. An optimized structure of state channel network to improve scalability of blockchain algorithms. In *2020 17th International ISC Conference on Information Security and Cryptology (IS- CISC)*, 73–76. IEEE.

Al-Qaseemi, Sarah A., et al. 2016. IoT architecture challenges and issues: Lack of standardization. In *2016 Future Technologies Conference (FTC)*, 731–738. https://doi.org/10.1109/FTC.2016.7821686.

Ali, Zainab H., et al. 2015. Internet of Things (IoT): definitions, challenges and recent research directions. *International Journal of Computer Applications* 128 (1), 37–47.

Alizadeh, Faezeh, and Amir Jalaly Bidgoly. 2023. Bit flipping attack detection in low power wide area networks using a deep learning approach. In *Peer-to-Peer Networking and Applications*, 1–11.

Aruna, K., and G. Pradeep. 2020. Performance and scalability improvement using IoT-based edge computing container technologies. *SN Computer Science* 1: 1–7.

Barua, Arup, et al. 2022. Security and privacy threats for bluetooth low energy in IoT and wearable devices: a comprehensive survey. *IEEE Open Journal of the Communications Society* 3: 251–281.

Ben Saad, Sabra, et al. 2022. A trust and explainable federated deep learning framework in zero touch B5G Networks. In *GLOBECOM 2022–2022 IEEE Global Communications Conference*, 1037–1042. https://doi.org/10.1109/GLOBECOM48099.2022.10001371.

Benites, Fernando, and Elena Sapozhnikova. 2017. Improving scalability of ART neural networks. *Neurocomputing* 230: 219–229. ISSN: 0925–2312. https://doi.org/10.1016/j.neucom.2016.12.022. https://www.sciencedirect.com/science/article/pii/S0925231216314977.

Bhat, Showkat Ahmad, et al. 2022. Agriculture-food supply chain management based on blockchain and IoT: A narrative on enterprise blockchain interoperability. *Agriculture* 12 (1). ISSN: 2077-0472. https://www.mdpi.com/2077-0472/12/1/40.

Buhalis, Dimitrios, and Rosanna Leung. 2018. Smart hospitality-Interconnectivity and interoperability towards an ecosystem. *International Journal of Hospitality Management* 71: 41–50. ISSN: 0278-4319. https://doi.org/10.1016/j.ijhm.2017.11.011. https://www.sciencedirect.com/science/article/pii/S0278431917301974.

Carvalho, Rodrigo, et al. 2021. Q-learning ADR agent for LoRaWAN optimization. In *2021 IEEE International Conference on Industry 4.0, Artificial Intelligence, and Communications Technology (IAICT)*, 104–108. https://doi.org/10.1109/IAICT52856.2021.9532518.

Caso, Giuseppe, et al. 2021. NB-IoT random access: data-driven analysis and ML-based enhancements. *IEEE Internet of Things Journal* 8 (14), 11384–11399. https://doi.org/10.1109/JIOT.2021.3051755.

Chauhan, Chetan, and Manoj Kumar Ramaiya. 2022. Advanced model for improving iot security using blockchain technology. In *2022 4th International Conference on Smart Systems and Inventive Technology (ICSSIT)*, 83–89. IEEE.

Chen, Mi, et al. 2023. Dynamic parameter allocation with reinforcement learning for LoRaWAN. *IEEE Internet of Things Journal* 10(12): 10250–10265. https://doi.org/10.1109/JIOT.2023.3239301.

El Soussi, Mohieddine, et al. 2018. Evaluating the performance of eMTC and NB-IoT for smart city applications. In *2018 IEEE International Conference on Communications (ICC)*, 1–7. IEEE.

Farhad, Arshad, Dae-Ho Kim, et al. 2022. Deep learning-based channel adaptive resource allocation in LoRaWAN. In *2022 International Conference on Electronics, Information, and Communication (ICEIC)*, 1–5. https://doi.org/10.1109/ICEIC54506.2022.9748580.

Farhad, Arshad, and Jae-Young Pyun. 2023. AI-ERA: Artificial intelligence-empowered resource allocation for LoRa-enabled IoT applications. *IEEE Transactions on Industrial Informatics*, 1–13. https://doi.org/10.1109/TII.2023.3248074.

Fu, Hua, et al. 2023. Deep learning based RF fingerprint identification with channel effects mitigation. *IEEE Open Journal of the Communications Society*, 1668–1681.

Hasan, Ayesha, and Bilal Muhammad Khan. 2023. Deep learning aided wireless interference identification for coexistence management in the ISM bands. *Wireless Networks*, 1–21.

Hazra, Abhishek, Mainak Adhikari, et al. Nov. 2021a. A comprehensive survey on interoperability for IIoT: taxonomy, standards, and future directions. *ACM Computing Surveys* 55 (1). ISSN: 0360-0300. https://doi.org/10.1145/3485130.

Hazra, Abhishek, Prakash Choudhary, et al. 2021b. Recent advances in deep learning techniques and its applications: an overview. In *Advances in Biomedical Engineering and Technology: Select Proceedings of ICBEST 2018*, 103–122.

Hazra, Abhishek, et al. (2022). Fog computing for energy-efficient data offloading of IoT applications in industrial sensor networks. *IEEE Sensors Journal* 22 (9): 8663–8671. https://doi.org/10.1109/JSEN.2022.3157863.

Hazra, Abhishek, Pradeep Rana, et al. 2023. Fog computing for next-generation Internet of Things: Fundamental, state-of-the-art and research challenges. *Computer Science Review* 48: 100549.

Huang, Yiwei, and Kwan-Wu Chin. 2023a. A hierarchical deep learning approach for optimizing CCA threshold and transmit power in WiFi networks. *IEEE Transactions on Cognitive Communications and Networking*, 1–1. https://doi.org/10.1109/TCCN.2023.3282984.

Huang, Yiwei, and Kwan-Wu Chin. 2023b. A three-tier deep learning based channel access method for WiFi networks. *IEEE Transactions on Machine Learning in Communications and Networking*, 90–106.

Iannizzotto, Giancarlo, et al. 2023. Improving BLE-based passive human sensing with deep learning. *Sensors* 23 (5): 2581.

IoT in 2023 and beyond (2023). Report. https://techinformed.com/iot-in-2023-and-beyond/.

Kherani, Arzad Alam, and Poonam Maurya. 2019. Improved packet detection in LoRa-like chirp spread spectrum systems. In *2019 IEEE International Conference on Advanced Networks and Telecommunications Systems (ANTS)*, 1–4. https://doi.org/10.1109/ANTS47819.2019.9118076.

Kurniawan, Agus, and Marcel Kyas. 2022. Machine learning models for LoRa Wan IoT anomaly detection. In *2022 International Conference on Advanced Computer Science and Information Systems (ICACSIS)*, 193–198. https://doi.org/10.1109/ICACSIS56558.2022.9923439.

Lee, Junhee, et al. 2018. A scheduling algorithm for improving scalability of LoRaWAN. In *2018 International Conference on Information and Communication Technology Convergence (ICTC)*, 1383–1388. IEEE.

Levchenko, Polina, et al. 2022. Performance comparison of NB-Fi, Sigfox, and LoRaWAN. *Sensors* 22 (24). ISSN: 1424-8220. https://www.mdpi.com/1424-8220/22/24/9633.
Li, Ang, et al. 2023. Secure UHF RFID authentication with smart devices. *IEEE Transactions on Wireless Communications* 22 (7), 4520–4533. https://doi.org/10.1109/TWC.2022.3226753.
Li, Aohan. 2022. Deep reinforcement learning based resource allocation for LoRaWAN. In *2022 IEEE 96th Vehicular Technology Conference (VTC2022-Fall)*, 1–4. https://doi.org/10.1109/VTC2022-Fall57202.2022.10012698.
LoRa and LoRaWAN: A Technical Overview (Dec. 2019). en. Technical Paper. https://lora-developers.semtech.com/documentation/tech-papers-and-guides/lora-and-lorawan/.
LoRAWAN Regional Parameters (Sept. 2022). en. Specification RP002-1.0.4. https://resources.lora-alliance.org/technical-specifications/rp002-1-0-4-regional-parameters. Fremont, United States.
Magaia, Naercio, et al. 2020. Industrial internet-of-things security enhanced with deep learning approaches for smart cities. *IEEE Internet of Things Journal* 8 (8): 6393–6405.
Mao, Wenliang, et al. 2021. Energy-efficient industrial internet of things: overview and open issues. *IEEE Transactions on Industrial Informatics* 17 (11): 7225–7237. https://doi.org/10.1109/TII.2021.3067026.
Maurya, Poonam, and Arzad Alam Kherani. 2020. Tracking performance in LoRaWAN-like systems and equivalence of a class of distributed learning algorithms. *IEEE Communications Letters* 24 (11): 2584–2588. https://doi.org/10.1109/LCOMM.2020.3012569.
Maurya, Poonam, Aatmjeet Singh, et al. 2022a. A review: spreading factor allocation schemes for LoRaWAN. *Telecommunication Systems* 80 (3): 449–468.
Maurya, Poonam, Aatmjeet Singh, et al. 2022b. Design LoRaWAN network for unbiased communication between nodes and gateway. In *2022 14th International Conference on COMmunication Systems & NETworkS (COMSNETS)*, 581–589. https://doi.org/10.1109/COMSNETS53615.2022.9668447.
Mayer, Philipp, et al. 2019. ZeroPowerTouch: zero-power smart receiver for touch communication and sensing in wearable applications. In *2019 Design, Automation & Test in Europe Conference & Exhibition (DATE)*, 944–947. https://doi.org/10.23919/DATE.2019.8715062.
Minhaj, Syed Usama, et al. 2023. Intelligent resource allocation in LoRaWAN using machine learning techniques. *IEEE Access* 11: 10092–10106. https://doi.org/10.1109/ACCESS.2023.3240308.
Misra, Sudip, et al. 2021. *Introduction to IoT*. Cambridge: Cambridge University Press.
Mocanu, Elena, et al. (2019). On-line building energy optimization using deep reinforcement learning. *IEEE Transactions on Smart Grid* 10 (4): 3698–3708. https://doi.org/10.1109/TSG.2018.2834219.
Mohammed, Chand Pasha, and Shakti Raj Chopra. 2023. Blockchain security implementation using Python with NB-IoT deployment in food supply chain. In *2023 International Conference on Emerging Smart Computing and Informatics (ESCI)*, 1–5. IEEE.
Najm, Ihab Ahmed, et al. 2019. Machine learning prediction approach to enhance congestion control in 5G IoT environment. *Electronics* 8 (6): 607.
Natarajan, Yuvaraj, et al. 2022. An IoT and machine learning-based routing protocol for reconfigurable engineering application. *IET Communications* 16 (5): 464–475.
Nilsson, Jacob, and Fredrik Sandin. 2018. Semantic interoperability in industry 4.0: survey of recent developments and outlook. In *2018 IEEE 16th International Conference on Industrial Informatics (INDIN)*, 127–132. https://doi.org/10.1109/INDIN.2018.8471971.
Omar, Hassan Aboubakr, et al. 2016. A survey on high efficiency wireless local area networks: Next generation WiFi. *IEEE Communications Surveys & Tutorials* 18 (4): 2315–2344.
Praveen Kumar, Donta, et al. 2023. Exploring the potential of distributed computing continuum systems. *Computers* 12: 198.
Rajab, Husam, et al. 2021. Reducing power requirement of LPWA networks via machine learning. *Pollack Periodica* 16 (2): 86–91.
Rajawat, Anand Singh, et al. 2021. Blockchain-based model for expanding IoT device data security. In *Advances in Applications of Data-Driven Computing*, 61–71.

Ramezanpour, Keyvan, et al. 2023. Security and privacy vulnerabilities of 5G/6G and WiFi 6: Survey and research directions from a coexistence perspective. *Computer Networks* 221: 109515.

Rana, Bharti, et al. 2021. A systematic survey on internet of things: Energy efficiency and interoperability perspective. *Transactions on Emerging Telecommunications Technologies* 32 (8): e4166.

Raval, Maulin, et al. 2021. Smart energy optimization for massive IoT using artificial intelligence. *Internet of Things* 13: 100354. ISSN: 2542-6605. https://doi.org/10.1016/j.iot.2020.100354. https://www.sciencedirect.com/science/article/pii/S2542660520301852.

Recommendation ITU-T Y.4480 (Nov. 2021). Low Power Protocol for Wide Area Wireless Networks. en. Recommendation ITU-T Y.4480. https://www.itu.int/rec/T-REC-Y.4480/. Geneva, Switcherland: Telecommunication Standardization Sector of ITU.

Reddy, Gogulamudi Pradeep, et al. 2022. Communication technologies for interoperable smart microgrids in urban energy community: a broad review of the state of the art, challenges, and research perspectives. *Sensors* 22 (15). https://www.mdpi.com/1424-8220/22/15/5881.

Ren, Rong, et al. 2023. Deep reinforcement learning for connection density maximization in NOMA-based NB-IoT networks. In *2023 8th International Conference on Computer and Communication Systems (ICCCS)*, 357–361. IEEE.

Sanjoyo, Danu Dwi, and Masahiro Mambo. 2022. Accountable bootstrapping based on attack resilient public key infrastructure and secure zero touch provisioning. *IEEE Access* 10: 134086–134112. https://doi.org/10.1109/ACCESS.2022.3231015.

Self-evolving intelligent algorithms for facilitating data interoperability in IoT environments (2018). *Future Generation Computer Systems* 86: 421–432. ISSN: 0167-739X.

Shahjalal, Md. et al. 2022. Implementation of a secure LoRaWAN system for industrial internet of things integrated with IPFS and blockchain. *IEEE Systems Journal* 16 (4): 5455–5464. https://doi.org/10.1109/JSYST.2022.3174157.

Sivaganesan, Dr. D. 2021. A data driven trust mechanism based on blockchain in IoT sensor networks for detection and mitigation of attacks. *Journal of Trends in Computer Science and Smart Technology* 3 (1): 59–69.

Sivanandam, Nishanth, and T. Ananthan. 2022. Intrusion detection system for bluetooth mesh networks using machine learning. In *2022 International Conference on Industry 4.0 Technology (I4Tech)*, 1–6. https://doi.org/10.1109/I4Tech55392.2022.9952758.

Sodhro, Ali Hassan, et al. 2019. A novel energy optimization approach for artificial intelligence-enabled massive internet of things. In *2019 International Symposium on Performance Evaluation of Computer and Telecommunication Systems (SPECTS)*, 1–6. https://doi.org/10.23919/SPECTS.2019.8823317.

Srirama, Satish Narayana. (2023). A decade of research in fog computing: Relevance, challenges, and future directions. *Software: Practice and Experience*. https://doi.org/10.1002/spe.3243. EPRINT: https://onlinelibrary.wiley.com/doi/pdf/10.1002/spe.3243. https://onlinelibrary.wiley.com/doi/abs/10.1002/spe.3243.

Strebel, Raphael, and Michele Magno. 2018. Poster abstract: zero-power receiver for touch communication and touch sensing. In *2018 17th ACM/IEEE International Conference on Information Processing in Sensor Networks (IPSN)*, 150–151. https://doi.org/10.1109/IPSN.2018.00038.

Sudharsan, Bharath, et al. 2022. RIS-IoT: towards resilient, interoperable, scalable IoT. In *2022 ACM/IEEE 13th International Conference on Cyber-Physical Systems (ICCPS)*, 296–297. https://doi.org/10.1109/ICCPS54341.2022.00039.

Suresh, Setti, and Geetha Chakaravarthi. 2022. RFID technology and its diverse applications: A brief exposition with a proposed Machine Learning approach. *Measurement* 195: 111197. ISSN: 0263-2241. https://doi.org/10.1016/j.measurement.2022.111197. https://www.sciencedirect.com/science/article/pii/S026322412200450X.

Tan, Sheng, et al. 2022. Commodity WiFi sensing in ten years: status, challenges, and opportunities. *IEEE Internet of Things Journal* 9 (18): 17832–17843. https://doi.org/10.1109/JIOT.2022.3164569.

Tellache, Amine, et al. 2022. Deep reinforcement learning based resource allocation in dense sliced LoRaWAN networks. In *2022 IEEE International Conference on Consumer Electronics (ICCE)*, 1–6. https://doi.org/10.1109/ICCE53296.2022.9730234.

Tu, Lam-Thanh, et al. (2022). Energy efficiency optimization in LoRa networks—a deep learning approach. *IEEE Transactions on Intelligent Transportation Systems*, 1–13. https://doi.org/10.1109/TITS.2022.3183073.

Wheelus, Charles, and Xingquan Zhu. 2020. IoT network security: Threats, risks, and a data-driven defense framework. *IoT 1.2*, 259–285.

Yoshino, Manabu, et al. 2020. Zero-touch multi-service provisioning with pluggable module-type OLT on access network virtualization testbed. In *2020 Opto-Electronics and Communications Conference (OECC)*, 1–3. https://doi.org/10.1109/OECC48412.2020.9273446.

Zeadally, Sherali, and Michail Tsikerdekis. 2020. Securing Internet of Things (IoT) with machine learning. *International Journal of Communication Systems* 33 (1): e4169.

Zhang, Jiansheng, et al. 2023. Secure blockchain-enabled internet of vehicles scheme with privacy protection. *Computers, Materials & Continua* 75 (3).

Zohourian, Alireza, et al. 2023. IoT Zigbee device security: A comprehensive review. *Internet of Things*, 100791.

Chapter 6
Towards Large-Scale IoT Deployments in Smart Cities: Requirements and Challenges

Naser Hossein Motlagh, Martha Arbayani Zaidan, Roberto Morabito, Petteri Nurmi, and Sasu Tarkoma

6.1 Introduction

The number of Internet of Things (IoT) devices has long since surpassed the number of people on Earth and is expected to continue growing with estimates suggesting nearly 30 billion devices will be deployed by 2030 (Melibari et al. 2023). Cities and urban areas are one of the main areas for these devices with examples ranging from smart home sensors to driverless cars, portable IoT devices, smart wearables, and different types of drones. Examples of these devices in operation within a smart city are shown in Fig. 6.1.

The characteristics of the IoT devices vary depending on the device designs and their intended applications, which in turn poses requirements for the infrastructure that is available in the city. For example, driverless cars require continuous and persistent network connections, whereas wearables typically require discontinuous and transient connections. Similarly, applications that target the immediate needs of citizens tend to require support for real-time computation and processing, whereas analytics and other more long-term services can operate without support for real-time processing. Besides the need for real-time responsiveness of the networks, some of these applications would be computationally demanding. Providing the necessary networking and computational support in an affordable, efficient, and scalable manner is highly challenging (Zeadally et al. 2020). Besides these overall infrastructure challenges, deploying the sensors can also be demanding. IoT devices that benefit the city mostly can be categorized into fixed sensors and mobile sensors. Fixed sensors require strategic planning for deployment and to ensure the necessary electricity, networking, computations, and security support are in place. Mobile

N. Hossein Motlagh (✉) · M. Arbayani Zaidan · R. Morabito · P. Nurmi · S. Tarkoma
Department of Computer Science, University of Helsinki, Helsinki, Finland
e-mail: naser.motlagh@helsinki.fi; martha.zaidan@helsinki.fi; roberto.morabito@helsinki.fi; petteri.nurmi@helsinki.fi; sasu.tarkoma@helsinki.fi

PraveenKumar Donta et al. (eds.), *Learning Techniques for the Internet of Things*,
https://doi.org/10.1007/978-3-031-50514-0_6

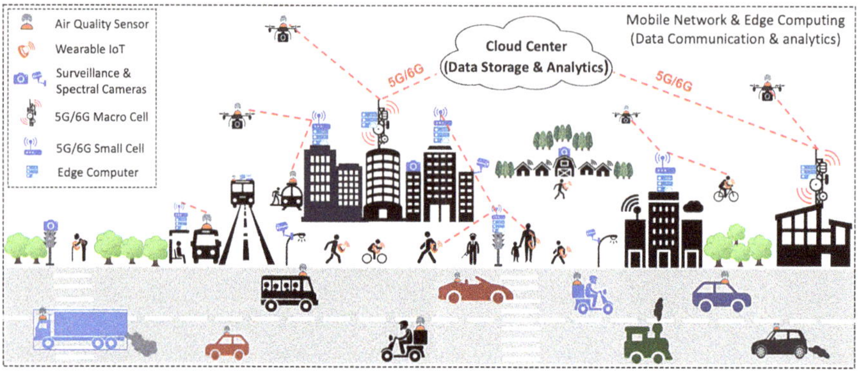

Fig. 6.1 Illustration of large-scale sensor deployment in a smart city

sensors, in turn, need to have sufficiently dense coverage and data quality may be an issue as certain locations or demographic groups may be overrepresented.

Taking all of the above into account, massive-scale deployments of IoT sensors in smart cities that meet the needs of citizens and applications are a highly challenging task. This chapter details these challenges, beginning from requirements (Sect. 6.2) and continuing to key challenges (Sect. 6.3). To highlight some of the potential benefits that can be obtained from IoT deployments in smart cities, in Sect. 6.4, we present a case study and results from a deployment of air quality sensors in the city of Helsinki. We further provide a discussion about the role of AI and emerging technologies in future smart cities in Sect. 6.5. Finally, we conclude the chapter in Sect. 6.6.

6.2 Requirements for IoT Deployment in Smart Cities

6.2.1 Reliable Network Connection

To ensure successful operations of deployed IoT devices in smart cities, it is mandatory to have robust and seamless network services in the cities. While some IoT applications would require ultralow latency services from the network, other applications may demand high bandwidth or may need to obtain massive connections (Jiang and Liu 2016). The following are examples of applications requiring different forms of services from the network.

Ultralow Latency: The driverless cars and drones are outstanding example applications that would need fast data processing in order to make precise decision-making, e.g., for avoiding obstacles and changing directions. To ensure the safe operations of the applications, thus, driverless cars and drones are expected to have stringent latency requirements (Zeadally et al. 2020; Gupta et al. 2021).

High Bandwidth: The surveillance cameras have been widely used in urban areas to monitor human activities as well as face recognition (Tripathi et al. 2018). To perform real-time image processing streamed from the cameras, therefore there is a need for enhanced bandwidth from the network such that it can support the transmission of tens of video frames every second, while each frame requires a few megabits of bandwidth from the network. Assuming a frame size of 20 KB and 30 fps be the standard frame rate, then the required bandwidth for a single frame would be 4.8 Mb/s (20 KB×30 fps×8 bits/byte). This bandwidth requirement is further enhanced with the frame transmission rate of the surveillance camera. Hyperspectral cameras that are widely used for environmental and pollution monitoring are the other prominent examples of IoT applications as they can produce images of 30300 MB in less than a second (Motlagh et al. 2020; Su et al. 2021). Therefore, for frame transmission, compared to the surveillance cameras, they require even higher bandwidth from the network.

Massive Connection: In addition to the example IoT devices mentioned earlier, the number of other types of IoT devices and applications rapidly increases which mandates obtaining ubiquitous and responsive network services in cities. Among many, examples of such applications include portable low-cost air quality sensors, smart homes, smart grids, smart metering, and different forms of wearables such as smartwatches and smart rings. The increasing number of IoT devices either mobile (carried by people or vehicles) or installed at fixed locations requires providing massive connections by the networks (Gupta et al. 2021).

6.2.2 Infrastructure Deployment

To provide effective network services and ensure successful operations of IoT, it is necessary to place first the network infrastructure such as base stations, IoT gateways, and edge computers in strategic locations (PoIs) to enable providing full network coverage in cities (Motlagh et al. 2022). This is needed to ensure support for mobile IoT devices (either vehicles, drones, or people) moving at various speeds.

Second, when deploying fixed sensors in urban areas, it is essential to install them in places that can easily connect to the network and maintain its connection. In addition, as IoT devices require power supplies as well as continuous maintenance, it is important to install the devices in locations with energy sources and easy access for inspection.

6.3 Key Aspects of Sensor Deployment and Data Management in Smart Cities

6.3.1 Sensor Deployment and Placement

Urban environments are complex systems as they consist of different urban elements such as residential areas, shopping centers, parks and green areas, and highways and streets (with high and low levels of traffic). These urban environments do not only span horizontally, but they grow vertically (such as tall buildings and skyscrapers) with the population growth in cities. Therefore, to optimally provide IoT services (Rashid and Rehmani 2016) and also better monitor the health of city infrastructures (Kim et al. 2007), there is a need for optimal sensor deployments and placement methods in order to cover the whole city environment.

In the existing methods, the solutions include "citizen-centric" sensor placement approach by (i) installing sensors near public places, e.g., schools and hospitals, (ii) providing local information by minimizing the distance between the sensors and the people, and (iii) placing and optimizing sensors on critical urban infrastructure, e.g., monitoring traffic emissions on roads with high traffic levels (Sun et al. 2019).

Furthermore, current sensor deployment and placement the most areas of a city are not covered. The areas that fall under a certain radius of a sensor are considered covered by sensing systems. Therefore, to cover the missing areas, the current approaches rely on interpolating data using the measurements of other sensor nodes in the same area. Indeed, the city environments because of their complex features and dynamics make sensor deployment challenging. Thus, sensor deployment and placement require new models that take into account the dynamics of the city blocks, urban infrastructure, building shapes, demographics, and the microenvironmental features of the regions.

In light of the challenges associated with sensor deployment and placement outlined in this section, it is crucial to consider the broader ecosystem in which these sensors operate. Effective sensor deployment is but the first step in a multifaceted process that ultimately leads to the delivery of valuable services and applications within smart cities.

Figure 6.2 provides an illustrative overview of this ecosystem, segmented into four primary layers: data collection, data transmission, data services, and

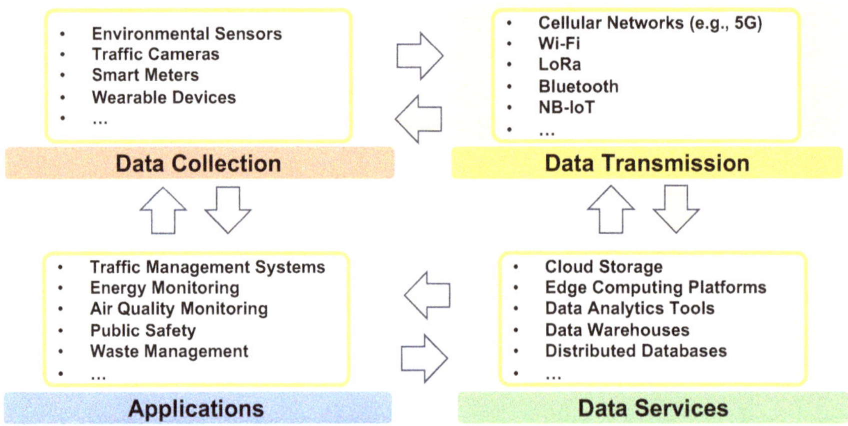

Fig. 6.2 Illustration of the four primary layers in smart city data management: data collection, data transmission, data services, and applications, each with representative examples

applications. Each layer represents a critical stage in the data life cycle, with its unique challenges and requirements.

Each of these layers is interconnected, collaboratively ensuring that data is effectively collected, transmitted, managed, and utilized to provide intelligent and responsive smart city applications.

Within this complex framework, security and ethical considerations permeate every layer. The process of data handling often involves sensitive or personally identifiable information, necessitating stringent ethical considerations and robust security measures. Techniques like data anonymization are implemented to protect privacy, while adherence to international and local legal frameworks, like the GDPR in Europe, guide the ethical collection and handling of data (Badii et al. 2020). Security considerations are equally crucial, involving the deployment of encryption technologies and access control mechanisms to safeguard data at rest and in transit, providing a secure environment for data storage and processing (Cui et al. 2018).

The following sections will delve deeper into the challenges and considerations associated with data collection, data transmission, and data services within this secured and ethically compliant framework. Then, in Sect. 6.4, we will explore a practical application of this layered framework through a case study on *air quality monitoring* with IoT for smart cities, offering real-world insights into how these layers function consistently to support smart city initiatives while upholding the highest standards of security and ethics.

6.3.2 Data Collection

Data collection is the foundational component in the IoT life cycle within smart city applications, requiring robust and efficient processes to ensure the efficacy of subsequent analytics and decision-making. In the realm of IoT, data collection entails gathering various types of data from devices like environmental sensors, traffic cameras, smart meters, wearable devices, and RFID tags, as illustrated in Fig. 6.2.

Each device plays a specific role in collecting different data types, which are essential for various applications in smart cities. For instance, environmental sensors gather crucial data on air quality, temperature, and humidity, providing real-time information necessary for monitoring and responding to changes in the urban environment.

To facilitate reliable and efficient data collection, adherence to established protocols and standards is crucial (Donta et al. 2022). Protocols like MQTT and CoAP (Mehmood et al. 2017), while also playing a role in the transmission, are fundamental at the collection stage for ensuring data is gathered and packaged correctly for transmission. MQTT is notable for its lightweight characteristics, making it ideal for scenarios with limited bandwidth, high latency, or unreliable networks. CoAP, used for devices in constrained environments, simplifies data transmission at the initial collection point.

Interoperability is another crucial factor at the data collection stage (Lee et al. 2021), ensuring that various devices can communicate and share data effectively. Interoperability not only considers the compatibility between different device types but also the protocols and standards they use, fostering a seamless and efficient data collection process (Hazra et al. 2021). Initiatives and efforts, such as those led by the Internet Engineering Task Force (IETF) and many other standardization bodies (e.g., 3GPP, IEEE, etc.), actively work toward standardization to ensure that different protocols, data formats, and devices can effectively interoperate with one another (Morabito and Jiménez 2020; Lee et al. 2021).

6.3.3 Data Transmission

Efficient data transmission is critical in the deployment of IoT systems within smart cities, as it acts as the bridge between data collection and data services. The significance of effective data transmission lies in the necessity for real-time (or near real-time), accurate, and secure transmission of data from myriad IoT devices to their respective endpoints.

The challenges in data transmission are multiple. Applications within smart cities necessitate the transmission of a wide and varied volume of data, requiring robust and adaptable networks (Javed et al. 2022). The latency in data transmission, or the delay in data transfer, becomes particularly significant for applications that mandate immediate or real-time responses. Limited bandwidth is another substantial challenge, often stressed in areas densely populated with devices simultaneously transmitting data.

Moreover, the heterogeneity of transmission technologies introduces complexity. Various technologies, including LoRa, Wi-Fi, Bluetooth, LTE-M, NB-IoT, and 5G, offer different advantages and challenges (Motlagh et al. 2023). For instance, while LoRa provides long-range connectivity and low-power consumption, it might not offer the high data rates required for some applications. Conversely, 5G provides high data rates and low latency, supporting applications with demanding throughput and responsiveness requirements.

Smart city applications can also be characterized by different requirements (Singh et al. 2023), aligning with the categorizations provided by 5G networks. Ultrareliable low latency communications (URLLC) is crucial for applications that require immediate responses with minimal delay. Enhanced mobile broadband (eMBB) caters to applications that need high data rates and bandwidth. Finally, massive machine-type communications (mMTC) is essential for supporting a massive number of connected devices, typically seen in densely populated urban areas.

To address these challenges, it is fundamental to deploy and use optimized data transmission protocols and technologies, ensuring each application's unique requirements are met. Techniques like data compression can be utilized to reduce the amount of data transmitted, saving bandwidth and improving transmission efficiency.

6.3.4 Data Services

Data services play a fundamental role in the framework of IoT within smart cities, offering a wide set of functionality that are essential for effectively managing and utilizing the data gathered. Within this landscape, we identify four main components belonging to data services: *data storage*, *data processing*, *data analytics*, and *data sharing and access*. These components are interconnected, each playing a critical role while collaboratively working to ensure that data flows seamlessly through the system from collection to actionable insight, ultimately serving as the backbone for various smart city applications.

Data storage and **data processing** are pivotal in the IoT life cycle within smart cities (Gharaibeh et al. 2017), serving as the repository and analysis mechanism for the vast data generated. Efficient and secure data storage solutions are essential due to the immense volume of data continuously produced by various IoT devices. These solutions must guarantee data integrity, swift retrieval times for real-time applications, and robust security to protect sensitive information from unauthorized access and potential breaches. On the processing end, transforming the raw data into actionable insights presents its challenges. First, there is a demand for substantial computational power to analyze and process the collected data efficiently. Quality control of the data is also paramount; ensuring accuracy is crucial for reliable analysis and insights. Strategies and technologies must be in place to handle incomplete or "noisy" data, requiring sophisticated data cleaning and validation processes. Additionally, for real-time applications, minimizing latency from data collection to insight generation is critical.

Several technologies and strategies have emerged to address the challenges associated with data storage and processing. Cloud computing (Pan and McElhannon 2017) offers a viable solution, providing scalable storage and computing resources. This technology is particularly well-suited for applications without stringent latency requirements. For applications demanding real-time data processing, edge computing (Hassan et al. 2018) offers a solution by processing data closer to its generation point, thereby reducing latency and conserving bandwidth. Data warehouses and distributed databases also play a crucial role (Diène et al. 2020). Data warehouses serve as centralized repositories that store integrated data from various sources, designed mainly for query and analysis. In contrast, distributed databases provide a framework for storing and processing large data volumes across a network of computers, offering scalability and fault tolerance.

Data analytics takes the processed data to the next level by employing advanced tools and algorithms to interpret and analyze it for patterns, trends, and hidden insights. While data processing prepares and refines the data, data analytics is concerned with drawing meaningful conclusions and providing foresight and understanding that inform decision-making processes. Within this framework, technologies like AI and ML play a significant role in providing deeper insights, offering predictive analytics and facilitating more informed and proactive decision-making and planning in the urban context. This process encompasses three main analytics types: descriptive, predictive, and prescriptive (Atitallah et al. 2020; Motlagh et al. 2023). *Descriptive analytics*, commonly utilized in business, measures and contextualizes past performance to aid decision-making. It brings out hidden patterns and insights from historical data but isn't primarily used for forecasting. *Predictive analytics*, on the other hand, goes beyond description, extracting information from raw data to identify patterns and relationships, thereby facilitating forecasts of behaviors and events. Using both historical and current data, predictive analytics provides valuable foresights. *Prescriptive analytics* advances further, quantifying the potential effects of future decisions to provide recommendations and insights on possible outcomes. This advanced analytics type supports decision-making by offering choices and suggestions based on data analysis, making it a crucial tool for planning and strategy in smart cities.

However, the integration of big data analytics necessitates a clear understanding of specific functional and nonfunctional requirements (Silva et al. 2013; Santana et al. 2017), given the diverse and dynamic nature of data sources and applications within smart cities. Functional requirements encompass aspects like interoperability, real-time monitoring, access to historical data, mobility, service composition, and integrated urban management. On the other hand, nonfunctional requirements include sustainability, availability, privacy considerations, social impact, and scalability. Addressing these requirements is imperative for developing robust and resilient smart city architectures that can seamlessly integrate and analyze data from heterogeneous sources, including IoT sensors, social media networks, and electronic medical records. Furthermore, the dynamic urban environment of smart cities demands attention to stream data analytics, enabling real-time services while also accommodating planning and decision-making processes through historical or batch data analytics. Essential characteristics that a big data analytics platform should embody to navigate the challenges of big data include scalability, fault tolerance, I/O performance, real-time processing capabilities, and support for iterative tasks.

Effective and secure **data sharing and access** is key to maximizing the utility of data in smart cities. This involves making collected data available to authorized entities, departments, or individuals who require it for various applications and analytics, always with robust data access policies and mechanisms in place to ensure both data sharing and privacy protection. Data sharing in the context of smart cities encompasses a set of technologies, practices, and frameworks aimed at facilitating secure and efficient data access among multiple stakeholders without compromising data integrity (What is data sharing?—Data sharing explained—AWS 2022). This process is integral to improving efficiency and fostering collaboration not only within city departments but also with external partners, vendors, and the community at large, all while being aware of and mitigating associated risks. There are at least two main factors that strengthen the importance of data sharing in smart cities. The first relates to the possibility of integrating data from different sources, which can possibly enhance the value and performance of dedicated services (Delicato et al. 2013). For instance, data sharing enables improved urban planning and transportation management by combining information from traffic cameras, sensors, and public feedback, leading to more effective and responsive city services. The second is linked to a more effective *data-driven decision-making*. Transparent information sharing facilitates improved analytics, enabling city officials and stakeholders to make informed and effective long-term decisions (Shahat and Elragal 2021). For example, integrating data from environmental sensors, healthcare institutions, and public service departments can provide a holistic view of city health and environmental conditions, aiding in timely decision-making and policy formulation.

However, the process of data sharing is not without challenges. Risks include potential privacy disclosure, where organizations must navigate legal and ethical obligations to protect customer data while sharing information responsibly. The process also opens up possibilities of data misinterpretation and issues related to data quality, including hidden biases in datasets (What is data sharing?—Data sharing explained—AWS 2022).

In mitigating risks and facilitating data sharing in smart cities, several technologies are essential. Among these, data warehousing is crucial for internal data sharing, serving as a repository for data from various departments and allowing isolated access to shared information (What is data sharing?—Data sharing explained—AWS 2022). Next, APIs play a key role by enabling fine-grained communication and controlled data sharing between software components. They precisely dictate accessible data and usage rules, ensuring structured and secure data exchange (Badii et al. 2017). Lastly, federated learning is transformative, allowing collaborative AI and ML development while maintaining data control and privacy for each contributor. This approach not only enhances data-driven insights but also ensures confidentiality, supporting robust and intelligent smart city applications (Jiang et al. 2020).

While data services provide the foundational support for various applications in smart cities, the effectiveness of these applications is highly dependent upon the quality of the data being collected, transmitted, and analyzed. The following section, therefore, will delve into the topic of data quality, exploring the challenges and considerations related to ensuring the accuracy, reliability, and validity of data in smart city ecosystems.

6.3.5 Data Quality

An IoT application may comprise hundreds or thousands of sensor devices that produce vast amounts of data. This data is rendered useless if it is riddled with errors as poor sensor data quality caused by the errors may lead to wrong decision-making results. In order to enable massive deployment, most IoT applications use low-cost sensor techniques, though at the expense of data quality. As a result, IoT often encounters soft faults (i.e., error) which are associated to outliers, bias, drifts, missing values, and uncertainty, which should be detected or quantified and removed or corrected in order to improve sensor data quality (Teh et al. 2020). Due to the diverse nature of IoT deployments and the likelihood of sensor failures in the wild, a key challenge in the design of IoT systems is ensuring the integrity, accuracy, and fidelity of sensor data (Chakraborty et al. 2018)

The error within an IoT application may take place for different reasons. For example, in a sensor network serving an IoT application, poor data quality may arise from congested and unstable wireless communication links and can cause data loss and corruption (Zhang et al. 2018). The other example pertains to the damage or exhaustion of battery in sensor devices that would cause the data quality to degrade, as toward the end of its battery life, sensors tend to produce unstable readings (Ye et al. 2016).

In addition, the role of external factors such as the hostile environment is not negligible on sensor readings and data quality. For example, air quality IoT devices that include aerosol, trace gases, and meteorological sensors are often placed outdoors and are subjected to extreme local weather conditions such as strong winds and snow, which might affect the operation of the sensor (Zaidan et al. 2022).

In IoT datasets, one of the most common data quality problems is called missing data (incomplete data) which indicates a portion of data that is missing from a time-series data (Wang and Strong 1996). In principle, the missing data may be caused by different factors such as unstable wireless connection due to network congestion; sensor device outages due to its limited battery life; environmental interferences, e.g., human blockage, walls, and weather conditions; and malicious attacks (Li and Parker 2014).

To cover missing data, one solution can be to retransmit the data. However, since most IoT applications are in real time, therefore, the data retransmission would not be effective as (i) rendering the data is not beneficial if there is a delay and (ii) the retransmission adds to the computation and energy costs. The latter is due to the fact that the sensor devices are usually limited in terms of battery, memory, and computational resources. However, to fill in the missing data, an alternative would be applying imputation based on Akima Cubic Hermite (Zaidan et al. 2020) and multiple segmented gap iteration (Liu et al. 2020) methods.

Another common problem that involves data quality is called outlier which can be in the forms of anomalies (Zaidan et al. 2022; Aggarwal 2017) and spikes (Ahmad et al. 2009; Bosman et al. 2017). An outlier takes place when sensor measurement values exceed thresholds or largely deviate from the normal behavior provided by the model. In other words, the outlier occurs when the sensor measurement value is significantly different from its previous and next observations or observations from neighboring sensor nodes (Rassam et al. 2014; Dereszynski and Dietterich 2011). In practice, outliers can be identified by applying anomaly detection methods based on adaptive Weibull distribution (Zaidan et al. 2022) and principal component analysis (PCA) (Zhao and Fu 2015; Harkat et al. 2000).

Besides the outliers, another common problem in IoT data quality is known as bias or offset (Ferrer-Cid et al. 2019), which occurs when the sensor measurement value is shifted in comparison with the normal behavior of a reference sensor. A drift is a specific type of bias that takes place when the sensor measurement values deviate from their true values over time. Drifts are usually caused by IoT device degradation, faulty sensors, or transmission problems (Rabatel et al. 2011).

In current solutions, the drifts caused by any reasons can be detected by comparing two types of Bayesian calibration models (Zaidan et al. 2023) or applying ensemble classifiers where each classifier will learn a normal behavior model and compare it with the current reading (Bosman et al. 2015). In order to correct the bias and drift, calibrations are usually required (Zaidan et al. 2023). For example, air quality low-cost sensors often experience bias and drift in the field due to the sensors' device quality and variations in environmental factors. The sensors can then be calibrated using machine learning (ML) models, such as nonlinear autoregressive network with exogenous inputs (NARX) and long short-term memory (LSTM), to improve data quality and meet the data quality of reference instruments (Zaidan et al. 2020).

With a clearer understanding of the importance of data quality analysis, and having navigated through the various challenges and solutions crucial to each aspect of the data life cycle in IoT as summarized in Table 6.1, we move forward to explore how the concepts and challenges discussed thus far manifest in real-world scenarios. The next section provides a practical perspective through a case study on *air quality monitoring with IoT for smart cities*. This case study offers a valuable understanding into the application of data collection, transmission, services, and quality principles in the development and implementation of smart city applications, serving as a tangible example of a theory translated into practice.

6.4 Case Study: Air Quality Monitoring with IoT for Smart Cities

This section presents a case study where IoT devices were used for an air quality monitoring network in Helsinki, Finland, a well-known smart city. Air pollution is known to be harmful to human health and the environment. According to the World Health Organization (WHO), air pollution causes approximately 7 million in deaths each year. Of this, an estimated 4.2 million deaths are due to outdoor exposure (World 2021). Official air quality monitoring stations have been established across many smart cities around the world. Unfortunately, these monitoring stations are sparsely located and consequently do not provide high-resolution spatiotemporal air quality information (Kortoçi et al. 2022). Thanks to advances in communication and networking technologies, and the Internet of Things (IoT), low-cost sensors have emerged as an alternative that can be deployed on a massive scale in cities (Zaidan et al. 2020). This deployment offers a high resolution of spatiotemporal air quality information (Motlagh et al. 2020). This case study demonstrates how air quality IoT devices benefit several aspects in terms of local pollution monitoring, traffic management, and urban planning.

Table 6.1 A summary of key challenges and solutions for deploying massive IoT in smart cities

Concern	Key challenge	Solution
Sensor deployment and placement	Identifying the key locations and finding optimal places that allow the most coverage	Current solutions use interpolations (of data collected from other nodes in the sensor network). The need for enhanced methods that consider the population, urban, and environmental factors
Data collection	Gathering various types of data that are reliable and efficient and interoperable	(i) Establishing protocols such as MQTT and CoAP, to ensure reliability and efficiency
		(ii) Establishing standardization to ensure that different protocols, data formats, and devices can effectively interoperate with one another
Data transmission	The latency in data transmission, limited bandwidth, limited connectivity, and heterogeneity of transmission technologies	(i) 5G networks also provide URLLC, eMBB, and mMTC to respond to the minimal delay, high bandwidth, and massive connection requirements, respectively
		(ii) Optimizing data transmission protocols and technologies, e.g., by applying data compression methods to reduce the amount of transmission data and save bandwidth
Data services	(i) Data storage and data processing to handle the immense volume of generated IoT data	(i) Cloud computing provides scalable storage and computing resources, and edge computing offers processing data closer to its generation point, data warehouses, and distributed databases facilitate storing and processing large data volumes
	(ii) Data analytics to draw meaningful conclusions and provide foresight for decision-making processes	(ii) Technologies like AI and ML can play a significant role in providing deeper insights, offering predictive analytics and facilitating more informed and proactive decision-making and planning
	(iii) Data sharing and access to maximize the utility of data	(iii) Integrating data from different sources can enhance the value and performance of dedicated services, and transparent data sharing can improve analytics and lead to more effective decision-making
Data quality	The occurrences and identification of poor data quality, e.g., missing data, outlier bias, and drift data	(i) Identification of anomalous and poor data quality through drift detection
		(ii) Correcting the data by applying imputation and calibration methods

6.4.1 IoT Installation

This subsection describes the experimental details including the sites, IoT devices, and the data collected from the experiments.

Experimental Sites

In this case study, two air quality IoT devices were installed at the following two different sites in the city of Helsinki, Finland. These sites include the following:

1. The Kumpula site is located at the Kumpula campus of the University of Helsinki in the front open yard and about 4 kilometers northeast of the Helsinki center. The site is also considered as an urban background that is situated at about 150 meters from a main street in Kumpula district in Helsinki (Järvi et al. 2009).
2. The Makelankatu site is known as a street canyon and is located just beside Makelankatu Street, which is one of the arterial roads and is lined with apartment buildings. The street consists of six lanes, two rows of trees, two tramlines, and two pavements, in a total of 42 meters of width. Every day, different types of vehicles including cars, buses, and trucks cross this street and thus cause frequent traffic congestion (Hietikko et al. 2018).

The map of both sites is presented on the left-hand side picture in Fig. 6.3. The Kumpula site is notated by **K**, whereas the Makelankatu site is notated by **M**. The distance between the two sites is 900 meters.

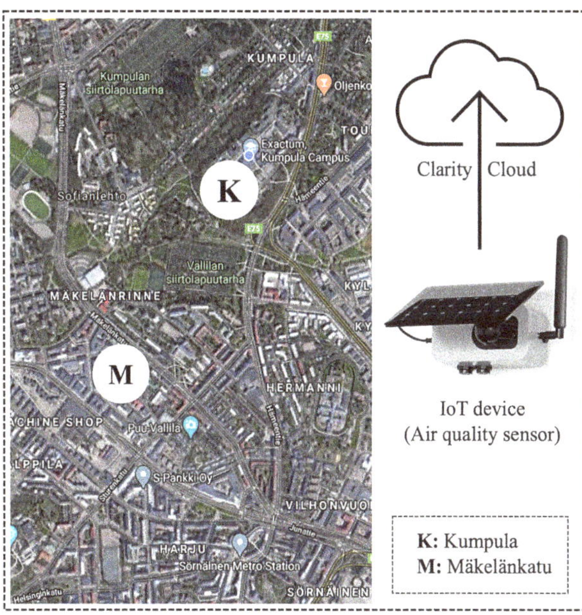

Fig. 6.3 The sites and the IoT devices used in the experiment

IoT Devices
Air quality IoT devices used in this experiment are developed by Clarity Corporation, a company that is based in Berkeley, California, USA. These IoT devices are shown on the right-hand side of Fig. 6.3. The weight of the device is 450 grams. The input power of the sensor is 5 volts. The sensor device is designed to operate by battery and has a battery lifetime of 15 days of continuous measurements. If the battery operates by harvesting solar power, its operation time extends to 1 to 2 years. In our experiment, we used grid electricity for the sensor's input power. The sensors offer sensing meteorological variables including the temperature (temp) which uses bandgap technology and relative humidity (RH) which uses capacitive technology. The sensors also measure particulate matter (PM) and CO_2 with laser light scattering technology and metal oxide semiconductor technologies, respectively.

The sensors underwent a laboratory calibration process, by the manufacturer, using federal reference method (FRM) instruments. The sensors are equipped with the LTE-4G communication module to transmit the measured data. The transmitted data is also stored in a cloud platform facilitated by Clarity.[1] The cloud platform allows access to the raw sensor and visualized data. The data can also be downloaded using a user interface accessible by SmartCity WebApp.[2] The measurement frequency of data varies around 16–23 minutes per data point. We installed one of these IoT devices on a container at the Kumpula site (**K**) about 2 meters from the ground level and another one at the Makelankatu site (**M**) on the top of a container about 4 meters above the ground level.

The Data
We collected the datasets from January 1 to December 31, 2018, from the two IoT devices. For our analysis, in this chapter, we use $PM_{2.5}$ and PM_{10}, and Air Quality Index (AQI) variables, extracted from the datasets. In our analysis, we process the data in an hour resolution. In practice, AQI is defined as the maximum of the indexes for six criterion pollutants, including PM_{10}, $PM_{2.5}$, CO, NO_2, O_3, and SO_2 (Fung et al. 2022).

6.4.2 *Air Quality IoT Monitoring for a Smart City*

This subsection explains how air quality IoT devices can benefit a smart city using the analysis extracted from the IoT experiments. These benefits include local air pollution monitoring, traffic management, and urban planning.

Local Air Pollution Monitoring
One of the key motivations for deploying dense air quality IoT devices in city districts is to provide local air pollution monitoring at fine-grained resolution. In

[1] smartcity.clarity.io.
[2] clarity.io/documents.

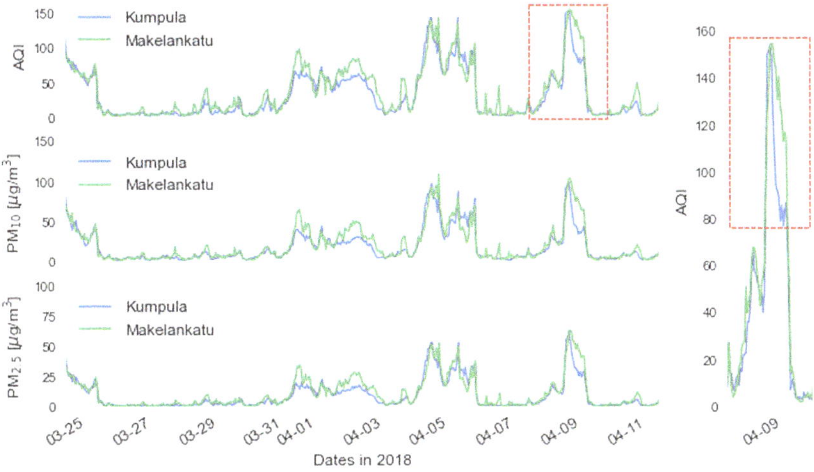

Fig. 6.4 Time-series data of AQI, PM_{10} and $PM_{2.5}$ concentrations (in $\mu g/m^3$) at Kumpula (K) and Makelankatu (M) sites

principle, in urban areas, the quality of air changes even at a few ten meters of distance. To show such a variation, we extract measurements of AQI, $PM_{2.5}$, and PM_{10} from our two IoT devices, between March 25 and April 11, 2018. Then, as illustrated in Fig. 6.4, we plot the time series of these variables. In the figure, the blue color presents the measurements from the Kumpula site, and the green color portrays the air quality captured at the Makelankatu site. In the figure, the top subfigure shows the AQI variations, and the middle and bottom subfigures depict the PM_{10} and $PM_{2.5}$ concentrations, respectively.

As shown in the plots, in general, both measurements have similar patterns. The green curves lie slightly above the blue curves most of the time, indicating that the pollution level in the Makelankatu site is higher than the Kumpula site. Between March 27 and 31, PM_{10} and $PM_{2.5}$ show relatively low pollution concentrations. These results are also confirmed by AQI which indicates overall low pollution levels for those dates. On April 1, all pollutant indexes fluctuate and show a slight increase and decrease. Then, we observe another fluctuation with a higher increase from April 5 to 7. Again, we observe another rapid fluctuation between April 9 and 10. Furthermore, by only considering the fluctuations in the air quality from April 9 to 10 (as zoomed in and shown on the right side of Fig. 6.4), we observe a large discrepancy between the pollution levels **K** and **M** with a difference of 80 $\mu g/m^3$.

As a result, the fluctuations shown for the period of the time-series plot, as well as the variations of the measurements in both sites **K** and **M**, call for the need for the deployment of air quality IoT devices separately at both sites in order to detect pollution hotspots and also monitor the air quality at fine-grained resolution in real time. Indeed, deploying dense air quality sensors in cities could provide more accurate information leading to more robust and reliable conclusions about air quality levels at higher resolution, even at a few meter distances. A

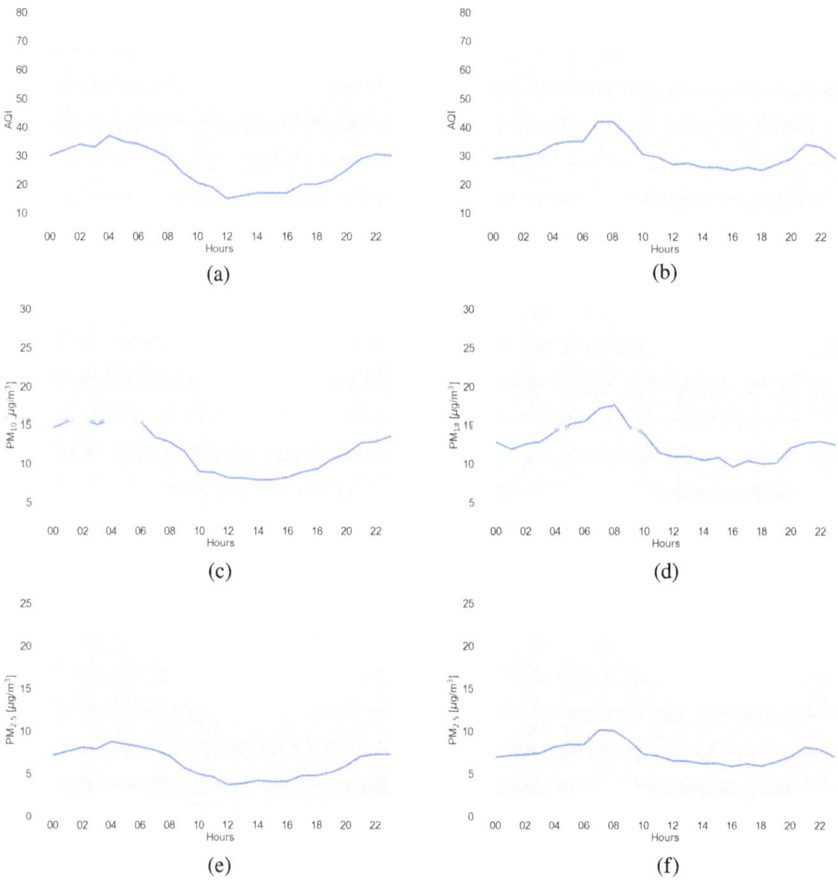

Fig. 6.5 Diurnal cycles for AQI, PM_{10}, and $PM_{2.5}$ in Kumpula (left) and Makelankatu (right) sites. (**a**) AQI at the Kumpula site. (**b**) AQI at the Makelankatu site. (**c**) PM_{10} at the Kumpula site. (**d**) PM_{10} at the Makelankatu site. (**e**) $PM_{2.5}$ at the Kumpula site. (**f**) $PM_{2.5}$ at the Makelankatu site

dense deployment can also assist in creating emission inventories of pollutants and detecting pollution sources, as well as allowing real-time exposure assessment for designing mitigation strategies (Kumar et al. 2015).

Traffic Management

Traffic is one of the main sources of outdoor air pollution in urban areas (Bigazzi and Rouleau 2017; Motlagh et al. 2021). The health effects of traffic-related air pollution continue to be of important public health risks (Boogaard et al. 2022). In order to carry out effective traffic management driven by the level of air pollution, it is important to have air quality IoT devices installed next to roads. Therefore, the patterns of air pollution can be observed in roads allowing designing appropriate traffic management strategies.

Figure 6.5 shows diurnal cycles of AQI, PM_{10}, and $PM_{2.5}$ at the sites of Kumpula (right) and Makelankatu (left). The x-axes show the 24-h time period, whereas the y-axes exhibit the levels of AQI and PM concentrations (in $\mu g/m^3$). The blue curves are the median of the data for each variable aggregated from one year of data, whereas the shaded areas represent the lower quartile (25%) and upper quartile (75%) of the data for each variable aggregated from one year of data (i.e., from January 1 to December 31, 2018).

As demonstrated in Fig. 6.5, on the Kumpula site (the left subfigures), the AQI, PM_{10}, and $PM_{2.5}$ do not increase during the peak hours (i.e., rush hours when people and vehicle movement is high). This is due to the fact that the Kumpula site is located in an urban background with less exposure to traffic emissions. However, on the Makelankatu site (the right subfigures), the AQI, PM_{10}, and $PM_{2.5}$ show an increase during peak hours, mainly between 8 AM and 10 AM. These patterns explain that Makelankatu street is a busy road during the rush hours, especially in the mornings.

As a result, these patterns and the pollution concentration levels can be used by authorities to study, for example, the traffic behaviors and types of vehicles and therefore devise possible interventions to reduce the amount of pollutants in the areas where the IoT devices are installed. For instance, $PM_{2.5}$ (that are known as fine particles) are predominantly emitted from combustion sources like vehicles, diesel engines, and industrial facilities; and PM_{10} (that are known as coarse particles) are directly emitted from activities that disturb the soil including travel on roads, construction, mining, open burning, or agricultural operations (Harrison et al. 2021). Hence, understanding the levels of PM_{10} and $PM_{2.5}$ concentrations at different locations enables planning appropriate interventions and designing effective traffic management strategies.

Urban Planning

Modern urban planning needs to consider environmental pollution and factors that threaten cities. Among many, AQI is known to be an important indicator that plays a vital role in urban life. Based on yearly AQI information, appropriate urban planning can be designed by considering the effects of different factors on air quality such as topography, buildings, roads, vegetation, and other external sources (e.g., traffic) (Falzone and Romain 2022). Thus, poor AQI levels may indicate areas that are unsuitable for certain types of land use. For instance, sensitive land uses like schools, hospitals, and residential areas can be kept away from major pollution sources like factories or highways.

Figure 6.6 presents different percentages of AQI levels in four different seasons for the two sites. The figure shows the whole data aggregated for a year (from January 1 to December 31, 2018). The AQI is divided into four levels including good (green), satisfactory (light green), fair (yellow), poor (orange), and very poor (red). For example, in the summer, the AQI levels in Kumpula (Fig. 6.6a) are better than in Makelankatu (Fig. 6.6b). This is because the Kumpula site is surrounded by vegetation and trees during the summertime. In wintertime, on the other hand, the Kumpula site is slightly more polluted than the Makelankatu site, as there is no vegetation and trees are without leaves, causing the Kumpula site to be exposed easily to air pollutants transported by nearby roads. The Kumpula area

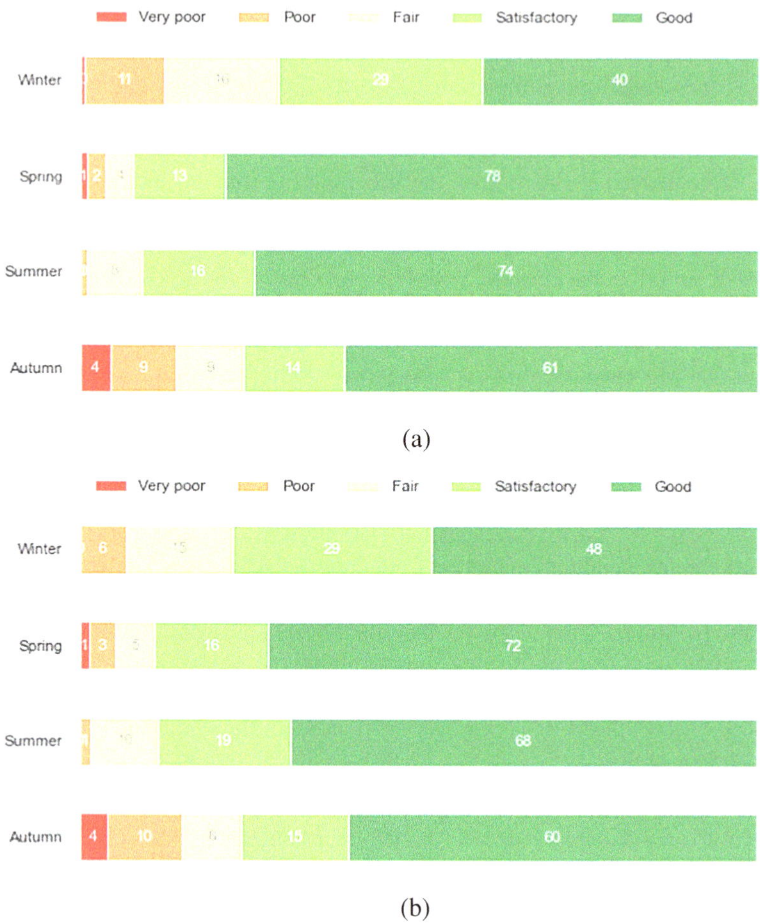

Fig. 6.6 Different AQI levels (%) in four different seasons at the two sites. (**a**) AQI at the Kumpula site. (**b**) AQI at the Makelankatu site

hosts residential buildings, university campuses, and a school; thus, to mitigate the air pollution effects, in this area, it is important for city planners to consider planting evergreen trees (He et al. 2020) such as Scots pine, Norway spruce, common juniper, and European yew.

In the Makelankatu site, on the other side, due to its proximity to the main road, the AQI levels are worse than the Kumpula site. Therefore, better traffic management strategies can be devised for the Makelankatu road. In general, air quality analysis based on AQI can provide information about prominent air pollution problems. Therefore, scientific assessments can be carried out in order to realize future development and planning for smart cities (Feng et al. 2019).

6.5 Role of AI and Emerging Technologies in Future Smart Cities

The convergence of AI and IoT—often defined as AIoT (Zhang and Tao 2020)—is not only expected but is already serving as a foundational element in the development of smart cities. With AI currently playing a key role in managing and interpreting the increasing volumes of data generated by a diverse array of IoT devices, it is evident that its significance will only amplify moving forward. As the data landscape continues to expand and AI methods undergo continuous refinement and innovation, there is growing potential for integrating newer, more efficient AI models and methodologies into key enabling technologies. Such integration can facilitate the creation of fully automated AI-enabled smart cities, and it also ensures that smart city ecosystems are equipped to adapt and respond to the ever-changing demands and challenges of evolving urban spaces.

Below, we outline a set of pivotal enabling technologies situated at the intersection of AI and IoT, each playing a crucial role in fostering the development of future smart cities. It is worth highlighting that the list presented is not exhaustive. Instead, it provides an illustrative snapshot of significant, emerging technological trends that are currently shaping the smart cities' landscape. These identified technologies are presented as key drivers facilitating the emergence of cities that are not only smarter but also more efficient and responsive. Each technology contributes its unique strengths and capabilities, offering varied solutions. Together, they equip smart cities with functional modules necessary for addressing the myriad challenges these complex ecosystems currently face and will encounter in the future.

Digital Twin Systems

Deploying IoT and sensor networks in urban areas provides the opportunity for the creation of digital twin systems in smart cities. For example, deploying a massive number of surveillance cameras in cities can enable real-time monitoring of the people and traffic flow in cities and learning patterns from the movements and moving directions, allowing better planning for the traffic design. Similarly, using the telecom infrastructure and wireless access points deployed in cities makes it possible to estimate the number of access requests by the users (even for specific IoT applications), and therefore planning better resource management and thus improving the quality of experiences by the users. Moreover, as highlighted earlier in this chapter, deploying air pollution sensors allows for capturing air pollution in real time and identifying hotspots in cities, leading to better planning for the cities. Using such massive deployments therefore enables the creation of digital twins, a powerful tool that provides the digital transformation of smart cities that enables real-time and remote monitoring of the physical elements (such as buildings and transportation systems) in cities and therefore enables effective decision-making by the policy makers (Deng et al. 2021).

On-Device Machine Learning
On-device ML, also known as TinyML, is pivotal in advancing Artificial Intelligence of Things (AIoT), offering substantial benefits in terms of efficiency, latency, and privacy (Dutta and Bharali 2021). TinyML enables devices to process and analyze data locally, reducing the need for constant connectivity and data transmission to centralized data centers, thereby decreasing latency and minimizing bandwidth usage. This approach makes AIoT applications more responsive and reliable while also enhancing privacy and security by keeping sensitive data on the device. In the specific context of smart cities, there are several application scenarios where TinyML can play a transformative role. For example, it fosters the development of smart and autonomous entities capable of making decentralized and quick decisions in applications like traffic and pollution monitoring, thereby contributing to the collective intelligence in smart cities. Such deployment simplicity of TinyML, coupled with its independence from the power grid, facilitates the establishment of smart spaces even in remote and disadvantaged areas, promoting their economic and technological revitalization (Sanchez-Iborra and Skarmeta 2020). With the impending surge in urban populations, and the consequent strain on city resources and infrastructure, the introduction of TinyML in smart spaces is crucial for efficient resource optimization and energy waste reduction. This is imperative not just for managing the increasing energy demands but is integral to meeting stringent carbon neutrality goals set for sustainable urban living (Rajapakse et al. 2022). Furthermore, practical applications of TinyML, such as deploying LSTM autoencoders on constrained devices for tasks like anomaly detection in urban noise sensor networks, showcase its potential and versatility in urban settings, paving the way for future explorations into on-device model training and trust management systems among sensor devices (Hammad et al. 2023). Each aspect of on-device ML represents a significant step toward decentralized, efficient, and intelligent urban planning and decision-making processes in smart cities.

6G Connectivity
Previous communication technologies including 4G, LTE-M, and NB-IoT as well as the current 5G technology have paved the way allowing large-scale deployment of IoT in smart cities. Indeed, the earlier 4G technology provided the communication resources that can support a large variety of IoT applications, and LTE-M and NB-IoT technologies were planned to specifically support machine-to-machine and IoT deployments. Later, the 5G improved the communication capabilities of the previous technologies by providing URLLC, eMBB, and mMTC. Currently, 5G paves the way for using AI for 6G, the next-generation communication technology (Strinati et al. 2019).

6G is expected not only to enhance communication capabilities (i.e., by URLLC+, eMBB+, and mMTC+) but also to offer AI and wireless sensing as new network services. In practice, 6G will see the physical objects through electromagnetic waves and will improve communication performance by providing high-resolution sensing, localization, imaging, and environment reconstruction capabilities. 6G will provide joint communication and sensing that will integrate localization, sensing, and communication and will facilitate edge intelligence and enable the transformation from connected things and people to connected intelligence (Rong 2021).

The edge intelligence will also offer intelligence at the edge and will enable the processing of large datasets for critical IoT applications and computations. The edge intelligence will thus provide swift replies with precise decisions for the requested services by the specific IoT applications. Moreover, 6G is expected to provide high-density IoT connections and support one million connections per square kilometer (Gupta et al. 2021). Benefiting from these advanced features, therefore, 6G will support a wide variety of IoT applications at very large scales and with very high dense deployments. Examples of such applications would include but are not limited to activity detection, gesture recognition, mobility identification, remote sensing, simultaneous localization and mapping, object tracking, and security screening (Jiang et al. 2023).

Blockchain

While AI-based technologies provide the intelligence required for insight generation and decision-making automation in smart cities, ensuring the security and integrity of the data utilized by AI algorithms is equally crucial. Blockchain emerges as a key enabler, safeguarding data collected and transmitted by AI-enabled smart city systems, providing a secure, reliable, and trustworthy environment (Li 2018).

Furthermore, blockchain not only enhances the security and efficiency of IoT-enabled smart city applications but also mitigates data vulnerability and addresses single-node failures inherent in cloud-based solutions (Kumari et al. 2021). Though cloud-based architectures are widely used, they are susceptible to cyberattacks, including data tampering and false data injection, and can experience reliability issues due to single-node failures (Wang et al. 2019). In this context, blockchain, with its decentralized distributed ledger technology (DLT), offers a robust and transformative alternative. It ensures transparency, data immutability, and integrity while providing pseudonymity. This technology is vital for smart cities, offering secure, resilient, and dynamic services across various sectors, including smart grids and intelligent transportation systems (ITS) (Kumari et al. 2021). Blockchain facilitates trust-free, peer-to-peer transactions without central authorities and protects users' identities through public pseudonymous addresses (Rahman et al. 2019).

The use of smart contracts, which is related to the *data sharing* case discussed in Sect. 6.3.4, automates transactions between parties, streamlining smart city operations seamlessly. The convergence of blockchain's security features, 6G connectivity, and AI intelligence is fundamental for the development of secure, resilient, and adaptable smart cities, ready to meet the evolving requirements of future applications (Javed et al. 2022).

6.6 Conclusion

Information and communication technologies are advancing rapidly, causing an increase in the deployed network infrastructure and fostering an increase in the variety and scale of IoT applications that support smart cities. This chapter addressed the requirements and challenges associated with large-scale IoT deployments in smart cities considering the advances in emerging communication and computing technologies. The chapter also highlighted the roles of AI and 5G beyond networks as well as the computing technologies that are needed to enable massive-scale IoT deployments in cities. To showcase the benefits of IoT deployments in cities, the chapter also presented the results obtained from a real-world case study of deploying two air quality IoT devices in the city of Helsinki, deployed at two separate locations. The results explain how these IoT devices can benefit decision-making by providing local air pollution monitoring, traffic management, and urban planning. Finally, the chapter explains the role of AI and emerging technologies by addressing the advances toward blockchain, digital twin systems, on-device machine learning, and 6G connectivity that would play a fundamental role in the creation of future smart cities.

References

Aggarwal, Charu C. 2017. *An introduction to outlier analysis*. New York: Springer.

Ahmad, Nor Faizah, et al. 2009. Robust preprocessing for health care monitoring framework. In *2009 11th International Conference on e-Health Networking, Applications and Services (Healthcom)*, 169–174. IEEE.

Atitallah, Safa Ben, et al. 2020. Leveraging deep learning and IoT big data analytics to support the smart cities development: Review and future directions. *Computer Science Review* 38: 100303.

Badii, Claudio, Pierfrancesco Bellini, Daniele Cenni, et al. 2017. Analysis and assessment of a knowledge based smart city architecture providing service APIs. *Future Generation Computer Systems* 75: 14–29.

Badii, Claudio, Pierfrancesco Bellini, Angelo Difino, et al. 2020. Smart city IoT platform respecting GDPR privacy and security aspects. *IEEE Access* 8: 23601–23623.

Bigazzi, Alexander York, and Mathieu Rouleau. 2017. Can traffic management strategies improve urban air quality? A review of the evidence. *Journal of Transport & Health* 7: 111–124.

Boogaard, H., et al. 2022. Long-term exposure to traffic-related air pollution and selected health outcomes: A systematic review and meta-analysis. *Environment International* 164: 107262.

Bosman, Hedde H.W.J., et al. 2015. Ensembles of incremental learners to detect anomalies in ad hoc sensor networks. *Ad Hoc Networks* 35: 14–36.

Bosman, Hedde H.W.J., et al. 2017. Spatial anomaly detection in sensor networks using neighborhood information. *Information Fusion* 33: 41–56.

Chakraborty, Tusher, et al. 2018. Sensor identification and fault detection in IoT systems. In *Proceedings of the 16th ACM Conference on Embedded Networked Sensor Systems*, 375–376.

Cui, Lei, et al. 2018. Security and privacy in smart cities: Challenges and opportunities. *IEEE Access* 6: 46134–46145.

Delicato, Flavia C., et al. 2013. Towards an IoT ecosystem. In *Proceedings of the First International Workshop on Software Engineering for Systems-of-Systems*, 25–28.

Deng, Tianhu, et al. 2021. A systematic review of a digital twin city: A new pattern of urban governance toward smart cities. *Journal of Management Science and Engineering* 6 (2): 125–134.

Dereszynski, Ethan W, and Thomas G. Dietterich. 2011. Spatiotemporal models for data-anomaly detection in dynamic environmental monitoring campaigns. *ACM Transactions on Sensor Networks (TOSN)* 8 (1): 1–36.

Diène, Bassirou, et al. 2020. Data management techniques for Internet of Things. *Mechanical Systems and Signal Processing* 138: 106564.

Donta, Praveen Kumar, et al. 2022. Survey on recent advances in IoT application layer protocols and machine learning scope for research directions. *Digital Communications and Networks* 8 (5): 727–744.

Dutta, Lachit, and Swapna Bharali. 2021. TinyML meets IoT: A comprehensive survey. *Internet of Things* 16: 100461.

Falzone, Claudia, and Anne-Claude Romain. 2022. Establishing an air quality index based on proxy data for urban planning part 1: methodological developments and preliminary tests. *Atmosphere* 13 (9): 1470.

Feng, Yueyi, et al. 2019. Defending blue sky in China: Effectiveness of the Air Pollution Prevention and Control Action Plan on air quality improvements from 2013 to 2017. *Journal of Environmental Management* 252: 109603.

Ferrer-Cid, Pau, et al. 2019. A comparative study of calibration methods for lowcost ozone sensors in IoT platforms. *IEEE Internet of Things Journal* 6 (6): 9563–9571.

Fung, Pak Lun, et al. 2022. Improving the current air quality index with new particulate indicators using a robust statistical approach. *Science of the Total Environment* 844: 157099.

Gharaibeh, Ammar, et al. 2017. Smart cities: A survey on data management, security, and enabling technologies. *IEEE Communications Surveys & Tutorials* 19 (4): 2456–2501.

Gupta, Rajesh, et al. 2021. 6G-enabled edge intelligence for ultra-reliable low latency applications: Vision and mission. *Computer Standards & Interfaces* 77: 103521.

Hammad, Sahibzada Saadoon, et al. 2023. An unsupervised TinyML approach applied to the detection of urban noise anomalies under the smart cities environment. *Internet of Things* 23: 100848.

Harkat, Mohamed-Faouzi, et al. 2000. Sensor failure detection of air quality monitoring network. *IFAC Proceedings Volumes* 33 (11): 529–534.

Harrison, Roy M., et al. 2021. Non-exhaust vehicle emissions of particulate matter and VOC from road traffic: A review. *Atmospheric Environment* 262: 118592.

Hassan, Najmul, et al. 2018. The role of edge computing in Internet of Things. *IEEE Communications Magazine* 56 (11): 110–115.

Hazra, Abhishek, et al. 2021. A comprehensive survey on interoperability for IoT: Taxonomy, standards, and future directions. *ACM Computing Surveys (CSUR)* 55 (1): 1–35.

He, Chen, et al. 2020. Particulate matter capturing capacity of roadside evergreen vegetation during the winter season. *Urban Forestry & Urban Greening* 48: 126510.

Hietikko, Riina, et al. 2018. Diurnal variation of nanocluster aerosol concentrations and emission factors in a street canyon. *Atmospheric Environment* 189: 98–106.

Järvi, Leena, et al. 2009. The urban measurement station SMEAR III: Continuous monitoring of air pollution and surface-atmosphere interactions in Helsinki, Finland. *Boreal Environment Research* 14: 86–109.

Javed, Abdul Rehman, et al. 2022. Future smart cities: Requirements, emerging technologies, applications, challenges, and future aspects. *Cities* 129: 103794.

Jiang, Dajie, and Guangyi Liu. 2016. An overview of 5G requirements. *5G Mobile Communications*, 3–26.

Jiang, Ji Chu, et al. 2020. Federated learning in smart city sensing: Challenges and opportunities. *Sensors* 20 (21): 6230.

Jiang, Wei, et al. 2023. Terahertz Communications and Sensing for 6G and Beyond: A Comprehensive View. Preprint. arXiv:2307.10321.

Kim, Sukun, et al. 2007. Health monitoring of civil infrastructures using wireless sensor networks. In *Proceedings of the 6th International Conference on Information Processing in Sensor Networks*, 254–263.

Kortoçi, Pranvera, et al. 2022. Air pollution exposure monitoring using portable low-cost air quality sensors. *Smart Health* 23: 100241.

Kumar, Prashant, et al. 2015. The rise of low-cost sensing for managing air pollution in cities. *Environment International* 75: 199–205.

Kumari, Aparna, et al. 2021. Amalgamation of blockchain and IoT for smart cities underlying 6G communication: A comprehensive review. *Computer Communications* 172: 102–118.

Lee, Euijong, et al. 2021. A survey on standards for interoperability and security in the Internet of Things. *IEEE Communications Surveys & Tutorials* 23 (2): 1020–1047.

Li, Shuling. 2018. Application of blockchain technology in smart city infrastructure. In *2018 IEEE international Conference on Smart Internet of Things (SmartIoT)*, 276–2766. IEEE.

Li, YuanYuan, and Lynne E. Parker. 2014. Nearest neighbor imputation using spatial-temporal correlations in wireless sensor networks. *Information Fusion* 15: 64–79.

Liu, Yuehua, et al. 2020. Missing value imputation for industrial IoT sensor data with large gaps. *IEEE Internet of Things Journal* 7 (8): 6855–6867.

Mehmood, Yasir, et al. 2017. Internet-of-Things-based smart cities: Recent advances and challenges. *IEEE Communications Magazine* 55 (9): 16–24.

Melibari, Wesal, et al. 2023. IoT-based smart cities beyond 2030: enabling technologies, challenges, and solutions. In *2023 1st International Conference on Advanced Innovations in Smart Cities (ICAISC)*, 1–6. IEEE.

Morabito, Roberto, and Jaime Jiménez. 2020. IETF protocol suite for the Internet of Things: overview and recent advancements. *IEEE Communications Standards Magazine* 4 (2): 41–49.

Motlagh, Naser Hossein, Eemil Lagerspetz, et al. 2020. Toward massive scale air quality monitoring. *IEEE Communications Magazine* 58 (2): 54–59.

Motlagh, Naser Hossein, Lauri Lovèn, et al. 2022. Edge computing: The computing infrastructure for the smart megacities of the future. *Computer* 55 (12): 54–64.

Motlagh, Naser Hossein, Martha A Zaidan, et al. 2021. Transit pollution exposure monitoring using low-cost wearable sensors. *Transportation Research Part D: Transport and Environment* 98: 102981.

Motlagh, Naser Hossein, Martha Arbayani Zaidan, et al. 2023. Digital twins for smart spaces—beyond IoT analytics. *IEEE Internet of Things Journal*. https://doi.org/10.1109/JIOT.2023.3287032.

Pan, Jianli, and James McElhannon. 2017. Future edge cloud and edge computing for Internet of Things applications. *IEEE Internet of Things Journal* 5 (1): 439–449.

Rabatel, Julien, et al. 2011. Anomaly detection in monitoring sensor data for preventive maintenance. *Expert Systems with Applications* 38 (6): 7003–7015.

Rahman, Md Abdur, et al. 2019. Blockchain and IoT-based cognitive edge framework for sharing economy services in a smart city. *IEEE Access* 7: 18611–18621.

Rajapakse, Visal, et al. 2022. Intelligence at the extreme edge: A survey on reformable TinyML. *ACM Computing Surveys*. https://doi.org/10.48550/arXiv.2204.00827.

Rashid, Bushra, and Mubashir Husain Rehmani. 2016. Applications of wireless sensor networks for urban areas: A survey. *Journal of Network and Computer Applications* 60: 192–219.

Rassam, Murad A., et al. 2014. Adaptive and online data anomaly detection for wireless sensor systems. *Knowledge-Based Systems* 60: 44–57.

Rong, Bo, 2021. 6G: The next horizon: From connected people and things to connected intelligence. *IEEE Wireless Communications* 28 (5): 8–8.

Sanchez-Iborra, Ramon, and Antonio F. Skarmeta. 2020. TinyML-enabled frugal smart objects: Challenges and opportunities. *IEEE Circuits and Systems Magazine* 20 (3): 4–18.

Santana, Eduardo Felipe Zambom, et al. 2017. Software platforms for smart cities: Concepts, requirements, challenges, and a unified reference architecture. *ACM Computing Surveys (Csur)* 50 (6): 1–37.

Shahat Osman, Ahmed M., and Ahmed Elragal. 2021. Smart cities and big data analytics: a data-driven decision-making use case. *Smart Cities* 4 (1): 286–313.

Silva, Welington M da, et al. 2013. Smart cities software architectures: a survey. In *Proceedings of the 28th Annual ACM Symposium on Applied Computing*, 1722–1727.

Singh, Prabhat Ranjan, et al. 2023. 6G networks for artificial intelligence-enabled smart cities applications: a scoping review. *Telematics and Informatics Reports*, 100044.

Strinati, Emilio Calvanese, et al. 2019. 6G: The next frontier: From holographic messaging to artificial intelligence using subterahertz and visible light communication. *IEEE Vehicular Technology Magazine* 14 (3): 42–50.

Su, Xiang, et al. 2021. Intelligent and scalable air quality monitoring with 5G edge. *IEEE Internet Computing* 25 (2): 35–44.

Sun, Chenxi, et al. 2019. Optimal citizen-centric sensor placement for air quality monitoring: a case study of city of Cambridge, the United Kingdom. *IEEE Access* 7: 47390–47400.

Teh, Hui Yie, et al. 2020. Sensor data quality: A systematic review. *Journal of Big Data* 7 (1): 1–49.

Tripathi, Rajesh Kumar, et al. 2018. Suspicious human activity recognition: a review. *Artificial Intelligence Review* 50: 283–339.

Wang, Richard Y., and Diane M. Strong. 1996. Beyond accuracy: What data quality means to data consumers. *Journal of Management Information Systems* 12 (4): 5–33.

Wang, Xu, et al. 2019. Survey on blockchain for Internet of Things. *Computer Communications* 136: 10–29.

What is Data Sharing?—Data Sharing Explained—AWS. 2022. Accessed: 11/October/2023. Amazon Web Services. https://aws.amazon.com/what-is/data-sharing/.

World Health Organization. 2021. *World Health Statistics 2019: Monitoring Health for the SDGs, Sustainable Development Goals.*

Ye, Juan, et al. 2016. Detecting abnormal events on binary sensors in smart home environments. *Pervasive and Mobile Computing* 33: 32–49.

Zaidan, Martha Arbayani, Naser Hossein Motlagh, Pak L. Fung, et al. 2020. Intelligent calibration and virtual sensing for integrated low-cost air quality sensors. *IEEE Sensors Journal* 20 (22): 13638–13652.

Zaidan, Martha Arbayani, Naser Hossein Motlagh, Pak Lun Fung, et al. 2023. Intelligent air pollution sensors calibration for extreme events and drifts monitoring. *IEEE Transactions on Industrial Informatics* 19 (2): 1366–1379.

Zaidan, Martha Arbayani, Yuning Xie, et al. 2022. Dense air quality sensor networks: validation, analysis, and benefits. *IEEE Sensors Journal* 22 (23): 23507–23520.

Zeadally, Sherali, et al. 2020. A tutorial survey on vehicle-to-vehicle communications. *Telecommunication Systems* 73: 469–489.

Zhang, Haibin, et al. 2018. A Bayesian network model for data losses and faults in medical body sensor networks. *Computer Networks* 143: 166–175.

Zhang, Jing, and Dacheng Tao. 2020. Empowering things with intelligence: a survey of the progress, challenges, and opportunities in artificial intelligence of things. *IEEE Internet of Things Journal* 8 (10): 7789–7817.

Zhao, Chunhui, and Yongji Fu. 2015. Statistical analysis based online sensor failure detection for continuous glucose monitoring in type I diabetes. *Chemometrics and Intelligent Laboratory Systems* 144: 128–137.

Chapter 7
Digital Twin and IoT for Smart City Monitoring

Shitharth Selvarajan and Hariprasath Manoharan

7.1 Introduction

The implementation of intelligent methods is imperative in the monitoring process of smart cities to optimize time management and foster economic development in developing regions. Future developments will require the implementation of multiple monitoring devices or processes to effectively monitor a greater number of areas. Therefore, it is imperative to develop additional twin technology to enable remote monitoring of every city, thereby reducing the average time required for observing essential processes. The concept of a digital twin involves the virtual replication of a physical object, enabling real-time observation that closely resembles that of a live monitoring system. Furthermore, it is noteworthy that approximately 90% of nations have already integrated Internet of Things (IoT) technology to oversee the functioning of intelligent urban areas. Hence, the utilization of digital twin technology in conjunction with an appropriate application platform presents a significant benefit for the optimization of the Internet of Things (IoT) operational framework. The application platform typically refers to a system that operates using a distinct address and is utilized for the purposes of facilitating identification and processing of data.

Furthermore, the generation of digital twins can be executed in either two-dimensional or three-dimensional formats, necessitating a dependable data connection for the creation of said digital twins. Therefore, it is possible to split the data into appropriate links by applying clustering techniques in order to create a secure

S. Selvarajan (✉)
School of Built Environment, Engineering and Computing, Leeds Beckett University, Leeds, UK
e-mail: s.selvarajan@leedsbeckett.ac.uk

H. Manoharan
Department of Electronics and Communication Engineering, Panimalar Engineering College, Chennai, Tamil Nadu, India

P. K. Donta et al. (eds.), *Learning Techniques for the Internet of Things*,
https://doi.org/10.1007/978-3-031-50514-0_7

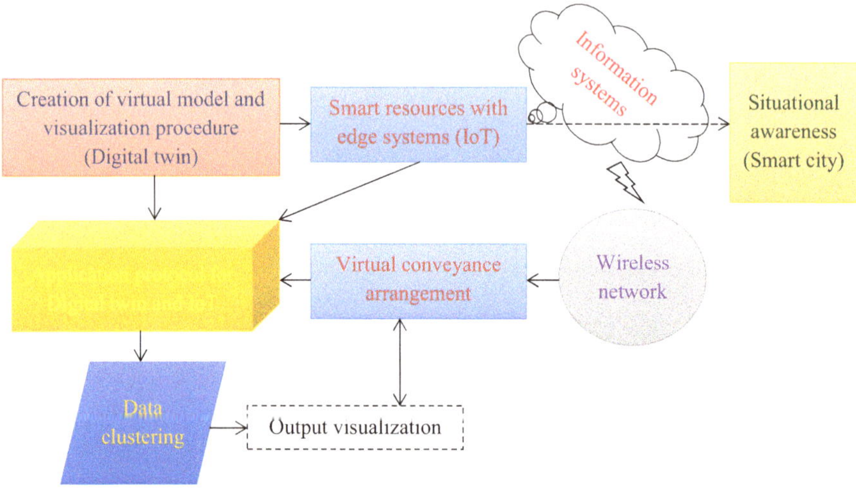

Fig. 7.1 Block diagram of visualization in smart cities with digital twins

data connection. The actual status of an item or other monitoring equivalents can be represented with low resource allocation technique by merging the developed digital twins with IoT, Constrained Application Platform (CoAP), and clustering algorithm. The block diagram of the suggested method, shown in Fig. 7.1, shows how a user-representative virtual model is built for visualization at the testing level. Using IoT modules, where information systems provide precise data processing techniques, resources can be allocated if the visualization duplicates the original twin's exact reproduction. The aforementioned procedure is carried out in smart cities, employing a wireless network to provide situational awareness and measure relevant metrics. Furthermore, the developed virtual platform is set up with a suitable data format, allowing the application platform (CoAP) to establish a direct link with clustered data. The output units for management, planning, and security are visualized after the conclusion of clustering and data connection with CoAP (Donta et al., 2022).

7.1.1 Background and Related Works

Prior to implementation, all existing works on digital twin technology must be evaluated since the chosen objective must be uniquely expressed and the processing system must react to the analytical representations. While other approaches, including IoT data transmission and sensor monitoring systems, are completely different, the process of creating digital twins is the same in all ways. Therefore, every aspect of the current methodology is carefully examined before implementing the new solution. Every difficulty that arises in creating a digital twin with a model perspective that adds significant value to twin capabilities is examined (Rasheed et al., 2020; Srirama, n.d.) in order to propose a basic form. The model viewpoint demonstrates that digital twins and virtual prototypes can be connected for all applications, but that data processing techniques must also be integrated to enable twin systems to make intelligent judgments. As a result, a digital twin concept based on real-time data collecting is put into practice, where the complete data is connected to a sensor and used to recognize a specific object (Jedermann et al., 2022). More than one life cycle is found in the digital twin during the object detection process, and it may even be seen with glasses. Therefore, live sensor monitoring data is not used, although twin monitoring systems are required for all IoT applications. All smart cities with a guaranteed sustainable transport network utilize an information processing technique utilizing the Internet of Things in addition to live sensor monitoring systems (Chen et al., 2021; Donta et al., 2023). However, the digital twin process generates more duplicate packets in a dynamic environment; as a result, it is considerably more challenging to depict the system as sustainable in networks' transfer mode.

Only a lightweight security model can be integrated for all effective networks since digital twins must be coupled in wireless representations, which are then followed by geographic network distribution (Kamruzzaman, 2021). Resource allocation awareness can be produced even when offloading techniques are used if a lightweight module is incorporated into the digital twin building process. But in order to keep the system operating sustainably, more security models must be integrated into the aforementioned process to advance technology. Even digital twins can be used in health-care monitoring systems where cloud infrastructure modules are used for both performance modeling and analysis (El Kafhali & Salah, 2019). The cost of installation can be decreased by connecting a person using a remote monitoring system during this procedure. Even so, a low-cost module for the remote monitoring system is available, and if it is used, the system's performance will be compromised. Numerous topologies and measuring techniques are used in smart cities to enhance the quality of service, and retransmission of packets in dynamic situations is ensured (Ali et al., 2022). Local and remote servers that connect the digital twin with IoT-based systems must deliver non-interruptible data since data processing techniques are involved in enhancing the quality of service to end users. Unreliable links must be identified and changed to reliable ones throughout the server connection procedure in order to prevent data in the generated twin from being impacted. A graphical model is developed to forecast whole-scale representations that are directly used in aerial mapping systems in order to assess the proportion of reliable and incorrect data (Kapteyn et al., 2021). Before calibrating control inputs, the digital twins created for aerial systems must be versatile in their architecture and give the appropriate state space representation. However, the sensor monitoring system's lack of flexibility makes the digital twin fail in this kind of implementation process.

Additionally, resilient services are offered for a variety of models where the end user platform and the digital twin method are working together (Lektauers et al., 2021). Since digital twins can provide real-time measurements through interactive simulations and other technologies, various data sharing processes can be developed based on need. When digital twins are generated, data shortages can be avoided since each twin carries unique data, reducing the likelihood that additional data will be needed during the connection process. The creation of a decision-based scheme utilizing mathematical techniques is then required so that a trustworthy network can be created using a twin representation system (Rathee et al., 2020). After detecting specific data, the deployment of such technologies enables the creation of a baseline strategy to address numerous issues linked to security threats and data breaches. Since the request procedure is not precise, it is still possible to transfer all of the information that is included in both sets of twins with great security, but specific protocols must be put in place for the data connection process. Digital twins and artificial intelligence algorithms are then combined to improve the autonomous process for creating zero-energy buildings under low impact (Garlik, 2022). Even after establishing digital twins, the implementation procedure is done in a very secure manner without having any negative effects on the environment. For digital twin representations, an intrusion-based system is also depicted, and every IoT layer is defined with a monitoring state (Elrawy et al., 2018). The possibility of involvement is significantly reduced with specific observations since all vulnerable twins are separated from wireless data representation modules. Table 7.1 compares comparable works in terms of their objective functions.

7.1.2 Research Gap and Motivation

The implementation of digital twins together with new processing approaches in existing methods, where the integration of the objective function is constrained, is

Table 7.1 Existing vs proposed

Reference	Methods/algorithms	Objectives			
		A	B	C	D
Wang (2022)	Energy-efficient smart city management	✓		✓	
Panteleeva and Borozdina (2019)	IoT-based smart city	✓	✓		
Wang (2021)	Spatial analytic geometry systems		✓		✓
Ganguli and Adhikari (2020)	Discrete dynamic system for smart cities	✓		✓	
Area et al. (2022)	Stieltjes derivative analysis for smart cities		✓	✓	
Segovia and Garcia-Alfaro (2022)	Digital twin for physical representations	✓			✓
Khalyutin et al. (2023)	Smart city energy process with digital twins		✓	✓	
Proposed	Digital twins, Constrained application protocol, and clustering optimization for smart cities	✓	✓	✓	✓

A: Minimum resource allocation; B: Reliable communication at active time periods; C: Reduction of inactive twins; D: Maximization of message transmissions evident from all previous research and Table 7.1. The majority of algorithms offer a basic illustration of smart city management employing dynamic depiction systems, but they do not fully examine the impact of the environment or other crucial factors like dependable communication during link active periods. Additionally, there are only so many messages that can be sent to end users, which results in a greater number of inactive twins and more untapped resources. Even with geometric design, the current representation does not follow a temporal step index for developing a digital twin.

Thus, by developing digital twins, an analytical representation is developed to address all the shortcomings of the current system, and data representation is accomplished via the Constrained Application Protocol (CoAP) (Donta et al., 2023). The data collected from smart cities are transferred with little resources during active time periods because every piece of data in the digital twin is clustered to many segments. Additionally, the link connections for digital twins are made during active time periods, guaranteeing a stable way of data transfer. With the proposed implementation procedure, more data that is represented in unique ways is conveyed, and the number of inactive twins in the design model is decreased.

7.1.3 Contributions

The primary contribution of the proposed work is to analyze how digital twins affect data management in smart cities using IoT and application protocols, with parametric analysis based on the following goals:

- To produce an original digital twin copy with a low error representation and time step index.
- Allocating specific reward functions to each cluster, which would represent the state model with sparse resources.
- To maximize the success rate of data transfer with produced digital twins by increasing the transmission of active message at a reduced.

7.2 Proposed System Model

Analytical equations must be used to depict the internals of the system model that was established for the digital twin in IoT operations. Because of this, representations of an analytical digital twin with IoT for smart city applications are offered in this part, along with time step demonstrations.

7.2.1 Twin Time Step

Moreover, digital twin creates a framework that is identical to original copy; thus, the noise constraints are formulated and it is represented using Eq. (7.1) as follows:

$$original_i = \min \sum_{i=1}^{n} E_1 + ... + E_n \tag{7.1}$$

where $E_1 + ... + E_n$ denotes total error that is observed at each time step.

7.2.2 Twin Reward Function

Economic development is improved whenever digital twins are used in smart city planning and talks; therefore, incentive functions are offered for normalized standard deviation data from connected wireless networks using IoT, as shown in Eq. (7.2):

$$reward_i = \max \sum_{i=1}^{n} \frac{original_i - reference_i}{SD_{sensor}} \tag{7.2}$$

where $reference_i$ indicates initial values from IoT devices and SD_{sensor} denotes connected sensor deviated values; Eq. (7.2) states that as each sensing unit is employed for a different detection method, the difference between the original and reference twins must be maximized.

7.2.3 Twin Representations

For all applications where the entire process is based on the time index as expressed in Eq. (7.3), the digital twin representation model is offered in a uniform manner. Intelligent system output is significantly influenced by the representation model:

$$DT_r = \sum_{i=1}^{n} Time_i + I_i + DT_c \tag{7.3}$$

where $Time_i$ denotes total time step index, I_i represents input values from IoT, and DT_c indicates the total number of components in the system. The output representation is degraded by the total of all the aforementioned variables, which is why the digital twins are arranged in a compound manner.

7.2.4 Twin-State Model

When digital twins are used to symbolize smart cities, less resource must be used in order to raise the caliber of improvements. Therefore, utilizing Eq. (7.4) to establish the probability values for the generated twin, the following quality developments can only be made:

$$resource_i = \min \sum_{i=1}^{n} [(v_i + ... + v_i) + (CS_1 + ... + CS_i)] \, b_i \tag{7.4}$$

where $v_1 + \cdots + v_i$ denotes the total number of created twin, $CS_1 + \cdots + CS_i$ indicates the current state of twin, and b_i represents the behavior of created twin. Equation (7.4) states that the behavior of digital twins can be used to obtain a minimum number of resources; hence, an intelligent behavior must be represented in order to distribute resources appropriately.

7.2.5 Twin Message Transmission

As IoT applications are connected to both wearable and non-wearable kinds, it is the responsibility of every user to check the data transmission process once the resources have been allotted. Data transmission must be secure and trustworthy to prevent broken links and the removal of duplicate twin packets from the system. Hence, Eq. (7.5) is designed to transmit messages with certain indicators:

$$MT_i = \max \sum_{i=1}^{n} (\rho_1 + ... + \rho_i) \times (R_1 + ... + R_i) m_t \tag{7.5}$$

where $\rho_1 + ... + \rho_i$ denotes the total number of twin transmitter, $R_1 + .. + R_i$ indicates the total number of twin receivers, and m_t describes data types. Equation (7.5) shows that as numerous types of data messages are conveyed, the number of transmitters and receivers must be maximized in order to establish a reliable connection for each specified data type.

7.2.6 Twin Communications

Only in systems with active data representations can digital twins communicate with one another. It is much more challenging to establish twin communications when the defined data type has few transmitters and receivers; hence, analytical representations for active data types are established using Eq. (7.6):

$$comm_i = \max \sum_{i=1}^{n} \alpha_m(i) \times \alpha_t(i) \tag{7.6}$$

where α_m, α_t represents active messages at active time periods. Every digital twin must transmit active messages instantly, according to Eq. (7.6), which reduces the overall time of all data.

7.2.7 Twin Transmission Delay

Digital twins will be taken into account during inactive time periods if active messages are not transmitted during active periods, which will have an impact on overall performance as shown in Eq. (7.7):

$$Inactive_i = \max \sum_{i=1}^{n} t_s - t_{start}(i) \tag{7.7}$$

where t_s denotes data reproduction time period and t_{start} indicates start time of data transmission.

7.2.8 Objective Function

The following are the parametric output criteria that make up the mathematical description of digital twin with IoT for smart cities:

$$obj_1 = \min \sum_{i=1}^{n} original_i, rersource_i, comm + i, Inactive_i \tag{7.8}$$

$$obj_2 = \max \sum_{i=1}^{n} reward_i, MT_i \tag{7.9}$$

According to the aforementioned objective functions, it is necessary to maximize transmission times while minimizing the resources provided for digital twin activities. As a result, the clustering algorithm described in Sect. 7.3 is integrated with the IoT Constrained Application Protocol (CoAP) for processing digital twin data.

7.3 Twin Protocol Integration

A set of guidelines for such an application protocol is required since digital twin uses IoT and needs to operate with a limited mechanism. Since CoAP is largely recommended for low-power networks, it is described in this part along with the integration of digital twin and IoT procedures. When using digital twin representations, the output form is constructed using low power, allowing both requests and responses to be driven without any issues related to practicality. The primary benefit of CoAP is the ability to construct a trustworthy system with message formats while just requiring a single datagram packet. Additionally, any asynchronous data can be communicated with the option to retransmit, preventing data transmission failures, and the ability to instantly regenerate the complete twin in response to requests. Due to the fact that CoAP operates by connecting two clients, the generated twin shares messages in an unidentified format. Tokens must be successfully received by both clients before they can be acknowledged; otherwise, the receiver must provide non-confirmable data. Additionally, related data transmitted by generated twins will be uniquely identified, preventing the transmission of duplicate data. According to the suggested system model, time series representations will be used if either of the aforementioned processes fails, after which a request for data must be made and its contents must be restored. The fundamental reason for selecting CoAP as the digital twin representation is that the selected smart city application will continue to operate in a constrained manner, necessitating the selection of resources like electricity, bandwidth, etc. in accordance with the produced twins. Therefore, CoAP is selected with minimal overhead to satisfy the needs of resource-constrained activities. Equation (7.10) is an analytical formulation of CoAP for a smart city application in the case of a digital twin and IoT for successful data transmission:

$$T_{success} = \max \sum_{i=1}^{n} \frac{1 - \delta_i}{\omega_i} \tag{7.10}$$

where δ_i, ω_i represents high and low data losses by created twins. Equation (7.10) shows that the success rate is maximized when high and low data are separated into cases with exact probabilities. Additionally, as shown in Eq. (7.11), the latency must be decreased prior to retransmission for unsuccessful data packets that are conveyed by generated twins:

$$T_{success} = \min \sum_{i=1}^{n} \gamma_e(i) \times TO_i \tag{7.11}$$

where γ_e denotes data exchange rate and TO_i represents data that is transferred after certain time bound

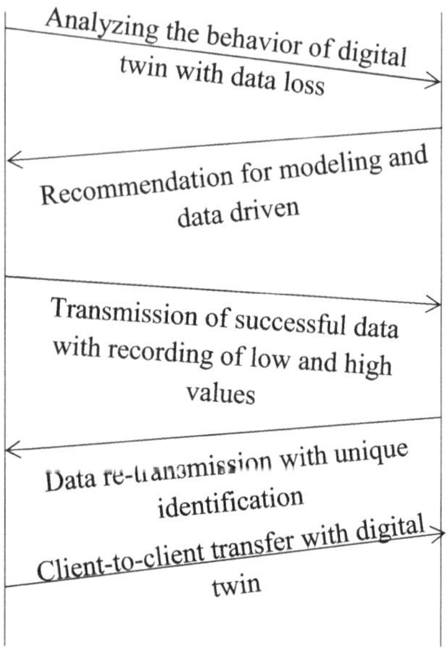

Fig. 7.2 CoAP for digital twin and IoT

Equation (7.11) states that data loss is directly decreased by IoT representation procedure since both data exchange rate and data transmitted after a specific time period must be minimized. Figure 7.2 depicts the protocol design for the digital twin and also includes the pseudo-code.

7.3.1 Optimization Algorithm

In order to process distinct data types, which must be distinguished from original data in applications involving digital twins and the Internet of Things, clustering is required. Any unlabeled data can be sorted using pre-defined clusters if data categories are clustered. However, if the data from the digital twins are not grouped, it will be considerably harder to recognize the data, which will result in inactive transmission throughout the monitoring process. The full transmission process can be completed without generating duplicate data from the produced twin if the processed data is present without any overlapping conditions. Additionally, the grid-based technology used to cluster the data from digital twins allows for the creation and transmission of independent data in a rapid mode environment. Without using a structured data set, inferences can be made since the data in processed digital twins is clustered. The ability to separate vast amounts of data even when digital twins are not structured in a similar manner is another significant benefit of clustering data in digital twin representation. As a result, every social network may be examined in smart cities along with the whole amount of value space that surrounds the complete background. Additionally, it is possible for data clustering to represent the data analysis phase by uncovering new information across the entire network, and digital twin representations allow for the extraction of more data with less resource restriction. One of the major benefits of clustering in digital twin data is that because there are so few interpretable data points, it is possible to reliably identify new patterns. Additionally, since clustering in the suggested method is done by looking at the closest data point, it is possible to combine all twin data into a single portioning system. Equation (7.12) illustrates the mathematical depiction of clustering in digital twins as follows:

$$C_{DT} = \sum i = 1^n (\varphi_1 + ... + \varphi_i) \times dis_i^2 \tag{7.12}$$

where $\varphi_1 + ... + \varphi_i$ represents total number of clustered data and dis_i^2 denotes clustered distance.

Equation (7.12) shows that the full set of data is categorized according to distance measurements. Since a distance separation is required to distinguish duplicate data from the digital twin, Eq. (7.13) is used to express the distance separation as follows:

$$dist_i = \min \sum i = 1^n DT_i - DT_1 \tag{7.13}$$

DT_i, DT_1 denotes the representation of original and reference twins.

Because Eq. (7.13) states that the difference between the original and reference twins must be minimized, the following observations are denoted by Eq. (7.14):

$$state_{DT} = \sum i = 1^n \frac{1}{z_i - y_i} \tag{7.14}$$

z_i and y_i denote inverse representation of digital twins to determine corresponding states. The pseudo-code for clustering in digital twin is provided with initial code representations, and block processing is included in Fig. 7.3.

7.4 Results and Discussions

In this phase, a digital twin is created and integrated with real-world circumstances in order to test the mathematical methodology that has been used. A digital replica is created, and data is gathered using wireless sensors, in order to mimic the smart city resources, which include a number of features. When the testing phase is initiated, various collected features are provided in the produced twin as test bed input, and it is important to supply the input data in the form of photos during this testing phase. Every input is stolen throughout the testing phase, and the end user the person who created the digital twin observes the traits and connections between twins and the outside world. Additionally, the monitoring procedure for smart cities starts with simulations that might be entirely based on real-world events. As a result, the

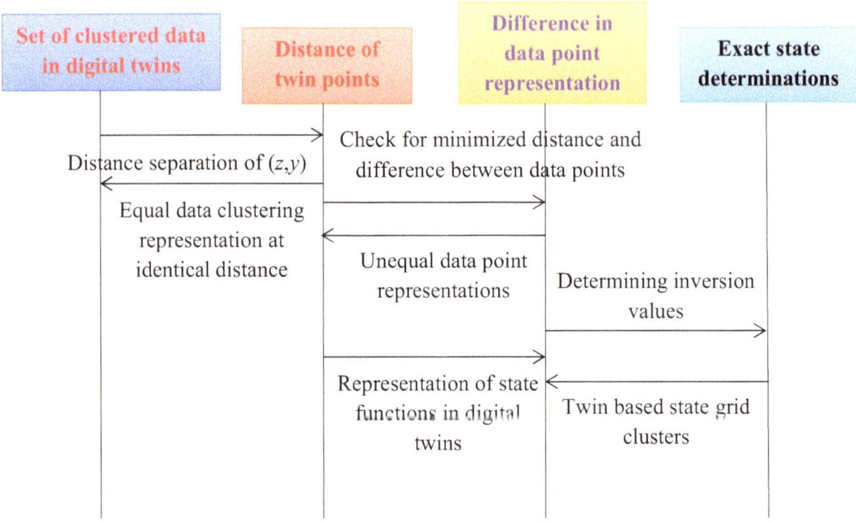

Fig. 7.3 Clustering procedure for digital twins in IoT

suggested solution takes into account information about current operating systems, and if something changes, the produced twin can be deleted from the network. Additionally, the data transmission approach is carried out utilizing a number of application protocols, with a restriction specified in the case of the suggested method employing CoAP. Losses are decreased for more successfully transmitted packets because the developed twin for monitoring the entities in smart cities adheres to the constraint principle. Additionally, practically hundreds of data are provided as input to the produced twins, making it difficult to identify the appropriate data. For this reason, it is crucial to group each data set into clusters by taking the monitoring system's distance into account. Since every cluster uses a time series representation, only the data that corresponds to that time period is given as input for the twin operation at that particular time. The fact that the precise status of the twins is known in the event of data loss in the connected network is another significant benefit of data clustering in digital twin operation. Five possibilities are taken into consideration based on the analytical representations in simulation studies to analyze the experimental data, and their significance is given in Table 7.2.

Scenario 1: Analysis of state model
Scenario 2: Twin communications
Scenario 3: Monitoring inactive twins
Scenario 4: Success rate
Scenario 5: Number of message transmissions

Table 7.2 Importance of scenario considerations

Scenario	Significance
Analysis of state model	Reduction of errors at every time step index
Twin communications	To determine number of active messages at given time index
Monitoring inactive twins	To provide retransmission for unsuccessful data
Success rate	To analyze total number of data loss by created twins
Number of message transmissions	To classify data types (Cluster) before transmitting a particular data

Table 7.3 Simulation parameters

Bounds	Requirement
Operating systems	Windows 8 and above
Platform	MATLAB and Autodesk
Version (MATLAB)	2015 and above
Version (Autodesk)	2.0 and above
Applications	Design of digital twins
Data sets	Wireless sensor networks with unique monitoring characteristics in the network

7.4.1 Discussions

As responses by digital twins according to the marked changes are determined without any external effect, all of the aforementioned scenarios are performed in real time to determine proper working functionality of the proposed method. As a result, in the future, the abrupt changes are taken into consideration, and efforts can be made for proper expulsion in the network. After connecting the digital twins to MATLAB, the simulation setup is created, and the code is then implemented utilizing the objective functions by sequentially developing the twin loop from i to n. The simulation parameters that are taken into account by the suggested method are shown in Table 7.3.

All the data set is integrated with autodesk; therefore, corresponding design is provided for creating digital twins. The detailed description of remaining scenarios is as follows.

7.4.1.1 Scenario 1: Analysis of State Model

This scenario uses time step techniques to track the precise state representation of digital twins, and every error that occurs at a specific time is tracked using clustered data. A reward function will be assigned to the appropriate twin if the associated clustered data is sent under low error conditions. By comparing original and reference values, which offers a percentage of difference and is directly separated by

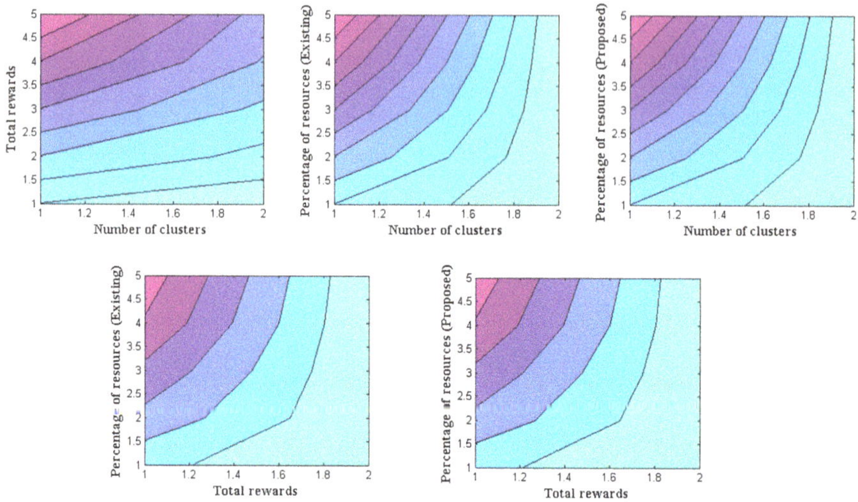

Fig. 7.4 Active time period measurement for dynamic message transmissions

tracked sensor value variations, the aforementioned procedure of awarding prizes is carried out. Furthermore, it is only true that digital twins in relation to physical representation have an exact state model when clustered data is given minimal resources. As a result, each twin's whole behavior is observed in order to allocate resources, and a distinct representation of the current state is discovered. The proposed and current approaches' state models are depicted in Fig. 7.4.

Figure 7.4 makes it clear that reward functions are maximized with minimal resources due to low inaccuracy in every input data. The overall number of clusters for various types of data is estimated to range from 1000 to 5000, as it is much simpler to monitor precise state models for large data clusters. The total reward functions for the aforementioned data clusters are 623, 1358, 1892, 2457, and 2791, respectively. If the data set in the form of images is not delivered to the input system in a proper manner, there will be more deviations. However, the monitored data set for IoT operation from wireless sensors is accurate since only detected values are taken into account, and adjustments can also be made manually, which does not indicate the precise method of monitoring systems. With the aforementioned rewards, the resources used for operating the digital twins stay at 40, 36, 30, 38, and 24 in the case of the current technique (Ali et al., 2022). In contrast, the proposed strategy minimizes the percentage of resources to 31, 26, 23, 20, and 18 accordingly.

7.4.1.2 Scenario 2: Twin Communications

As twin communication is only analyzed if transmissions are present in this scenario, the overall number of active messages is tracked in each cluster. In the

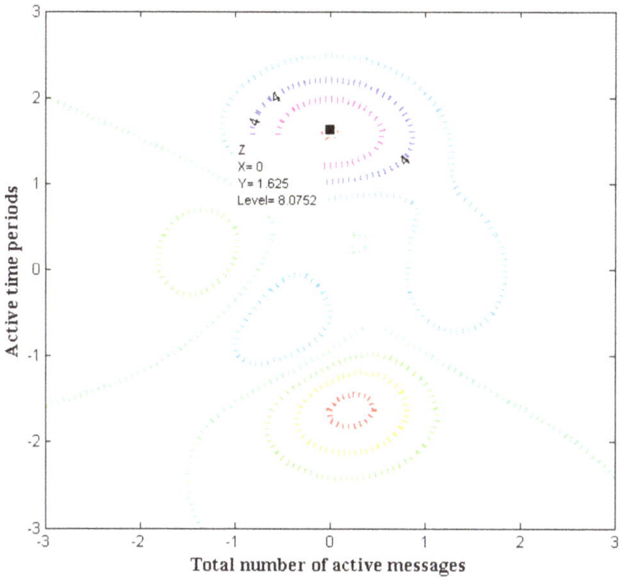

Fig. 7.5 Active time period measurement for dynamic message transmissions

communication province with high response transmitters and receivers, the twin generation process itself involves some active signals that must be transferred. Additionally, it is looked at whether it is possible to establish a proper link for delivering active messages in the situation of digital twins with the placement of low communication transmitters and receivers. However, at the same time, because less resources are allocated to low-processing transmitters and receivers, some data types are thought to be automatically replaced. Additionally, twin communication must occur at low-delay locations in order for the network to be able to make suitable connections at the given time index. Figure 7.5 depicts the number of active messages that is present in existing and proposed methods.

It is clear from Fig. 7.5 that the suggested method effectively transmits active signals when compared to the current method (Ali et al., 2022). To demonstrate the comparative example, the total number of active messages is taken to be 300, 500, 700, 900, and 1100, respectively, with the active time period of transmissions being limited to maximum values of 12, 17, 22, 24, and 28. Twin message transmission factor is minimized when the active messages are transmitted within the allotted time frame. In contrast, if there is a longer transmission time, twin communication will require more time to communicate the most recent signals, which should be avoided. In addition, the existing approach has a communication percentage of 22, 20, 16, 15, and 13, whereas the proposed method has a communication percentage of 17, 12, 10, 7, and 4, which is found to be the lowest. Therefore, the suggested technique offers adequate message transmissions within the proper time periods in the event of proposed digital twin formation for monitoring smart cities.

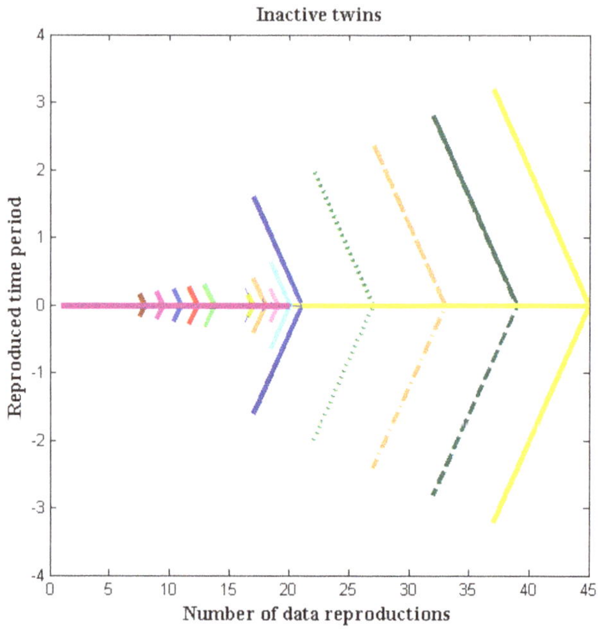

Fig. 7.6 Data reproductions for inactive twin discoveries

7.4.1.3 Scenario 3: Monitoring Inactive Twins

When data is provided as input to digital twins in the majority of smart city monitoring systems, it must remain active for a considerable amount of time until the most recent data set is received by end users, at which point a comparison with the relevant data set must be made. Thus, in this situation, inactive twins are monitored by taking into account the initial time period of transmission, during which every piece of information is sent in a clustered fashion. Additionally, since some data can be repeated, it is necessary to prohibit the reproduction of active states with inactive ones during active times. As a result, conspicuous points on the display where the reproduction period is substantially shorter for twin representations indicate the transformation in the data set (pictures). The inactive twins will be removed from the network in accordance with the reduction of starting and reproduction time periods, resulting in the high performance of the complete data process. In the example of an inactive twin monitoring system, simulation output is shown in Fig. 7.6 for both the current and proposed methods.

According to Fig. 7.6's observations, the proposed method has a lower number of inactive twins than the current method (Ali et al., 2022). The application protocol with CoAP monitors the reproduction time period, and it is deemed to be 1.98, 2.45, 2.96, 3.38, and 3.79 accordingly. This is done to verify the test results with clustered data. The percentage of twins that remain in inactive states is decreased in both the existing and suggested methods, and is displayed for the following data with values of 20, 25, 30, 35, and 40. The proposed strategy decreases the number of inactive twins since the data is delivered in the form of clusters and more active messages (outcomes from scenario 2) are represented. As a result, the percentage of inactive twins in the suggested method is 3, 1, 0.6, 0.2, and 0.1, whereas it is 10, 8, 5, 3, and 2 for the present strategy. As a result, more data is transferred to end users for smart city monitoring systems when there are fewer inactive twins.

7.4.1.4 Scenario 4: Success Rate

By calculating the success rate, the quantity of digital twin representations that successfully sent is seen. Smart city data, which are encoded as pictures and sent under high and low loss conditions, are lowered if the success rate of a twin is significantly higher. Additionally, the success rate is determined using CoAP, where the separation of low losses lowers the chance of difference in high loss scenarios. The success rate of data that is represented by adhering to the CoAP protocol is thus determined in the projected model by the ratio of the aforementioned conditions. Additionally, if a data transfer fails, the rate of retransmission increases after a predetermined time limit, minimizing the pace at which each data is exchanged. Contrarily, if the packets are swapped throughout the digital twin process, there is a chance that more data will remain in a duplicate condition, and this cannot be avoided. The percentage of success rate for suggested and current solutions is shown in Fig. 7.7.

Figure 7.7 shows that, in comparison with the current method, the success rate of data following the creation of digital twins for smart cities is maximized. Both data losses are taken into account within the illustrated bounds, where the data exchange rate is minimized over a predetermined time period, to test the success rate of packets. The success rate of the data is maximized to 87% and 97% in the case of the existing and new approaches, respectively, in the simulation outcome exchange rate of data is assumed as 2400, 2600, 2800, 3000, and 3200, respectively. While data success rates are maximized to 100% when exchange rates are higher, the proposed method's exchange rate is higher due to some loss causes. However, it is noted that at low exchange rates below 2400, there is higher loss, and in twin creation, it is not avoided because of poor data representations. Consequently, with considered loss factors, the proposed system's success rate is maximized in comparison with the current approach.

7.4.1.5 Scenario 5: Number of Message Transmissions

Different data types are defined in this scenario to reflect the calculation of the total number of messages sent via digital twins. With wireless technology, end users transmit additional messages, which are distinguished by identifying the specified data kinds. Additionally, all data communicated using digital twins is encoded in an

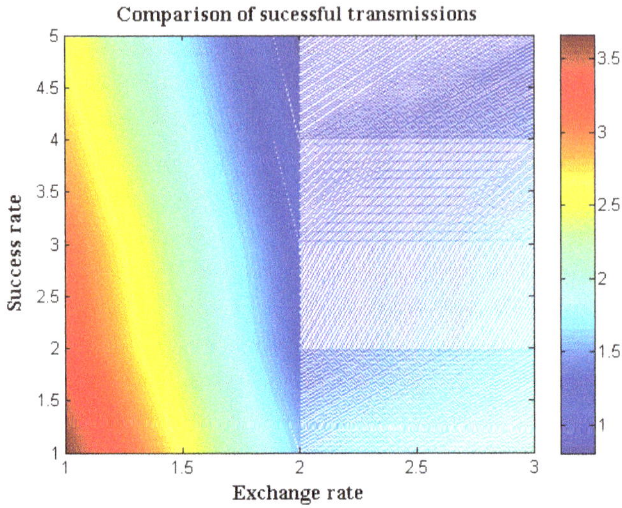

Fig. 7.7 Number of successful transmissions at minimized exchange rate

unknowable format, making it impossible for other users to decode it using more specialized data. Since the overall number of transmitters and receivers is also kept to a minimum, only designated data types are given the top priority. In the event that a specified data type has identifying problems, the twin immediately discards the data without sending it again. In order to remove failure connections or duplicate data, the data format must be robust and easily identifiable within the allotted time limits. This strengthens the security of digital twin transmissions. Figure 7.8 compares the message transmission for the suggested and current approaches.

Figure 7.8 shows that the total number of messages transmitted under the suggested method is higher than under the current methodology (Ali et al., 2022). Transceivers in digital twins are taken into account in a three-step factor, ranging from 6 to 18, to verify the total number of message transmissions. For each message transmission, new data kinds are established. The identified data types continue to be 68, 75, 79, 84, and 89 in the suggested method, while the total numbers of messages transmitted in the aforementioned scenario are 23, 27, 33, 35, and 38 in the case of the current methodology. However, under the suggested manner, the number of messages transmitted remains at 31, 36, 44, 48, and 53 due to accurate identification types. Therefore, monitoring smart cities is achievable with the total amount of messages transmitted, and if reference data changes, current state values can be indicated, boosting the effectiveness of the suggested digital twins in the proposed technique.

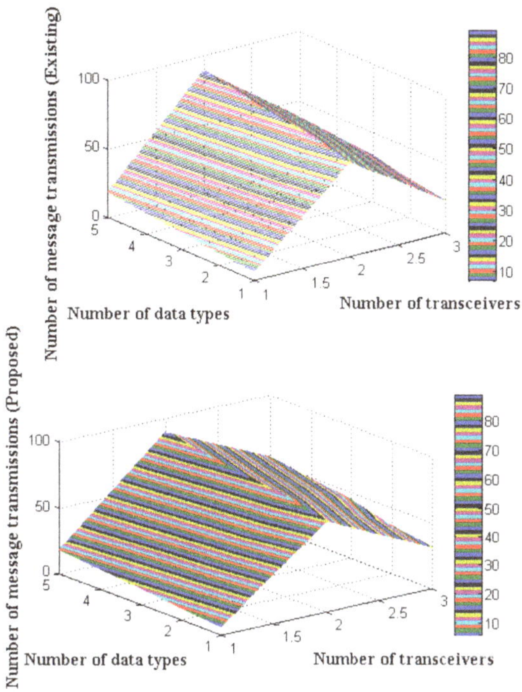

Fig. 7.8 Message transmissions for different data types

7.5 Conclusion

The proposed method uses the time series factor to carry out the process of digital twin representations via IoT and CoAP. The complete twin representations are based on specific reward functions since in the projected model error measurements are represented as data is created in particular clusters. The data is implemented using a unique image set since digital twins are also used in smart cities to assess the changing effects in relation to reference values. By recognizing the behavior of each twin, the resources allocated for twin representations are further minimized in this manner. As less resources are allotted, the maximum transmission period for each data is shortened, and the designed system fully eliminates inactive periods. Additionally, the active time periods are extended due to the low loss factor, resulting in a significant increase in active message transmissions, which is associated with connection to CoAP. Different data categories are identified specifically, and each data point is clustered into many data points. As a result, data success rates are raised and data exchange rates between digital twins are decreased. Additionally, each twin's physical representation includes high and low loss factors that specify the precise state of the system and enable efficient data transfer. Five situations are taken into consideration in accordance with the decided mathematical model, where state models are built precisely with precise twin communications, to test the impact of the suggested strategy. The proposed technique is seen to minimize inactive periods by maximizing success rates greater than 95% in the comparator case study. Future extensions of the suggested digital twin paradigm that include direct connection representations to other application platforms could boost societal economic growth.

References

Ali, Jawad, et al. 2022. Mathematical modeling and validation of retransmission-based mutant MQTT for improving quality of service in developing smart cities. *Sensors* 22 (24): 9751.

Area, Iván, et al. 2022. Concept and solution of digital twin based on a Stieltjes differential equation. *Mathematical Methods in the Applied Sciences* 45 (12): 7451–7465.

Chen, Xiangcong, et al. 2021. IoT cloud platform for information processing in smart city. *Computational Intelligence* 37 (3): 1428–1444.

Donta, Praveen Kumar, Boris Sedlak, et al. 2023. Governance and sustainability of distributed continuum systems: A big data approach. *Journal of Big Data* 10 (1): 1–31.

Donta, Praveen Kumar, Satish Narayana Srirama, et al. 2022. Survey on recent advances in IoT application layer protocols and machine learning scope for research directions. *Digital Communications and Networks* 8 (5): 727–744.

Donta, Praveen Kumar, Satish Narayana Srirama, et al. 2023. iCoCoA: Intelligent congestion control algorithm for CoAP using deep reinforcement learning. *Journal of Ambient Intelligence and Humanized Computing* 14 (3): 2951–2966.

El Kafhali, Said, and Khaled Salah. 2019. Performance modelling and analysis of Internet of Things enabled healthcare monitoring systems. *IET Networks* 8 (1): 48–58.

Elrawy, Mohamed Faisal, et al. 2018. Intrusion detection systems for IoT-based smart environments: a survey. *Journal of Cloud Computing* 7 (1): 1–20.

Ganguli, R., and Sondipon Adhikari. 2020. The digital twin of discrete dynamic systems: Initial approaches and future challenges. *Applied Mathematical Modelling* 77: 1110–1128.

Garlik, Bohumir. 2022. Energy centers in a smart city as a platform for the application of artificial intelligence and the Internet of Things. *Applied Sciences* 12 (7): 3386.

Jedermann, Reiner, et al. 2022. Digital twins for flexible linking of live sensor data with real-time models. *Sensors and Measuring Systems; 21th ITG/GMA-Symposium*, 1–7. VDE.

Kamruzzaman, MM. (2021). 6G- Enabled smart city networking model using lightweight security module. 10.21203/rs.3.rs-954242/v1.

Kapteyn, Michael G., et al. 2021. A probabilistic graphical model foundation for enabling predictive digital twins at scale. *Nature Computational Science* 1 (5): 337–347.

Khalyutin, Sergey, et al. 2023. Generalized method of mathematical prototyping of energy processes for digital twins development. *Energies* 16 (4): 1933.

Lektauers, Arnis, et al. 2021. A multi-model approach for simulation-based digital twin in resilient services. *WSEAS Transactions on Systems and Control* 16: 133–145.

Panteleeva, Margarita, and Svetlana Borozdina. 2019. Mathematical model of evaluating the quality of "smart city" transport interchanges functioning. In *E3S Web of Conferences*. Vol. 97, 01006. Les Ulis: EDP Sciences.

Rasheed, Adil, et al. 2020. Digital twin: Values, challenges and enablers from a modeling perspective. *IEEE Access* 8: 21980–22012.

Rathee, Geetanjali, et al. 2020. A trusted social network using hypothetical mathematical model and decision-based scheme. *IEEE Access* 9: 4223–4232.

Segovia, Mariana, and Joaquin Garcia-Alfaro. 2022. Design, modeling and implementation of digital twins. *Sensors* 22 (14): 5396.

Srirama, Satish Narayana. n.d. A decade of research in fog computing: Relevance, challenges, and future directions. *Software: Practice and Experience* 54: 3–23. https://doi.org/10.1002/spe.

3243. eprint: https://onlinelibrary.wiley.com/doi/pdf/10.1002/spe.3243. https://onlinelibrary.wiley.com/doi/abs/10.1002/spe.3243.

Wang, Jing. 2022. Optimized mathematical model for energy efficient construction management in smart cities using building information modeling. *Strategic Planning for Energy and the Environment*, 61–80.

Wang, Kaiqi. 2021. A theoretical analysis method of spatial analytic geometry and mathematics under digital twins. *Advances in Civil Engineering* 2021: 1–14.

Chapter 8
Multi-Objective and Constrained Reinforcement Learning for IoT

Shubham Vaishnav and Sindri Magnússon

8.1 Introduction

In recent years, the process of digitization in society has brought about significant advancements in wireless and related networks. The expansion of the Internet of Things (IoT) ecosystem, with its ever-growing number of interconnected devices, has presented various challenges, including bandwidth limitations and latency demands (Yousefpour et al. 2019). To tackle these issues, emerging paradigms like fog and edge computing have gained widespread popularity (Rao et al. 2011). These approaches enhance processing efficiency by enabling data computation to be conducted in closer proximity to the devices that generate it.

Optimization plays a critical role in the IoT as it helps enhance network performance metrics like bandwidth and latency. However, traditional optimization methods face challenges due to unknown parameters, making them less effective in practice. However, the advent of machine learning has unlocked tremendous potential, as it enables us to optimize network performance based on data-driven learning. Unlike conventional methods, machine learning allows us to adaptively learn and improve without explicit programming, making it highly promising for enhancing network efficiency. In particular, among various machine learning paradigms, reinforcement learning (RL) stands out due to its unique feature of learning through interactions with the environment (Geddes et al. 1992). This attribute makes RL particularly suitable for dynamic and adaptive IoT networks.

A key challenge in learning to optimize network performance lies in the inherent multi-objective nature of IoT networks. Balancing competing objectives such as energy lifetime, bandwidth, and latency presents intricate complexities.

S. Vaishnav (✉) · S. Magnússon
Department of Computer and Systems Science, Stockholm University, Stockholm, Sweden
e-mail: shubham.vaishnav@dsv.su.se; sindri.magnusson@dsv.su.se

P. K. Donta et al. (eds.), *Learning Techniques for the Internet of Things*,
https://doi.org/10.1007/978-3-031-50514-0_8

For instance, when optimizing for the energy lifetime of a network, if we do not consider other objectives, then a simple solution would be to simply shut down the network. However, this approach is clearly impractical, as it would render the network nonfunctional and defeat its intended purpose. To ensure a fully operational network, we must carefully consider and optimize other critical objectives such as bandwidth and latency in conjunction with energy conservation. Dealing with these multiple conflicting objectives presents complex challenges in the algorithm design of RL algorithms, demanding innovative and efficient approaches to strike a balance between the diverse goals and constraints in IoT optimization.

In this chapter, we will review the key problems, challenges, and algorithms for multi-objective RL in IoT networks. Section 8.2 discusses the common optimization problems in IoT and the objectives considered for those problems. In Sect. 8.3, we discuss the fundamentals of multi-objective optimization, followed by the fundamentals of RL in Sect. 8.4. Section 8.5 discusses the different existing approaches for MORL and their applicability in IoT networks. In Sect. 8.6, we explore the future scopes related to the improvisation of the existing MORL algorithms, and we also discuss the challenges in applying MORL in IoT.

8.2 Objectives and Problems in IoT Networks

In IoT networks, there are multiple metrics or objectives that we might like to optimize for. These objectives frequently exhibit conflicts, necessitating a skillful balancing act. Simultaneously, certain objectives may take on the role of constraints, where meeting them to a satisfactory level is acceptable. To navigate this complex landscape, intelligent strategies are required to prioritize objectives, allocate resources judiciously, and achieve an optimal trade-off that aligns with the specific needs and challenges of the IoT environment.

Optimizing energy consumption stands as a crucial objective in IoT networks, given that many IoT devices are deployed in fields with limited, unreliable, and intermittent power sources (Sarangi et al. 2018). However, IoT optimization encompasses a broader spectrum of objectives that demand attention. Among these, reducing latency to enhance real-time responsiveness, managing costs to ensure efficient resource allocation, fortifying security to safeguard sensitive data, addressing mobility challenges, and improving scalability to accommodate the ever-growing number of IoT devices are paramount (Yousefpour et al. 2019). Regardless of the specific IoT scenario, the pursuit of enhanced performance typically revolves around optimizing a subset of these objectives. Striking the right balance among these diverse objectives poses a significant challenge, requiring intelligent decision-making and resource allocation to tailor solutions that best fit the unique requirements and constraints of each IoT application.

The primary aim of most IoT optimization problems is to optimize a subset of these objectives. The following list outlines the main IoT optimization problems commonly encountered in practical applications:

1. **Routing:** Intelligent routing protocols are essential for the connectivity and functioning of IoT networks. Several objectives need to be considered for designing routing protocols. The nodes consume a lot of energy while transmitting packets, receiving packets, processing locally, or being active. The network's total energy consumption needs to be reduced. This aids but doesn't necessarily ensure the optimization of the network lifetime—another important objective. The routing protocols should also be able to direct the packets on paths with less delay or latency. This could sometimes conflict with the most energy-efficient paths. Moreover, due to the open exposure of IoT networks, they're susceptible to several security attacks. The nodes which have been attacked are known as malicious nodes. The routing protocols should be able to avoid routing paths involving malicious nodes, which could further conflict with other objectives.
2. **Task Scheduling:** IoT nodes are heterogeneous and involve different amounts of processing times and energy consumption. Further, the transmission time and energy also have to be accounted for. Thus, proper scheduling of tasks to different nodes is important for network optimization. This complex decision-making problem may involve trade-offs between objectives like delay and energy and spatial-temporal sampling rates.
3. **Efficient Communication:** IoT devices collect much data through embedded sensors. All nodes in a network have information to be conveyed to some other nodes for further processing. However, communicating all information to other nodes is inefficient and involves many costs like energy consumption, network congestion, etc. However, communicating very less information can be detrimental as well. Thus, the node has to find a trade-off between the two conflicting objectives of communication cost and the value of the information it has (Vaishnav et al. 2023).
4. **Data Aggregation:** Data transmission costs energy and increases bandwidth use and network congestion. Thus, minimizing the amount of data transmission is important to improve the average sensor lifetime and overall bandwidth utilization. Summarizing and aggregating sensor data to reduce data transmission in the network is called data aggregation (Ozdemir & Xiao 2009). As stated earlier, IoT networks are prone to security attacks. Hence, network security has to be ensured while designing data aggregation algorithms.
5. **Network slicing:** The partitioning of a physical network into multiple virtual networks is called network slicing. This helps customize and optimize each network for a particular application (Zhang 2019). The shared network resources can be dynamically scheduled to the different logical network slices using a demand-driven approach. There can be a conflict of interest between users. While some may focus on minimizing latency, others may focus on minimizing the energy and installation costs.
6. **Deployment:** IoT networks comprise interconnected IoT devices. However, for efficient collection, analysis, and data processing, the interconnectivity of IoT nodes and sensing coverage are very important. Deployment algorithms focus on ensuring this coverage and connectivity. However, other potentially conflicting objectives also need to be optimized. These are energy, latency, and throughput.

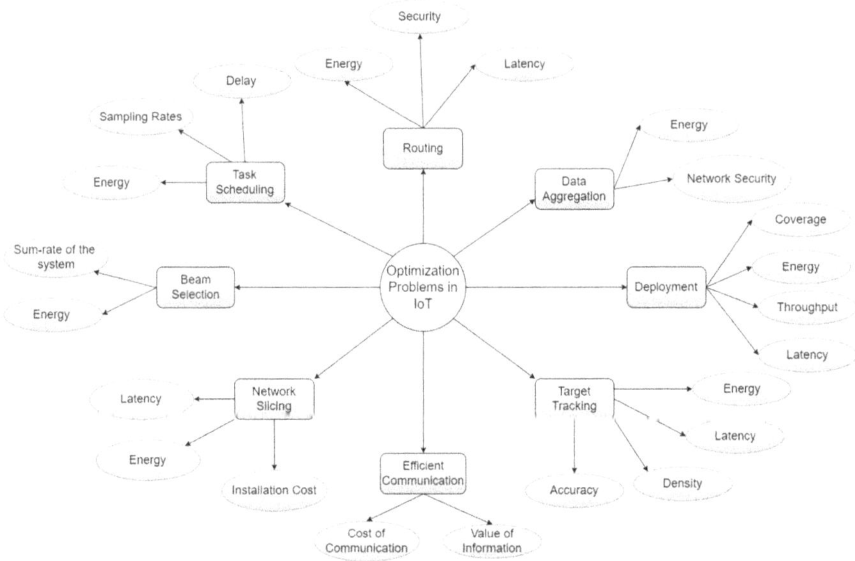

Fig. 8.1 Objectives in some common optimization problems in IoT

7. **Target Tracking:** Target tracking is the localization and tracking of mobile targets, like the localization of firemen in an IoT-enabled building under fire. Ensuring localization accuracy in some applications is critical. However, based on applications, there could also be several other important objectives like energy, latency, and density.
8. **Beam Selection**: Beam selection refers to selecting the best beam pair for the transmitting and receiving antennas in wireless communications. Energy efficiency and the sum rate of the system are often two objectives to be optimized in such wireless communication scenarios.

Figure 8.1 visually illustrates these objectives, providing a comprehensive overview of the diverse challenges faced in IoT optimization. When addressing these optimization problems, striking a balance among various objectives is essential to arrive at solutions that best align with each IoT application's specific requirements and constraints.

8.3 Multi-Objective Optimization

Optimization is a branch of applied mathematics that aims to find the particular values of the associated variables, which result in either the minimum or the maximum values of a single objective function or multiple objective functions (Adnan et al. 2013). Most of the problems we face daily can be characterized as multi-objective problems (MOPs). They have more than one aspect to consider (Fei et al. 2016). The same can be said for problems in wireless sensor networks.

A typical MOP involves multiple objectives to be simultaneously optimized under certain constraints. As an example, a multi-objective problem with n objectives, m variables, and one constraint can be formulated as

$$\begin{aligned} &\text{minimize} \quad f(\mathbf{x}) := \text{minimize} \quad [f_1(\mathbf{x}), f_2(\mathbf{x}), \ldots, f_n(\mathbf{x})], \\ &\text{subject to} \quad g(\mathbf{x}) \leq M \end{aligned}$$

where $\mathbf{x} \in \mathbb{R}^m$ and $f(\mathbf{x}) \in \mathbb{R}^n$, with $\mathbb{R}^m$ and $\mathbb{R}^n$ representing the decision and objective space, respectively.

The objectives can be mutually conflicting. In multi-objective optimization problems where objectives conflict with each other, a single solution does not necessarily exist that maximizes the rewards (or minimizes the loss) for each objective simultaneously (Kochenderfer & Wheeler 2019). Depending on the weight given to each objective during the optimization process, each identified solution will potentially differ in the total reward gained for each objective.

8.3.1 *Pareto Front*

For a finite solution space with conflicting objectives, there will be several solutions for which it is impossible to improve the total reward for one objective without decreasing the reward for the other. Each solution with this characteristic is considered Pareto efficient, and all solutions with this characteristic make up the Pareto frontier. In MORL with conflicting objectives, the learning process aims to approximate the policy which leads to these Pareto-efficient solutions. The optimal solutions for a MOP comprise the Pareto front; see Fig. 8.2. Mathematically, let $X \subseteq \mathbb{R}^n$ be the set of feasible solutions for the optimization problem described above. A feasible solution $x_1 \in X$ is said to (Pareto) dominate another solution $x_2 \in X$ if:

1. $\forall i \in \{1, 2, \ldots n\}$, $f_i(x_1) \leq f_i(x_2)$,
2. $\exists i \in \{1, 2, \ldots n\}$, $f_i(x_1) < f_i(x_2)$

A solution $x^* \in X$ and the corresponding solution $f(x^*)$ is considered Pareto optimal if another solution does not exist that dominates it. The set of Pareto optimal solutions, denoted by X^*, is called the Pareto front, Pareto frontier, or Pareto boundary.

8.3.2 *Preference Vector*

In MOO, the goal is to find optimal solutions that balance these current objectives' trade-offs. Fei et al. (2016) highlight the importance of utilizing MOO in wireless

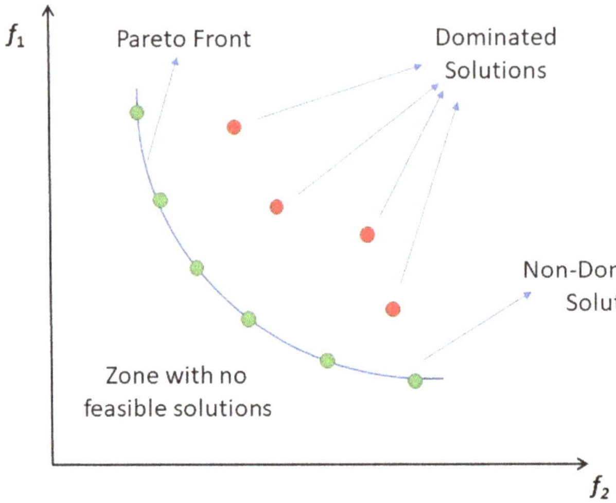

Fig. 8.2 Pareto front in multi-objective optimization

sensor networks by arguing that it is more realistic than single-objective optimization (SOO). The reasoning is that the importance of the single metric chosen as the objective may often be overestimated. They present categories into which a majority of MOO approaches can be divided, two examples being RL and scalarization. A shared characteristic of the scalarization methods is that they convert the MOP into a single-objective problem (SOP). The problem under consideration can be formulated as a linearly scalarized SOP as follows:

$$\text{minimize} \qquad f(\mathbf{x}) := \text{minimize}[\lambda_1 f_1(\mathbf{x}) + \lambda_2 f_2(\mathbf{x}) + \ldots + \lambda_n f_n(\mathbf{x})],$$

where $\lambda_i \in [0, 1]$ is a parameter indicating how much weight we put on the different objectives. The vector $\boldsymbol{\lambda}$ is thus called the preference vector. It is well-known from multi-objective optimization that by varying $\boldsymbol{\lambda}$, we capture all Pareto-efficient solutions for the n objectives (Van Moffaert & Nowé 2014). This can be seen in contrast with heuristic algorithms, which aim to find an approximate solution to a problem (Kokash 2005). Some standard scalarization methods are linear weighted sum, ϵ-constraints, goal programming, and Chebyshev.

IoT networks are known for their dynamic network characteristics and priorities. In such situations, a single utility for the optimizer might not be sufficient to describe the real objectives involved in sequential decision-making. A natural approach for handling such cases is optimizing one objective with constraints on others (Altman 2021). This allows us to understand the trade-off between the various objectives.

8.3.3 Traditional Approaches for MOO in IoT

Several approaches for MOO have been traditionally used in wireless networks. Linear programming and integer programming are two of them. Linear programming (LP) is a mathematical technique for maximizing or minimizing a linear function of several variables, such as output and cost. An integer programming (IP) problem is a special category of linear programming (LP) problem in which the decision variables are further constrained to take integer values. Many optimization problems in IoT either are NP-hard and cannot be solved by any polynomial time algorithms or cannot be expressed as a linear programming problem. Heuristic-based algorithms involve decision-making based on certain heuristic values or functions designed by a decision-maker. There is no well-defined process to design a good heuristic. Moreover, these approaches usually do not guarantee convergence to optimal solutions.

Fuzzy logic-based algorithms have also been popularly used. Fuzzy logic is a logic system that can describe to what degree something is true. Fuzzy membership functions are used to compute the membership of a variable. The fuzzy membership functions can be multi-objective, similar to the scalarized functions described in Sect. 8.3. However, fuzzy logic can be impractical for IoT if the logic is complex or if the IoT network is large and dynamic. Evolutionary algorithms have been applied in many scenarios in wireless networks to obtain near-optimal solutions. Evolutionary algorithms are nature-inspired population-based metaheuristic algorithms. Some of these are multi-objective genetic algorithms (MOGA), multi-objective particle swarm optimization (MOPSO), and multi-objective ant colony optimization (MOACO). However, they do not necessarily result in the Pareto front. ML-based algorithms are increasingly used to approach many optimization problems in IoT, including NP-hard problems. The evolution of IoT has paved roads for small devices to be autonomous decision-makers. This is being made possible using ML. Many supervised and semi-supervised ML algorithms have been used in IoT; however, they have a drawback. These algorithms need training data before deployment. However, no such training data is available in many scenarios before training.

8.4 Reinforcement Learning

In the future of IoT networks, decision-making policies will require dynamic adaptation based on incoming data due to the highly dynamic nature of these networks. This ongoing evolution presents challenges that cannot be fully addressed by traditional optimization methods alone. While traditional optimization techniques can be effective in static or slowly changing environments, they may struggle to cope with the rapid and unpredictable changes characteristic of IoT networks (Fig. 8.3).

In contrast, machine learning, particularly RL, emerges as a powerful alternative for IoT networks. RL enables devices and systems to learn from experience and

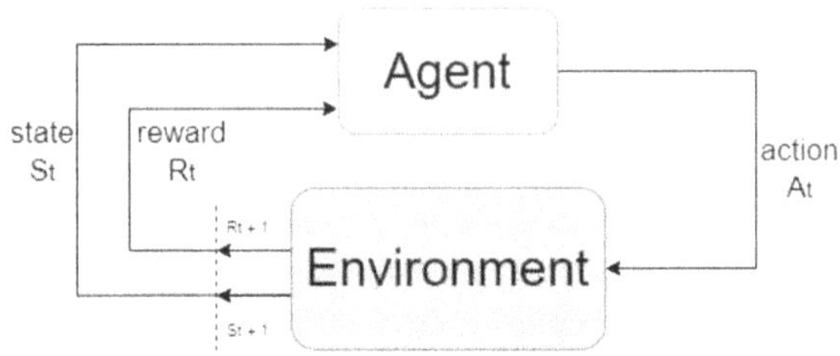

Fig. 8.3 Reinforcement learning cycle (Sutton & Barto 2018)

interactions with the environment. This capability allows them to adapt and make decisions in real time without relying on pre-existing training data. RL's learning-by-interaction approach aligns well with the evolving nature of IoT networks, where decisions need to be made dynamically based on changing conditions and incoming data. By combining RL with IoT networks, devices and systems can learn from their environment, identify patterns, and optimize decision-making processes to achieve their objectives efficiently. This autonomous and adaptive nature of RL empowers IoT networks to handle uncertainties, make data-driven decisions, and optimize performance in complex and rapidly changing scenarios.

RL problems are typically formulated using Markov Decision Processes (MDPs) (Sutton & Barto 2018). An MDP is a mathematical framework that represents sequential decision-making problems. It consists of a tuple (S, A, P, R, γ), where:

- $\mathcal{S}$ is the set of states representing the possible conditions of the environment.
- $\mathcal{A}$ is the set of actions that the agent can take to interact with the environment.
- P is the state transition probability function, denoting the likelihood of transitioning from one state to another when the agent takes a particular action.
- R is the reward function, which specifies the immediate reward the agent receives for performing an action in a given state.
- γ is the discount factor, representing the agent's preference for immediate rewards over future rewards.

In an MDP, the agent starts in an initial state s_0, and at each time step t, it chooses an action a_t based on its current state s_t. The environment then transitions to a new state s_{t+1} with a corresponding reward r_{t+1}, and the process continues over a series of time steps. The agent's goal in an MDP is to learn an optimal policy $\pi : S \to A$, which is a mapping from states to actions that maximizes the expected cumulative reward, known as the return. The return is defined as the sum of discounted rewards over time:

$$G_t = \sum_{k=0}^{\infty} \gamma^k r_{t+k+1}$$

The optimal policy π^* is the policy that maximizes the value function $V^{\pi}(s)$, which represents the expected cumulative reward from being in state s and following policy π:

$$V^{\pi}(s) = \mathbb{E}_{\pi}\left[G_t \mid s_t = s\right]$$

In addition to the value function, it is often convenient to work with the action-value function, commonly known as the Q-function. The Q-function, denoted as $Q(s, a)$, represents the expected cumulative reward an agent can achieve by taking action a in state s and following the optimal policy thereafter. Mathematically, this can be expressed as

$$Q^{\pi}(s) = \mathbb{E}_{\pi}\left[G_t \mid s_t = s, a_t = a\right],$$

i.e., the sum of discounted rewards from time step t onward.

Reinforcement learning algorithms aim to find the optimal policy π^* or the optimal value function $Q^*(s, a)$, which represents the expected cumulative reward from taking action a in state s and following the optimal policy thereafter. One of the key algorithms in RL is Q-Learning. Q-Learning is a model-free algorithm that learns the optimal action-value function $Q^*(s, a)$ directly from experience. The algorithm iteratively updates the Q-function based on the rewards received for actions taken in different states. The updates are performed using the following formula:

$$Q(s, a) \leftarrow Q(s, a) + \alpha \left[R_a(s, s') + \gamma \max_{a' \in \mathcal{A}}(Q(s', a')) - Q(s, a) \right]$$

where $Q(s, a)$ is the current estimate of the optimal Q-table and α is the learning rate. Q-learning is guaranteed to converge to the optimal Q-table provided that all state and action pairs are explored sufficiently often under nonsummable diminishing learning rate. Once the optimal Q-table is learned, then optimal policy can be obtained by simply looking up in the Q-table.

The Q-function returns the reward of a specific action and state mapping. The mapping could be represented as a table or a neural network. Tabular Q-Learning is typically used in low-dimensional state spaces with a small number of discrete actions. However, in high-dimensional state spaces with a large number of continuous actions, Q-Learning may become infeasible due to the curse of dimensionality. DRL is more common in advanced implementations because the table must be unrealistically large to fit all state and action mapping. The usage of a neural network is what characterizes deep reinforcement learning (DRL).

8.5 Multi-Objective and Constrained Reinforcement Learning in IoT Networks

In RL, the reward formulation captures the optimization goal. Multi-objective reinforcement learning (MORL) can be defined as a generalization of RL where the reward signals are extended to multiple feedback signals, one for each objective

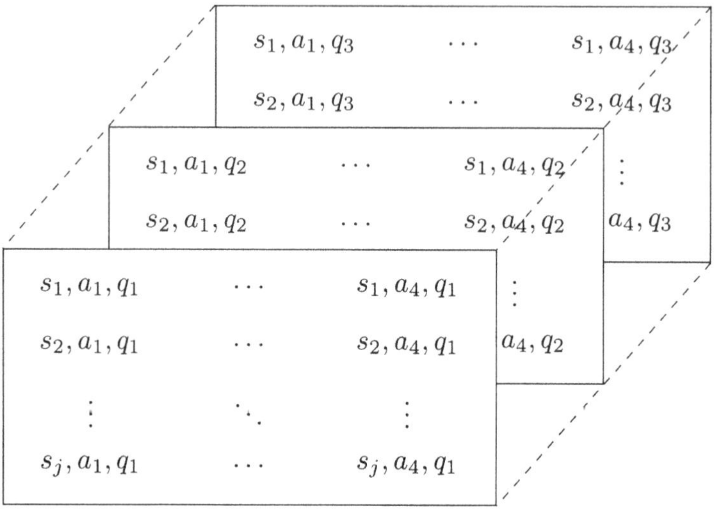

Fig. 8.4 A visualization of the Q-table approximated by the multi-objective Q-Learning algorithm corresponding to three preference vectors

(Van Moffaert & Nowé 2014). Thus, an RL algorithm can approach MOO by altering the reward signal into a scalarized multi-objective function, as described in Sect. 8.3. The vector $\boldsymbol{\lambda}$, known as the preference vector, defines the weightage given to the different objectives. In Q-Learning, each value of λ would correspond to a unique Q-table (or policy), giving the optimal decision for the selected preference vector. Figure 8.4 gives a visualization of the Q-tables approximated by the multi-objective Q-Learning algorithm corresponding to three preference vectors. In the last decade, several approaches for MORL have been proposed, which have the potential to be used in IoT. These are discussed below.

8.5.1 Single-Policy Approaches

In MORL based on a single-policy approach, the multidimensional reward vector is converted into a scalar value. This mapping of a multidimensional reward vector into a single scalar value can be done by several approaches. Some of these are the weighted sum approach, W-learning (Humphrys 1996), the analytical hierarchy process (AHP) (Zhao et al. 2010), the ranking approach (Gábor et al. 1998), and the geometric approach (Mannor & Shimkin 2004). The weighted sum approach, as discussed in Sect. 8.3, is the most commonly used one. MORL is then used just like any single-objective RL algorithm to find the optimal policy giving the Pareto front for the chosen preference vector. Table 8.1 reviews some of these recent works. We can see that in most of the MORL approaches in IoT domain (Wang et al. 2019;

Table 8.1 MOO problems in IoT, proposed MORL approaches, and their objectives

MORL in IoT			
Ref.	Problem/application	MORL approach	Objectives
Wang et al. (2019)	Workflow scheduling	Manually designed reward	Workflow completion time, Cost of Virtual Machines
Ren et al. (2021)	IoT-based Canal-Control	A reward network learns the preference vector by interacting with the environment. The learned reward function is fed to a Deep Q-Network (DQN)	Speed, Safety, Efficiency
Kruekaew and Kimpan (2022)	Task Scheduling	Preference vector is manually tuned using hyperparameter optimization	Task execution time, Processing cost, Resource Utilization
Caviglione et al. (2021)	Placement of Virtual Machines	Preference vector is manually designed	Deployment reliability, Co-location interference, Power consumption
Ghasemi and Toroghi Haghighat (2020)	Placement of Virtual Machines	Tabular Q-Learning with manually designed preference vector	Balancing loads of CPU, memory and bandwidth of different host machines and ensuring intra-host machine balance
Cui et al. (2021)	Resource Allocation for Internet of Vehicles (IoV)	Tabular Q-Learning with manually designed reward	Reliability, Latency
Peng et al. (2020)	Cloud Resource Scheduling	DRL is used, and the preference vector is manually tuned using hyperparameter optimization	Energy, Quality of Service
Yu et al. (2021)	Optimization for Unmanned Aerial Vehicle (UAV)-Assisted IoT Networks	An extended deep deterministic policy gradient (DDPG) algorithm is used. The preference vector is manually set	Sum data rate, Total harvested energy, and UAV's energy consumption over a particular mission period
Kaur et al. (2021)	Routing	Three Deep Q-networks are used for three objectives. The results from the three are aggregated based on a user-defined preference vector	Delay, Network Lifetime, Throughput
Vaishnav and Magnússon (2023)	Datastream processing and offloading	R-Learning is used and preference vector is chosen by the decision-maker	Energy, Delay

Kruekaew & Kimpan 2022; Caviglione et al. 2021; Ghasemi & Toroghi Haghighat 2020; Cui et al. 2021; Peng et al. 2020; Yu et al. 2021; Kaur et al. 2021; Vaishnav & Magnússon 2023), the preference vector for the scalarized reward is manually decided. Thus, most of the existing MORL approaches in IoT rely either on the decision-maker's judgment or extensive hyperparameter optimization to decide the preference vector. However, this is a highly inefficient approach for the dynamic IoT scenario.

Another approach based on linear scalarization in MORL trains a separate Q-table for each objective. Each Q-table is considered a vector, and the scalarized Q-table is then formed by taking the dot product of Q-vectors with the preference vector:

$$\hat{Q}(s, a) := \boldsymbol{\lambda}.\mathbf{Q}(s, a)$$

The scalarized Q-table $\hat{Q}(s, a)$ is then used for decision-making. This seems promising for dynamically changing preference vectors of the IoT scenario since no retraining of the RL agent would be required whenever the preference changes. However, the biggest limitation of this approach is that it gives only the solutions lying in the complex regions of the Pareto front (Vamplew et al. 2008). Apart from this, Vamplew et al. have proposed other variations of single-policy approaches like W-Steering and Q-Steering (Vamplew et al. 2015). However, most of these approaches have the limitation of relying on the decision-maker to choose a preference vector without learning from interaction with the environment. An attempt has been made in the IoT domain to decouple this dependence by introducing a separate network that learns the preference vector while interacting with the environment (Ren et al. 2021).

8.5.2 Multiple-Policy Approaches

In this approach to MORL, the dimensions of the objective space are not reduced. Thus, the RL agent must learn several optimal policies simultaneously or iteratively (Oliveira et al. 2021). The Convex Hull Value Iteration (CHVI) algorithm exemplifies this approach (Barrett & Narayanan 2008). CHVI algorithm is capable of simultaneously learning optimal policies for multiple assigned preference vectors. Other algorithms follow a multiple-policy approach by iteratively learning multiple policies, each customized for a particular preference vector. A linear scalarization is then performed to get the policy for the current preference vector. However, these algorithms are quite inefficient. A modification of the Q-Learning algorithm has been proposed by (Vamplew et al. 2011). This is suitable for multi-objective optimization problems. To convert the regular Q-Learning algorithm to support multiple objectives, two alterations are necessary:

1. The values learned by the algorithm must be in vector form, where each element in the vector corresponds to a single objective in the environment.
2. Greedy action selection is performed by applying a weighting function to the vector values of the Q-table.

With these alterations, a multi-objective Q-Learning algorithm can update multiple state values during a single reward iteration, converging toward the approximate optimal policies for multiple objectives. However, scalability issues will be faced in such an approach in most IoT applications, which have huge state and action spaces. Moreover, the post-training preference changes, which are quite common in IoT networks, still pose a challenge to the widespread applicability of the approaches discussed so far.

8.5.3 Approaches Based on Dynamic Preferences

In IoT networks, both the network characteristics and the objective preferences may change over time. For example, consider an IoT device that receives a stream of incoming tasks, as shown in Fig. 8.5. It can process some or all of the tasks locally or offload some portions to an edge node for processing. Each incoming task usually is associated with a deadline under which the processing must be done. Based on changing applications, the nature of tasks may vary. Some tasks may be delay-sensitive and must be processed within short deadlines. Other tasks could be more computationally intensive. While processing delay-sensitive tasks, first preference should be given to minimizing the delay, often done by processing them locally. However, while processing computation-intensive tasks, reducing the delay may not be the first preference. Rather, preference should be given to energy conservation. Thus, an intelligent offloading algorithm should be able to adapt according to changing preferences. MORL approaches that perform well in environments with dynamic preferences have been proposed. There is great potential for utilizing these approaches for intelligent decision-making in IoT:

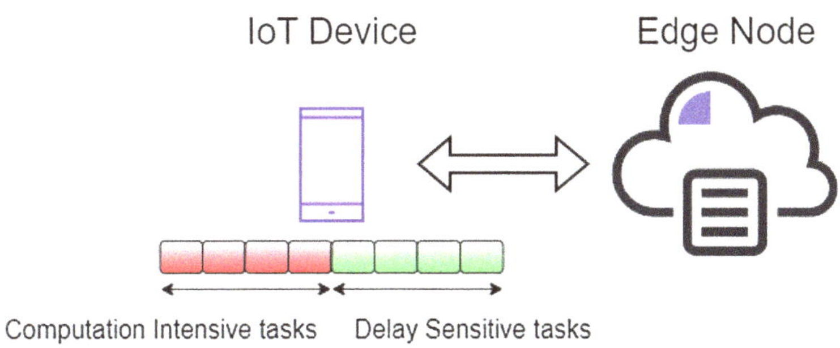

Fig. 8.5 An IoT-edge offloading scenario

1. Tabular setting: A dynamic preference-based approach was proposed for tabular Q-Learning in 2005 (Natarajan & Tadepalli 2005). In this approach, when the objective preference changes online, the agent learns from the best policy it has learned so far rather than starting the learning process from scratch for the new preference vector. Further, it was found that after learning a couple of policies, the agent need not learn much later because the existing policies form a good relevant coverage set. However, this approach is based on R-Learning. R-Learning is a variant of Q-Learning where average rewards are considered instead of discounted rewards. There is a possibility of designing frameworks such that R-Learning can be applied in IoT. One such framework has been proposed recently for data stream processing in IoT (Vaishnav & Magnússon 2023). Frameworks like this open possibility of utilizing this R-Learning-based (Natarajan & Tadepalli 2005) dynamic preference adaptability in IoT ecosystems. However, R-Learning might often be slower to converge as compared to Q-Learning. Moreover, tabular methods often fall short in scenarios with large and continuous action and state spaces. DRL is preferable in such scenarios.
2. DRL: A DRL approach for dynamic weights has also been proposed recently (Abels et al. 2019). In this approach, a Q-network (known as a conditioned network) generates preference-dependent Q-value vectors. This network has two inputs—the current environment state and a preference vector. Training is done by taking samples from both current preferences and some preferences previously encountered. Thus, previous knowledge is not forgotten, and the network doesn't have to train everything from scratch. This approach has been applied to design a multi-objective offloading framework in multi-access edge computing (Song et al. 2022). However, it can also be applied in applications of IoT. However, this approach is still not quick enough to transfer knowledge from old preferences to new ones.
3. Online policy transfer: A very recent work (Alegre et al. 2022) proposes a MORL algorithm using a combination of two existing concepts: successor features and optimal policy transfer. In the training phase, the successor features form a minimum convex coverage set of policies. These policies can be used in the execution phase. During execution, a new policy can be learned for any new preference vector without interacting with the environment. There is a very quick adaptation from old preferences to new ones using an optimal policy transfer algorithm. The transfer learning achieved in online approaches like this has the potential to aid dynamic decision-making in IoT.

8.6 Future Scope and Challenges in MORL

The last few decades have witnessed an upsurge in research carried out in wireless networks and IoT. Numerous algorithms and approaches have been proposed to solve optimization problems in IoT. Among ML-based approaches, RL has gathered a lot of attention because of the ability to learn by interacting with the environment without much prior information. However, one drawback of many proposed RL-based approaches is that they come at a cost. RL-based approaches may be computation-intensive and energy-consuming. Many existing works emphasize the improvements gained through RL without concerning themselves with the overall resources used in the network and the decision-making systems. This sometimes shows great results but doesn't show the overall picture of total resource usage, including that done for training the RL agent. Moreover, the resources available may also vary from one network to another, and from one time to another. When we speak of resources, we refer to the computation capacity, battery capacity, channel bandwidth capacity, etc. There is an increasing need to design algorithms that don't just optimize certain objectives but are also adaptive to the changing and limited network resources.

Selecting the right preference vector in a MORL algorithm is another challenge. Existing MORL algorithms proposed for IoT networks depend on hyperparameter optimization before applying the RL policy in the real-world scenario. However, the training data is rarely available at this phase. Better than hyperparameter optimization is the online plotting of Pareto fronts by learning multiple policies simultaneously while interacting with the environment. This can be accomplished using reward-free reinforcement learning. Reward-free reinforcement learning is suitable for scenarios where the agent does not have access to a reward function during exploration but must propose a near-optimal policy for an arbitrary reward function revealed only after exploring (Wagenmaker et al. 2022).

In highly dynamic IoT networks, the network characteristics, constraints, and the user's preferences are dynamically changing. Thus, there is a need to utilize transfer learning to adapt to new policies by transferring knowledge from old policies. Optimal policy transfer can provide solutions to this problem. The RL-based approaches for dynamic preferences are discussed in Sect. 8.4. However, there can be scenarios of variable constraints which are largely unexplored. Again, consider the IoT-edge offloading scenario shown in Fig. 8.5. Apart from the delay objective, energy consumption could be a constraint. For instance, if a mobile phone is put in power-saving mode, there is a constraint on the energy consumption the device can afford per unit of time. But the energy-constraint value may change if the mobile is put back to normal mode. In normal mode, the device can afford more energy consumption per unit of time. There is a need to study MORL algorithms that can adapt according to changing constraints as well.

Before the evolution of ML and RL, many simple heuristic-based approaches were used to solve optimization problems in IoT. They're simple and not so computation-intensive but do not guarantee convergence to the optimal solution. RL usually begins with an agent randomly making decisions and exploring as to which decisions are more rewarding. It takes time for RL-based approaches to converge to optimal solutions. There is an untapped potential in IoT to begin RL exploration using some existing simple heuristics, which are more rewarding than random exploration. It has been shown that such an exploration can help the RL agents converge faster to the optimal solutions (Beikmohammadi & Magnússon 2023).

8.7 Conclusion

In this chapter, we first discussed the nature of optimization problems in IoT networks and the multiple objectives involved. We highlighted the importance of efficient MOO algorithms for such problems. We discussed why MORL is more suitable than traditional and other ML approaches. We explored the challenges of implementing simple RL algorithms in IoT ecosystems. We then presented the existing MORL approaches and reflected on their applicability in IoT. We also highlighted the potential for utilizing advanced MORL algorithms for IoT networks. Finally, in Sect. 8.6, we suggested various avenues where existing MORL can be further improvised and applied for intelligent decision-making in IoT networks. Herein, we also reflect on the challenges that could be faced in the future.

References

Abels, Axel. et al. 2019. Dynamic weights in multi-objective deep reinforcement learning. In *International Conference on Machine Learning*, 11–20. PMLR.

Adnan, Md Akhtaruzzaman, et al. 2013. Bio-mimic optimization strategies in wireless sensor networks: A survey. *Sensors* 14 (1): 299–345.

Alegre, Lucas Nunes, et al. 2022. Optimistic linear support and successor features as a basis for optimal policy transfer. In *International Conference on Machine Learning*, 394–413. PMLR.

Altman, Eitan. 2021. *Constrained Markov Decision Processes*. Milton Park: Routledge.

Barrett, Leon, and Srini Narayanan. 2008. Learning all optimal policies with multiple criteria. In *Proceedings of the 25th International Conference on Machine Learning*, 41–47.

Beikmohammadi, Ali, and Sindri Magnússon. 2023. TA-Explore: Teacher-assisted exploration for facilitating fast reinforcement learning. In *Proceedings of the 2023 International Conference on Autonomous Agents and Multiagent Systems*, 2412–2414.

Caviglione, Luca, et al. 2021. Deep reinforcement learning for multi-objective placement of virtual machines in cloud datacenters. *Soft Computing* 25 (19): 12569–12588.

Cui, Yaping, et al. 2021. Reinforcement learning for joint optimization of communication and computation in vehicular networks. *IEEE Transactions on Vehicular Technology* 70 (12): 13062–13072. https://doi.org/10.1109/TVT.2021.3125109.

Fei, Zesong, et al. 2016. A survey of multi-objective optimization in wireless sensor networks: Metrics, algorithms, and open problems. *IEEE Communications Surveys & Tutorials* 19 (1): 550–586.

Gábor, Zoltán, et al. 1998. Multi-criteria reinforcement learning. In *ICML*. Vol. 98, 197–205.

Geddes, K. O. et al. 1992. *Algorithms for Computer Algebra*. Boston: Kluwer.

Ghasemi, Arezoo, and Abolfazl Toroghi Haghighat. 2020. A multi-objective load balancing algorithm for virtual machine placement in cloud data centers based on machine learning. *Computing* 102: 2049–2072.

Humphrys, Mark. 1996. Action selection methods using reinforcement learning. *From Animals to Animats* 4: 135–144.

Kaur, Gagandeep, et al. 2021. Energy-efficient intelligent routing scheme for IoT-enabled WSNs. *IEEE Internet of Things Journal* 8 (14): 11440–11449. https://doi.org/10.1109/JIOT.2021.3051768.

Kochenderfer, Mykel J., and Tim A. Wheeler. 2019. *Algorithms for Optimization*. Cambridge: MIT Press.

Kokash, Natallia. 2005. An introduction to heuristic algorithms. In *Department of Informatics and Telecommunications*, 1–8.

Kruekaew, Boonhatai, and Warangkhana Kimpan. 2022. Multi-objective task scheduling optimization for load balancing in cloud computing environment using hybrid artificial bee colony algorithm with reinforcement learning. *IEEE Access* 10: 17803–17818. https://doi.org/10.1109/ACCESS.2022.3149955.

Mannor, Shie, and Nahum Shimkin. 2004. A geometric approach to multi-criterion reinforcement learning. *The Journal of Machine Learning Research* 5: 325–360.

Natarajan, Sriraam, and Prasad Tadepalli. 2005. Dynamic preferences in multi-criteria reinforcement learning. In *Proceedings of the 22nd International Conference on Machine Learning*, 601–608.

Oliveira, Thiago, Henrique Freire de et al. 2021. Q-Managed: A new algorithm for a multiobjective reinforcement learning. *Expert Systems with Applications* 168: 114228.

Ozdemir, Suat, and Yang Xiao. 2009. Secure data aggregation in wireless sensor networks: A comprehensive overview. *Computer Networks* 53 (12): 2022–2037.

Peng, Zhiping, et al. 2020. A multi-objective trade-off framework for cloud resource scheduling based on the deep Q-network algorithm. *Cluster Computing* 23: 2753–2767.

Rao, Lei, et al. 2011. Coordinated energy cost management of distributed internet data centers in smart grid. *IEEE Transactions on Smart Grid* 3 (1): 50–58.

Ren, Tao, et al. 2021. An application of multi-objective reinforcement learning for efficient model-free control of canals deployed with IoT networks. *Journal of Network and Computer Applications* 182: 103049.

Sarangi, Smruti R., et al. 2018. Energy efficient scheduling in IoT networks. In *Proceedings of the 33rd Annual ACM Symposium on Applied Computing*, 733–740.

Song, Fuhong, et al. 2022. Offloading dependent tasks in multi-access edge computing: A multi-objective reinforcement learning approach. *Future Generation Computer Systems* 128: 333–348.

Sutton, Richard S., and Andrew G. Barto. 2018. *Reinforcement Learning: An Introduction*. Cambridge: MIT Press.

Vaishnav, Shubham, and Sindri Magnússon. 2023. Intelligent processing of data streams on the edge using reinforcement learning. In *2023 IEEE International Conference on Communications Workshops (ICC Workshops)*.

Vaishnav, Shubham, Efthymiou Maria, et al. 2023. Energy-efficient and adaptive gradient sparsification for federated learning. In *ICC 2023 - IEEE International Conference on Communications*.

Vamplew, Peter, John Yearwood, et al. 2008. On the limitations of scalarisation for multi-objective reinforcement learning of pareto fronts. In *AI 2008: Advances in Artificial Intelligence: 21st Australasian Joint Conference on Artificial Intelligence Auckland, New Zealand, December 1–5, 2008. Proceedings 21*, 372–378. Berlin: Springer.

Vamplew, Peter, Richard Dazeley, et al. 2011. Empirical evaluation methods for multiobjective reinforcement learning algorithms. *Machine Learning* 84: 51–80.

Vamplew, Peter, Rustam Issabekov, et al. 2015. Reinforcement learning of Pareto-optimal multiobjective policies using steering. In *AI 2015: Advances in Artificial Intelligence: 28th Australasian Joint Conference, Canberra, ACT, Australia, November 30–December 4, 2015, Proceedings 28*, 596–608. Berlin: Springer.

Van Moffaert, Kristof, and Ann Nowé. 2014. Multi-objective reinforcement learning using sets of pareto dominating policies. *The Journal of Machine Learning Research* 15 (1): 3483–3512.

Wagenmaker, Andrew J., et al. 2022. Reward-free RL is no harder than reward-aware RL in linear Markov decision processes. In *International Conference on Machine Learning*, 22430–22456. PMLR.

Wang, Yuandou, et al. 2019. Multi-objective workflow scheduling with deep-Q-network-based multi-agent reinforcement learning. *IEEE Access* 7: 39974–39982. https://doi.org/10.1109/ACCESS.2019.2902846.

Yousefpour, Ashkan, et al. 2019. All one needs to know about fog computing and related edge computing paradigms: A complete survey. *Journal of Systems Architecture* 98: 289–330.

Yu, Yu, et al. 2021. Multi-objective optimization for UAV-assisted wireless powered IoT networks based on extended DDPG algorithm. *IEEE Transactions on Communications* 69 (9): 6361–6374. https://doi.org/10.1109/TCOMM.2021.3089476.

Zhang, Shunliang. 2019. An overview of network slicing for 5G. *IEEE Wireless Communications* 26 (3): 111–117. https://doi.org/10.1109/MWC.2019.1800234.

Zhao, Yun, et al. 2010. Multi-objective reinforcement learning algorithm for MOS-DMP in unknown environment. In *2010 8th World Congress on Intelligent Control and Automation*, 3190–3194. Piscataway: IEEE.

Chapter 9
Intelligence Inference on IoT Devices

Qiyang Zhang, Ying Li, Dingge Zhang, Ilir Murturi, Victor Casamayor Pujol, Schahram Dustdar, and Shangguang Wang

9.1 Introduction

IoT devices (including smartphones and smart tablets) have gained significant popularity and have become the primary gateway to the Internet (Xu et al. 2019, 2020). Meanwhile, the exceptional performance of deep learning (DL) models in computer vision over the past decade has led to an increased reliance on deep neural networks (DNNs) for cloud-based visual analyses. These DNNs are utilized for diverse tasks such as inference and prediction after deployment. This integration of DNNs and cloud-based visual analyses has facilitated the realization of various applications, including object detection (Girshick et al. 2015), vehicle and person reidentification (Liu et al. 2016), pedestrian detection (Sun et al. 2014), and landmark retrieval (Wang et al. 2017), etc.

Q. Zhang (✉)
State Key Laboratory of Network and Switching, Beijing University of Posts and Telecommunications, Beijing, China

Distributed Systems Group, TU Wien, Vienna, Austria
e-mail: qyzhang@bupt.edu.cn

Y. Li
College of Computer Science and Engineering, Northeastern University, Shenyang, China

Distributed Systems Group, TU Wien, Vienna, Austria
e-mail: liying1771@163.com

D. Zhang · S. Wang
State Key Laboratory of Network and Switching, Beijing University of Posts and Telecommunications, Beijing, China
e-mail: zdg@bupt.edu.cn; sgwang@bupt.edu.cn

I. Murturi · V. C. Pujol · S. Dustdar
Distributed Systems Group, TU Wien, Vienna, Austria
e-mail: imurturi@dsg.tuwien.ac.at; v.casamayor@dsg.tuwien.ac.at; dustdar@dsg.tuwien.ac.at

P. K. Donta et al. (eds.), *Learning Techniques for the Internet of Things*,
https://doi.org/10.1007/978-3-031-50514-0_9

Developers are actively exploring the integration of DL into mobile applications to enhance intelligence and improve the user experience on IoT devices. While DL on the cloud has received significant research attention, the study of DL on IoT devices remains relatively limited. There is a lack of comprehensive understanding regarding the key challenges and issues associated with DL on IoT devices. Therefore, further research is needed to gain insights into this domain.

Supporting a diverse range of devices while maintaining high performance with a single convolutional neural network (CNN) presents a significant challenge. However, the development of models specifically designed for IoT devices has greatly improved the capabilities of AI. These models are optimized for performance and efficiency, considering the limited computational resources, power constraints, and memory limitations of IoT devices. These models are typically lightweight and compact, enabling fast and efficient inference without sacrificing accuracy. They incorporate techniques such as model compression (Choudhary et al. 2020), quantization (Polino et al. 2018), and efficient network architectures (Iandola et al. 2016; Zhang et al. 2018) to minimize computational and memory requirements while achieving high performance.

Device vendors have responded to the demand for efficient CNNs on IoT devices by introducing System-on-Chips (SoCs) and inference libraries that incorporate specialized units for CNN acceleration. These SoCs are equipped with high-performance CPU/GPU units, and dedicated accelerators designed for machine learning (ML) and image processing tasks. While these accelerators enable on-device processing, developers still face the challenge of supporting the diverse array of devices available in the market. The advancements in hardware have significantly enhanced the overall performance and capabilities of IoT devices, allowing them to handle computationally intensive tasks and deliver enhanced user experiences. Moreover, software solutions play a crucial role in accelerating on-device DL inference alongside hardware advancements. For instance, fine-tuned implementation can achieve up to a 62,806$\times$ performance improvement compared to vanilla implementations (Leiserson et al. 2020; Zhang et al. 2022). Inference libraries provide developers with the necessary tools and runtime environments to optimize inference on resource-constrained devices. These libraries enable real-time and on-device inference for a wide range of applications, further enhancing the efficiency and effectiveness of DL on IoT devices.

Recently, there has been a notable increase in the adoption of cloud-based visual analysis, driven by advancements in network infrastructure. To achieve SOTA performance and ensure compatibility with a wide range of IoT devices, developers often choose to offload computational tasks, either partially or entirely, to high-computing-power infrastructures such as cloud servers. The rapid development of 5G communication technology has further facilitated offloading, allowing applications with stringent latency requirements like to be supported effectively. While offloading offers benefits such as improved inference latency and the ability to handle device diversity, it also comes with certain challenges. One of these challenges is the high operational costs associated with maintaining and utilizing cloud resources. Additionally, remote execution raises concerns regarding privacy and security, and the user experience can be affected by variations in networking conditions. To address these challenges, researchers have been exploring collaborative approaches that leverage both local and cloud resources for CNN inference (Huang et al. 2020; Laskaridis et al. 2020). These approaches aim to strike a balance between leveraging the computing power of cloud and utilizing local resources to enhance performance and reduce latency. By distributing the computational workload and optimizing resource utilization, these collaborative methods offer potential solutions to the limitations of cloud-based visual analysis, paving the way for more efficient and effective AI applications.

In summary, IoT devices have made computing pervasive, accessible, and personalized, enriching our daily lives and opening up new possibilities for applications and services in various domains. The remainder of this work is structured as follows: Sect. 9.2 introduces the preliminary work on inference. Section 9.3 explores the diverse applications of inference in IoT, highlighting the range of domains where inference finds utility. Sections 9.4–9.6 present comprehensive reviews of commodity hardware, model optimization, and inference libraries, focusing on their relevance and effectiveness in IoTs, respectively. Section 9.7 reviews the current inference system in edge computing. Section 9.8 presents the research challenges and future opportunities. Lastly, Sect. 9.9 concludes the paper.

9.2 Inference on IoT Devices: Preliminaries

DL model deployment involves two main stages: model training and inference (Xu et al. 2022; Wang et al. 2022; Li et al. 2023). During the training stage, a significant volume of training data is utilized, and the backpropagation algorithm is employed to determine the optimal model parameter values. This process necessitates substantial computing resources and is typically conducted offline. On the other hand, model inference involves utilizing a trained model to process individual or continuous input data. The results of these computations often require real-time feedback to users, making factors such as computing time and system overhead (e.g., memory usage, energy consumption) crucial considerations. This two-stage deployment methodology allows for efficient utilization of computing resources during the training stage and facilitates real-time inference on IoT devices.

Inference refers to the execution of data analysis, decision-making procedures, and related tasks directly on edge devices or servers situated within a decentralized computing infrastructure, thus mitigating the exclusive dependence on cloud-based computing systems. This strategy facilitates timely, context-aware decision-making processes near the network edge, in closer proximity to the data source, proffering numerous advantages and opportunities for IoT devices:

- Real time: Inference empowers devices to make immediate decisions and take action without relying on cloud connectivity. By processing data locally, proximate to the data source, devices can provide real-time responsiveness.

- Reduced Bandwidth: Inference enables IoT devices to locally process large data volumes, thereby reducing latency and conserving bandwidth, proving advantageous in situations with limited network connectivity.
- Privacy Enhancement: Inference bolsters data privacy by limiting sensitive data transmission to the cloud. Conducted locally, it ensures sensitive information remains confined to IoT devices, minimizing potential risks and reinforcing privacy.
- Context-aware Decision-making: Inference utilizes contextual data from IoT devices, such as sensor readings and device-specific information, to enhance result accuracy and relevance. This facilitates intelligent, environment-specific decisions, leading to heightened operational efficiency and effectiveness.

Overall, inference underpins the autonomy, responsiveness, and the capacity to handle intricate tasks of IoT devices. It facilitates the evolution and potentiality of the IoT ecosystem, endowing devices with the ability to exploit their computational prowess and make judicious decisions.

9.3 Promising Intelligence Applications

AI applications, due to their complexity and high computational requirements, are housed in cloud centers. However, this computing paradigm struggles to deliver real-time services like analytics and smart manufacturing. Therefore, situating AI applications on IoT devices widens the application scope of AI models. As shown in Fig. 9.1, DL models can execute on edge devices (i.e., IoT devices and edge servers) or depend on cloud centers. In this section, we spotlight several notable AI applications and their merits.

9.3.1 Real-Time Video Analytic

Video analytics, integral to VR/AR, necessitates considerable computational power and extensive storage resources (Xu et al. 2021). Performing such tasks in the cloud often leads to unexpected latency and high bandwidth usage. However, the progression of edge computing allows for the migration of video analytics closer to the data source, mitigating these issues (Dustdar & Murturi 2021).

Video analysis applications, such as face recognition and object detection, benefit from various effective DL algorithms, including artificial neural networks and histogram analysis (Bajrami et al. 2018). Nonetheless, utilizing a singular model for analysis without noise reduction and feature extraction proves challenging. Thus, integrating multiple models often results in enhanced video analytic performance. For instance, face recognition entails several steps, each addressed by a different

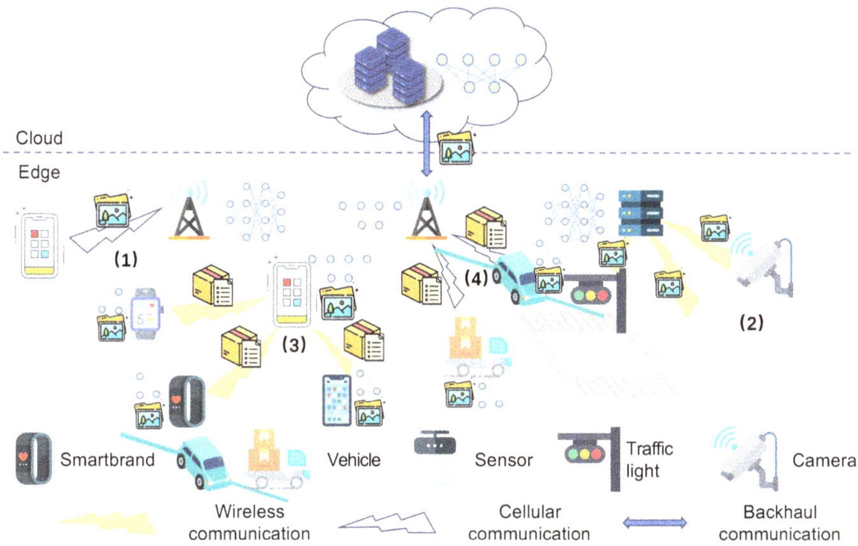

Fig. 9.1 Deep learning models can execute on edge devices (i.e., IoT devices and edge servers) or depend on cloud centers

model: AdaBoost (Viola & Jones 2001) for face detection, a nonlinear SVM classifier for gender and age classification, and a basic algorithm (Lucas & Kanade 1981; Zhou et al. 2022) for face tracking to calculate optimal flow and depict pixel trajectories.

9.3.2 *Autonomous Driving*

Autonomous vehicles, equipped with a plethora of sensors, generate a vast amount of data necessitating swift processing. The interconnectivity of these vehicles enhances safety, streamlines efficiency, and mitigates traffic congestion. Notably, autonomous driving aims to deliver services characterized by low latency, high-speed communication, and rapid response. ML- and DL-based solutions present potential for optimizing the complex operations intrinsic to autonomous vehicles. For example, ML algorithms deployed in self-driving vehicles extract features from raw data to discern real-time road conditions, facilitating informed decision-making. Similarly, for demanding tasks in autonomous driving—such as sensing, perception, and decision-making—DL algorithms process raw data through sensing to reach final decisions (Liu et al. 2019).

9.3.3 Smart Manufacturing

Smart manufacturing fundamentally hinges on automation and data analysis—the former being the primary goal, and the latter serving as an invaluable tool (Li et al. 2018). Ensuring low latency, privacy protection, and risk control is paramount to adhering to these principles. Within the realm of a smart factory, intelligent inference proves beneficial, bolstering computational resources and facilitating resource scheduling and data processing throughout the manufacturing process. Given the exponential proliferation, remote management of DL models and their continuous evaluation have emerged as pressing imperatives. To tackle these challenges, (Soto et al. 2016) pave the way for the development of real-time applications. Furthermore, DL algorithms are set to be significant catalysts propelling the industry's advancement by transforming all stages of the product lifecycle—from design and manufacturing to service—thereby driving substantial productivity enhancements.

9.3.4 Smart City and Home

The proliferation of IoT devices has sparked the emergence of intelligent services in various aspects of home lifestyles, encompassing appliances like smart TV and air conditioners (Kounoudes et al. 2021; Ain et al. 2018). Furthermore, the deployment of multiple IoT sensors and controllers in smart homes has become a prerequisite. Edge computing-based inference assumes a crucial role in optimizing indoor systems, aiming for low latency and high accuracy, thereby enhancing the capabilities and diversity of services. Moreover, extending edge computing beyond individual homes to encompass communities or cities holds significant potential. The inherent characteristic of geographically distributed data sources in urban environments enables location awareness, latency-sensitive monitoring, and intelligent control. For instance, we integrate large-scale ML algorithms, such as data mining combined with semantic learning, to extract advanced insights and patterns from the voluminous data generated by smart homes and cities (Mohammadi & Al-Fuqaha 2018).

9.4 Commodity Hardware for IoT Devices

With advancements in hardware, low-power IoT devices have the capability to independently handle AI tasks without relying on cloud communication. For instance, commodity CPUs, which are widely available in these devices, serve as the primary hardware for executing inference. CPUs play a crucial role in the inference, supported by toolchains and software libraries that facilitate practical inference. These CPUs share similar microarchitectures, allowing for the effective utilization of optimization techniques. However, performing computationally intensive tasks still poses challenges. For instance, processing a single image using the common VGG model (Sengupta et al. 2019), which consists of 13 CNN layers and 3 fully connected neural network (FCNN) layers, may take hundreds of seconds on devices like the Samsung Galaxy S7 (Xiang & Kim 2019).

Mobile GPUs have revolutionized high-dimensional matrix operations, including matrix decomposition and multiplications in CNN (Owens et al. 2008). Notably, GPUs have emerged as a standout option for edge computing, as they consume less power compared to traditional desktop and server GPUs. In particular, the Jetson family of GPUs, including the latest Jetson Nano, showcases a 128-core affordable GPU module that NVIDIA has successfully introduced. Additionally, the concept of caching computation results in CNN has sparked optimizations in frameworks like DeepMon (Huynh et al. 2017). DeepMon implements a range of optimizations specifically designed for processing convolution layers on mobile GPUs, resulting in significantly reduced inference time.

Due to power and cost constraints on devices, traditional CPU- and GPU-based solutions are not always viable. Moreover, devices often need to handle multiple application requests simultaneously, making the use of CPU- and GPU-based solutions impractical. As a result, hardware integrated with FPGA has gained attention for Edge AI applications. FPGA-based solutions offer several advantages in terms of latency and energy efficiency compared to CPUs and GPUs. However, one challenge is that developing efficient algorithms for FPGA is unfamiliar to most programmers, as it requires the transplantation of models programmed for GPUs into the FPGA platform.

There are also AI accelerators specifically designed for inference that have been introduced by several manufacturers. One notable example is the Myriad VPU (Leon et al. 2022), developed by Movidius, which is optimized for computer vision tasks. It can be easily integrated with devices like Raspberry Pi to perform inference. However, these AI accelerators are not widely available on all devices, limiting their accessibility. Additionally, the ecosystem surrounding these accelerators is still in its early stages and tends to be closed due to their black box structure and proprietary inference frameworks. This creates barriers for widespread adoption and usage. For instance, the Edge TPU, currently found only in Google Pixel smartphones, is limited to running models built with TensorFlow (Developers 2022).

Looking ahead, AI accelerators are expected to play a crucial role in IoT devices. With the introduction of powerful AI SoCs, there is potential for significant improvements in inference performance. As hardware accelerators and software frameworks continue to evolve and upgrade, more AI applications will be able to execute directly on IoT devices.

9.5 Model Optimization for IoT Devices

The limited computing resources on IoT devices necessitate developers to make trade-offs between model accuracy and real-time performance requirements, leading to the inability to deploy SOTA models. A fundamental challenge of this trend is the constrained resources of devices. Therefore, performance optimization has been a primary research direction for both academia and industry.

9.5.1 Lightweight Model Design

To optimize the computational overhead of DL inference, one approach is to ensure the lightweight nature of the DL models themselves. This can be achieved through the design of a lightweight model or the compression of a trained model. For example, SqueezeNet demonstrates such optimization by achieving comparable accuracy to AlexNet while utilizing only 2% of the parameters (Iandola et al. 2016).

The key innovation of SqueezeNet lies in its novel convolution method and the introduction of a fire module. As shown in Fig. 9.2, the fire module consists of a squeeze layer and an expand layer. The squeeze layer employs a 1×1 convolution kernel to alter the number of channels while maintaining the resolution (H×W) of the feature map to achieve compression. The subsequent expand layer utilizes 1×1 and 3×3 convolutional layers, whose outputs are combined to obtain the fire module's output. SqueezeNet follows a similar network design concept as VGG, utilizing stacked convolutional operations, with the difference being the incorporation of the fire module. Furthermore, ShuffleNet integrates group convolution to significantly decrease the number of parameters and computational complexity (Zhang et al. 2018). Group convolution divides channels into subgroups, where the output of each subgroup depends solely on the corresponding input subgroup. To address potential issues caused by group convolution, ShuffleNet (Zhang et al. 2018) introduces the shuffle operation, which rearranges the channels within each part to create a new feature map. The architecture of ShuffleNet is inspired by ResNet (Targ et al. 2016), transitioning from the basic ResNet bottleneck unit to the ShuffleNet bottleneck unit and stacking multiple ShuffleNet bottleneck units to form the complete model.

9.5.2 Model Pruning

Numerous researchers have explored techniques to reduce model complexity in DL models through parameter sharing and pruning. In many neural networks, the computationally intensive matrix multiplication in fully connected layers leads to a large number of model parameters and computing. To overcome this challenge, circulant

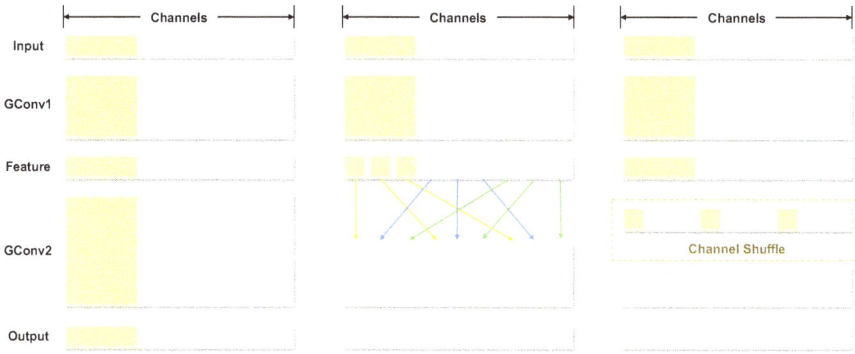

Fig. 9.2 Grouping convolution used in the lightweight model ShuffleNet

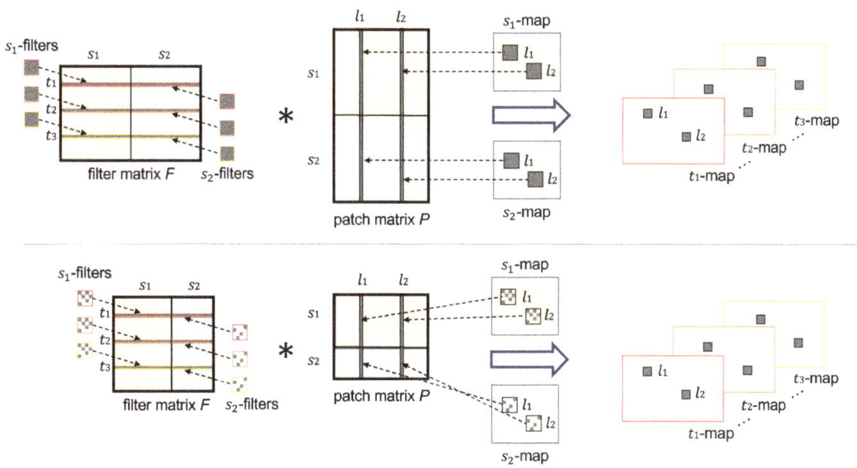

Fig. 9.3 Fast ConvNets reduce operations by pruning convolution kernels

projections have been proposed as a method to accelerate the computation of fully connected layers. As shown in Fig. 9.3, by employing a weight matrix with circulant projections (Cheng et al. 2015), memory requirement is reduced from $O(d^2)$ to $O(d)$ for a matrix of size d × d. Furthermore, the multiplication of rotation matrices can be accelerated using fast Fourier transform (FFT). This technique reduces the computational complexity from $O(d^2)$ to $O(dlogd)$ for the multiplication of a 1 × d vector and a d × d matrix. Given the significant role of CNN models in mobile vision, various approaches have been proposed for parameter pruning and sharing algorithms specifically tailored to CNNs. Fast ConvNets achieves computational reduction by pruning convolution kernels (Lebedev & Lempitsky 2016).

The commonly used technique for implementing convolution operations in DL libraries such as *TensorFlow* and *Caffe* is referred to as *im2col*. This process involves three steps: (1) During convolution, the input image is transformed into a matrix by sliding the convolution kernel. Each column in the matrix represents the information of a small window processed by the kernel. Rows in the matrix correspond to the product of the kernel's height, width, and number of input channels, while the column represents the product of the height and width of the single-channel image output by the convolution layer, representing the overall processing of the small window. (2) By reshaping the convolution kernel, a matrix is obtained where the rows correspond to the number of output image channels and the columns match the row values of the previous matrix. (3) The *im2col* operation converts complex convolution operations into matrix multiplications. The process involves performing matrix multiplication between two matrices and reshaping the resulting matrix into the final output. By leveraging existing optimization algorithms and libraries for efficient matrix operations, such as the BLAS algebraic operation library, *im2col* benefits from optimized matrix operations. Techniques like parameter pruning in Fast ConvNets (Lebedev & Lempitsky 2016) further reduce the matrix dimension after expansion, leading to accelerated computational workload for matrix multiplication.

9.5.3 Model Quantization

Quantization is a technique used to compress DL models by reducing the number of bits required for each parameter. In traditional DL models, parameters are typically represented using 16-bit floating-point numbers. However, experimental studies have shown that using lower-precision representations can significantly reduce memory consumption and computation time without compromising precision. In some cases, researchers have even employed 1-bit representations for storing parameters during both training and inference, achieving comparable results by using values of 1 or −1 (Rastegari et al. 2016). Additionally, other studies have investigated the use of vector quantization and product quantization techniques to compress models and further improve their efficiency.

Vector quantization and product quantization are widely used data compression techniques that involve grouping scalar data into vectors and quantizing them as a whole in the vector space, resulting in lossless data compression. By applying product quantization to the connection layer and the convolution layer, it is possible to achieve benefits such as reduced model size and improved operation time. These techniques are effective in optimizing DL models for improved efficiency and performance (Wu et al. 2016). As illustrated in Fig. 9.4, the main concept involves partitioning the input space into M equally sized subspaces, where each subspace of the weight matrix is assigned a sub-codebook obtained through a clustering algorithm. A codebook consists of multiple codewords, and the core idea of the algorithm is to approximate all subvectors of the same dimension in the space using a limited number of codewords. During inference, the input vector is divided into M sub-vectors, which are then multiplied only with the codewords in their respective subspaces. The final output can then be obtained based on the index of the pre-

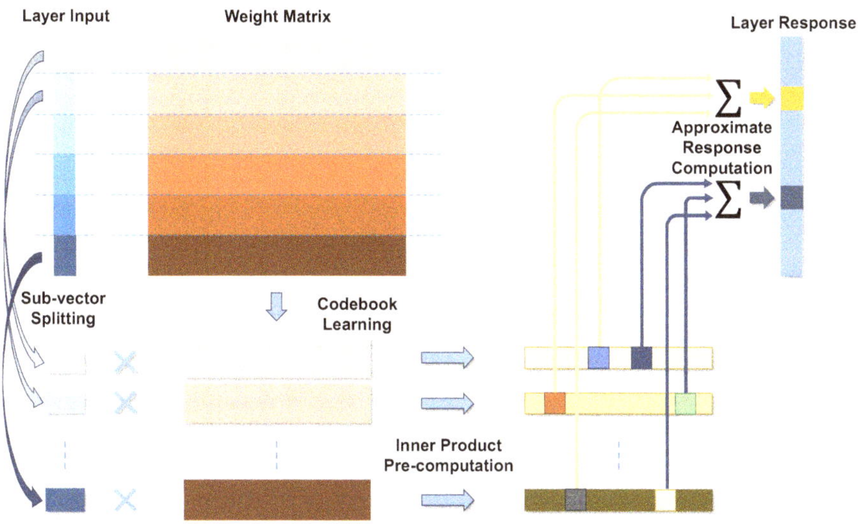

Fig. 9.4 Model compression method based on product quantization

calculated codebook mapped to the output matrix. Consequently, the computational complexity is reduced from the original $O(C_s C_t)$ to $O(C_s K + C_t M)$, where C_s and C_t denote the input and output dimensions, respectively. K represents the number of codewords in each codebook. Experimental results demonstrate that the algorithm can achieve a 4–6$\times$ increase in computational speed and reduce the model size by 15–20$\times$ with only a marginal 1% loss of accuracy.

9.5.4 *Knowledge Distillation*

Knowledge distillation (KD) is an effective algorithm widely used for compressing DL models. When employing small models for classification inference, relying solely on one-hot encoding in the training set is insufficient. This encoding method treats categories as independent entities and fails to capture the relationships between them. However, by allowing the small model to learn from the probability distribution generated by a larger model, additional supervision signals are provided, and similarity information between categories is incorporated, facilitating easier learning. For example, in the recognition of handwritten digits, certain images labeled as 3 may bear a resemblance to 8 or 2. One-hot encoding is unable to capture such nuances, whereas a pre-trained large model can provide this information. As a result, researchers modify the loss function to align the small model with the probability distribution outputted by the large model, a process known as KD training.

The effectiveness of KD training has been demonstrated in two datasets: handwriting recognition and speech recognition. However, (Romero et al. 2014) argue that directly mimicking the outputs of large models poses challenges for small models. Additionally, as the model depth increases, emulating the large model becomes more difficult as the supervision signal from the final layer needs to propagate to the earlier layers. To address this challenge, the researchers propose Fitnets, which involve incorporating supervisory signals in the middle layers. By comparing and minimizing the discrepancy between the outputs of the intermediate layers in both the large and small models, the small model can learn from the larger model during an intermediate step of prediction. Here, the "small" refers to the width of the layers rather than the depth. This training approach, known as hint training, involves pre-training the parameters of the first half of the small model using hint training. Subsequently, KD training is employed to train all parameters, enabling the small model to better emulate the knowledge of each layer in the larger model. However, it is important to note that this more active learning method may not be universally applicable due to the significant capacity gap between large and small models.

Building upon the work of (Yim et al. 2017), researchers have extended the concepts and applications of KD. Instead of having the small model directly fit the output of the large model, the focus is on aligning the relationships between the layers of the two models. These relationships are defined by the inner product between layers. *A* matrix of size $M \times N$ is constructed to represent this relationship, where each element (i, j) corresponds to the inner product between the $i - th$ channel of layer A and the $j - th$ channel of layer B. Yim et al. propose a two-stage method: first, adjusting the parameters of the small model based on the feature similarity preservation (FSP) matrix of the large model to align the layer relationships; then, continuing fine-tuning the small model parameters using the original loss function, such as cross-entropy. This approach aims to preserve the feature similarity between the two models while maintaining the original learning objective (Yim et al. 2017).

9.6 Inference Library for IoT Devices

The inference performance of on-device models is influenced by multiple factors, including hardware, models, and software, such as DL execution engines or libraries. DL libraries aim to enable on-device inference, and several major vendors have developed their own DL libraries, including TFLite (Haris et al. 2022), Core ML (Deng 2019), NCNN (Courville & Nia 2019), MNN (Jiang et al. 2020; Zhang et al. 2023), etc. TensorFlow and Caffe have been deprecated and replaced by their lightweight implementations, TFLite and PyTorchMobile, respectively. This work provides a summary of popular DL libraries such as TFLite (Haris et al. 2022), PyTorchMobile, NCNN (Courville & Nia 2019), MNN (Jiang et al. 2020), MACE (Lebedev & Belecky 2021), and SNPE (Zhang et al. 2022). Table 9.1 presents

Table 9.1 A comparison of representative DL libraries on mobile devices

Library	Developer	CPU FP32	CPU INT8	GPU FP32	GPU INT8	DSP INT8
TFlite	Google	✓	✓	✓	✓	✓
Pytorch Mobile	Facebook	✓	✓			
NCNN	Tencent	✓	✓	✓	✓	
MNN	Alibaba	✓	✓	✓	✓	
MACE	Xiaomi	✓	✓	✓	✓	
SNPE	Qualcomm	✓	✓	✓	✓	✓

a comparison of these DL libraries, considering their support for various model precision and hardware configurations.

Current DL libraries often do not fully leverage the capabilities of different hardware platforms. Each DL library typically supports at least one hardware platform, such as CPU, although PyTorchMobile lacks GPU acceleration support (Courville & Nia 2019). Interestingly, the DL library that achieves the best performance for a given model can vary depending on the specific hardware used. This difference in inference performance can be attributed to two main factors. Firstly, the hardware ecosystem exhibits a high level of fragmentation due to variations in architecture, such as Big. Little Core, cache size, GPU capacity, etc. Secondly, the heterogeneity of model structures also plays a role. The implementation of depth-wise convolution operators, for example, differs significantly from that of traditional convolution operators, as they have distinct cache access patterns (Zhang et al. 2022).

Even when using the same GPU, DL libraries provide different backend options. For example, MNN offers three backends: Vulkan, OpenGL, and OpenCL (Jiang et al. 2020). Interestingly, different backend choices can be more suitable for different models and devices. This may seem unexpected since MNN's Vulkan backend is primarily designed for cross-platform compatibility, including desktop, while OpenGL and OpenCL are mobile-specific programming interfaces that are highly optimized for mobile devices. This phenomenon can be attributed to both the underlying design of these backends and how DL developers implement the DL operators on top of them.

TFLite and SNPE provide acceleration capabilities for INT8 models running on DSP. For example, Qualcomm DSP is equipped with AI capabilities, such as HTA and HTP (Zhang et al. 2022), which are integrated with Hexagon vector extension (HVX) acceleration. The Winograd algorithm is also utilized to accelerate convolution calculations on the DSP. Furthermore, the energy-saving benefits of the DSP are particularly significant compared to the speed of inference.

To achieve optimal performance when executing models on devices, developers often need to integrate multiple DL libraries and dynamically select the appropriate one based on the current model and hardware platform. However, this approach is seldom implemented in practice due to the substantial overhead in terms of software complexity and development efforts. There is a need for a more lightweight system that can efficiently leverage the best performance from different DL libraries.

9.7 Inference Systems for IoT Devices

A considerable amount of research has been dedicated to exploring the collaborative utilization of local and cloud resources for inference. In contrast to the traditional completely offloading tasks to the server, these studies leverage the inherent characteristics of CNNs to optimize the offloading process (Donta & Dustdar 2022). Existing literature addresses the offloading of CNN inference to various destinations, including IoT devices within the local network (Xu et al. 2020; Mao et al. 2017), third-party IoT devices that possess the CNN computational graph (Almeida et al. 2022), and a choice between devices and servers through model selection (Han et al. 2016). Although these research endeavors share close relationships, the systems developed in these studies often have distinct requirements, such as the inclusion of multiple devices in a local area network, diverse optimization goals, such as distributing computations in a share-nothing setup, or significant overhead associated with maintaining multiple models.

9.7.1 Edge Cache-Based Inference

DL applications typically depend on real-time data provided by IoT devices, demonstrating specific similarity traits: (1) Temporal similarity: sequential frames in video streams captured by cameras frequently reveal similarities, such as consistent background or scene elements. (2) Spatial similarity: individuals' daily movement patterns, such as recurring travel between places like a laboratory and a restaurant, often show a high degree of repetition. Although variations in the captured images can occur due to alterations in lighting or background, image feature extraction algorithms like Speeded Up Robust Features (SURF) can capture these disparities (Xu et al. 2021). Note that despite the pronounced similarity between frames, they are not identical. The recurrent use of identical data can lead to a decrease in model accuracy. Consequently, the effective use of caching strategies is crucial to maintaining a balance between accuracy and efficiency. In this regard, there are some representative properties as follows.

Starfish enables the execution of multiple mobile vision applications on wearable devices while effectively managing computing resources and shared memory (LiKamWa & Zhong 2015). Its workflow is shown in Fig. 9.5. Researchers initially identified the need to parallelize multiple vision applications on existing wearable devices for simultaneous recognition and prediction tasks. However, these algorithms often rely on common data preprocessing steps such as grayscale processing, size adjustment, and feature extraction. Executing these algorithms separately on the same input image by different applications leads to redundant operations and memory consumption. To address this, Starfish decouples the CV library from the application, running it as an independent process (Core). API calls are transformed into cross-process calls, allowing the Core process to handle CV library calls

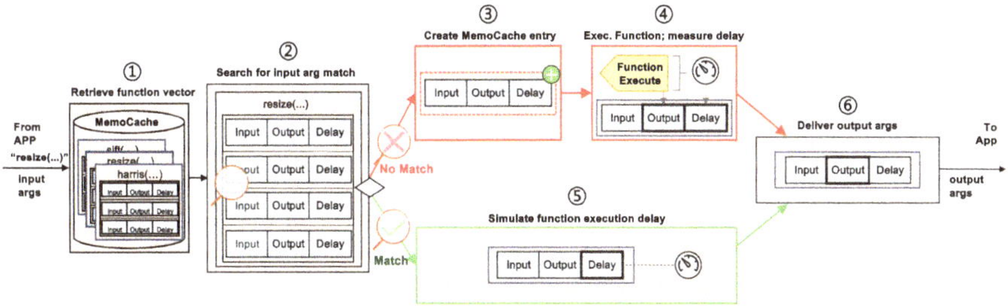

Fig. 9.5 The overview of Starfish framework

from all vision applications. A cache mechanism is employed to reduce redundant calculations and storage, with the API called by the CV library being key cached elements. Starfish, built on Android and optimized for the OpenCV (Bradski et al. 2000) library, effectively mitigates redundant operations and enhances resource utilization.

DeepMon leverages caching in mobile CV to enhance the performance of CNN inference (Huynh et al. 2017). The camera captures a series of video streams, and each frame is processed to convert it into an information stream that is then outputted to the users. The authors observe that consecutive frames in the video stream captured by IoT devices, such as smartphones and smart glasses, often contain significant pixel similarities, as vision applications typically require the camera to remain focused on a specific scene (e.g., object recognition, selfies) for a certain duration. Exploiting this pixel similarity, the author proposes using it as a cache to reduce the computation required by the CNN. However, traditional CNNs are considered a black box, where the output is obtained directly from the input image. The nature of convolution operations necessitates the dependence of output results on each frame's input image. To address this challenge, the authors redesign the forward algorithm of each CNN layer, enabling the (partial) computation results from the previous frame to be reused in the intermediate calculation process.

Guo and Hu (2018) incorporate a cross-application caching mechanism to minimize redundant calculations on the same or similar sensor input data, such as images. For instance, when both an image recognition application and an AR application utilize the same CNN model for image recognition, the output of CNN models can be cached. The cache lookup key is a variable-length feature vector that developers can specify, such as SIFT, SURF, etc. Furthermore, Potluck assigns an importance value to each cache item, indicating its frequency of use and potential time-saving benefits. Given the limited storage capacity, priority is given to caching the more important items. To find the most suitable cache entry, Potluck employs a nearest neighbor algorithm. The authors implement Potluck as a background service on the Android system and demonstrate its ability to reduce computing delay by up to 2.5–10× when used in applications.

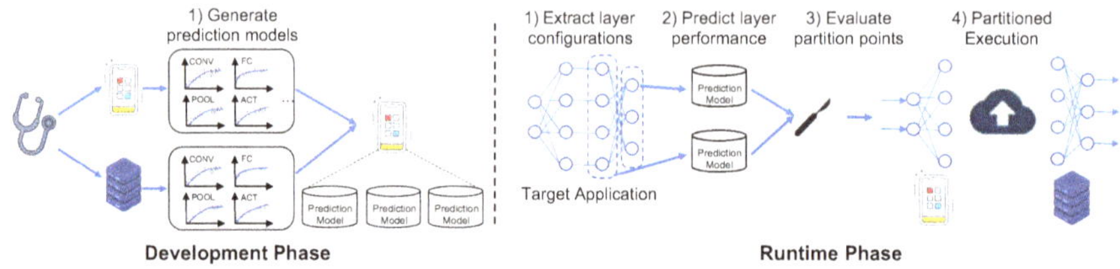

Fig. 9.6 The Neurosurgeon (Kang et al. 2017) framework tailored for computing offloading-based framework

9.7.2 Computing Offloading-Based Inference

Table 9.2 presents a comprehensive comparison of inference systems based on computing offloading. Note that these systems are specifically designed for a particular class of CNNs that are optimized for tasks such as classification and object detection. Among these systems, one notable work in this area is Neurosurgeon (Kang et al. 2017), as illustrated in Fig. 9.6, a framework that focuses on selecting an optimal split point to offload models between the devices and servers, with the aim of minimizing latency or energy consumption. However, the evaluation of Neurosurgeon primarily involves simple sequential CNN models, and the offloading decisions tend to be polarized, either offloading nothing or offloading everything, depending on the network conditions. Another relevant work is Hu et al. (2019), which introduces a scheduling scheme for partitioning DNNs under different network conditions to minimize either overall latency or throughput. However, it should be noted that the proposed scheduler lacks support for SLO deadlines, which are important in real-time applications. MCDNN (Han et al. 2016) is a framework that enables parallel computing for multiple applications. It uses a shared feature extraction layer and dynamically selects smaller, faster models to optimize accuracy and efficiency. The model selection is based on model catalogs, allowing for flexible adaptation to task requirements and available resources.

In terms of the compression of transferred data, JALAD (Li et al. 2018), SPINN (Laskaridis et al. 2020), and DynO (Almeida et al. 2022) incorporate the quantization scheme. However, it should be noted that SPINN utilizes a fixed 8-bit quantization level that is uniform across the split layers, without considering the dynamic range of the data or the resilience of each layer to quantization. DynO's compression method includes the compression method used in SPINN. DynO offers greater adaptability by dynamically selecting the optimal combination of bitwidth and split points based on performance targets and networking conditions. On the other hand, JALAD utilizes a decoupled DL model to make offloading decisions using a joint accuracy- and latency-aware execution framework (Li et al. 2018). However, JALAD is associated with a significant accuracy drop to achieve performance improvements. Additionally, JALAD only provides static configurations and lacks the ability to adapt to dynamic network conditions during runtime, limiting its efficiency on resource-constrained devices.

Table 9.2 Comparison of the existing inference systems

Work	Model	Network	Offloading granularity	Communication optimization	Scheduler	Decision variable(s)	Objectives
Neurosurgeon (Kang et al. 2017)	AlexNet, VGG, Deepcace	WiFi, LTE, 3G	Layer	×	Dynamic, SO, Exhaustive	Split point	Latency, Energy
JALAD (Li et al. 2018)	VGG, ResNet	LAN	Layer	Dynamic bitwidth	Static, SO, ILP	Split point, bitwidth	Latency
DADS (Hu et al. 2019)	YOLO, Inception	3G, 4G, WiFi	Layer	×	Dynamic, SO, Heuristic	Split point	Latency, Throughput
MoDNN (Mao et al. 2017)	VGG16	WLAN	neurons	×	Static, SO, Heuristic	Neuron partition	Latency
DeepThings (Zhao et al. 2018)	YOLOv2	WLAN	Cross-layer tile	×	Static, SO, Manual	Tile size, no. of layers	Latency, Throughput
MCDNN (Han et al. 2016)	VGG16	WLAN	Model	×	Dynamic, MO, Heuristic	Model variant, cloud or device	Latency, Energy
Clio (Huang et al. 2020)	MobileNetV2, VGG, ResNet	LORA, ZigBee, BLE, WiFi	Layer	Dynamic width	Dynamic, SO, Exhaustive	Split point, cloud-model width	Latency
ELF (Xu et al. 2020)	FastRCNN	WiFi	Image patch	×	Dynamic, SO, Multi-server	Patch packing, server allocation	Latency
IONN (Jeong et al. 2018)	Alexnet, Inception, ResNet	WLAN	Layer	×	Dynamic, SO, Heuristic	Split point	Latency
Edgnt (Li et al. 2018)	ALexNet	4G,LTE	Layer	EE	Dynamic, SO, Exhaustive	Split point, model depth	Latency
SPINN (Laskaridis et al. 2020)	Resnet50, Resnet56, mobileNetV2	4G,LTE	Layer	Fixed(8-bit) + EE	Dynamic, MO, Exhaustive	Split point, EE-polict	Latency, Throughput, Accuracy
DynO (Almeida et al. 2022)	mobileNetV2, ResNet152, InceptionV3	4G,WiFi	Layer	ISQuant + BitShuffling + LZ4	Dynamic, MO, Exhaustive	Split point, bitwidth	Server cost, Latency, Throughput, Accuracy

Clio (Huang et al. 2020) and SPINN (Laskaridis et al. 2020) focus on different aspects of model offloading: Clio considers the width of models, while SPINN focuses on the depth. However, these approaches require additional training for early classifiers or slicing-aware schemes, leading to increased computational overhead for pre-trained models. In contrast, DynO can directly target any pre-trained models without incurring additional costs. DynO proposes a distributed CNN inference framework that splits the computation between the client device and a more powerful remote end. It utilizes an online scheduler to optimize latency and throughput and minimize cloud workload and associated costs through device-offloading policies.

Early-Exit (EE)-based inference is a strategy that allows for accelerated inference by implementing early exits from specific branches within a model, capitalizing on the observation that early layers of models often capture significant features. One example of this approach is BranchyNet, which introduces supplementary side branches in addition to the main branch of the model (Teerapittayanon et al. 2016). BranchyNet enables the early termination of the inference process at an earlier layer when certain conditions are satisfied, resulting in substantial computational savings. It dynamically selects the branch that achieves the shortest inference time while maintaining a specified level of accuracy. By incorporating additional side branch classifiers, EE-based inference allows for early termination when processing easier samples with high confidence, while more challenging samples utilize more or all layers to ensure accurate predictions. This adaptive approach optimizes both inference speed and accuracy based on the characteristics of the input data. Another work, Edgent (Li et al. 2018), integrates BranchyNet to resize DNNs and enhance the efficiency of the inference process. By reducing the latency requirement, Edgent dynamically adjusts the optimal exit point in BranchyNet, resulting in improved accuracy. Additionally, Edgent utilizes adaptive partitioning, enabling collaborative and on-demand co-inference of DNNs.

In addition, another approach in this field focuses on exploiting the variability in the difficulty of different inputs to adapt the computations. Various works have been done in this area, including dynamic DNNs that adjust the depth of models (Panda et al. 2016; Kouris et al. 2022; Laskaridis et al. 2020; Panda et al. 2016), dynamic channel pruning (Jayakodi et al. 2020), or progressive inference schemes for generative adversarial network-based image generation (Jayakodi et al. 2020). These approaches offer flexibility in tuning the trade-off between accuracy and efficiency in the inference system.

9.8 Challenges and Opportunities of Inference

Despite the aforementioned benefits, the implementation of inference for IoTs still encounters various challenges and presents opportunities, as outlined below:

Model Optimization. With the advent of large models, how to run large sizes such as transformer-based models on IoT devices is an interesting direction. The acceleration of quantized models in inference is not universal and depends on the involved hardware (Zhang et al. 2022). However, there is significant potential to improve the inference speed of quantized models through software optimizations. By automating the compression, researchers can explore algorithms and strategies to effectively balance the trade-off between model size reduction and inference performance. On the other hand, the development trend for lightweight models on IoT devices is to achieve a similar trade-off in inference.

Algorithm-Hardware Codesign. The lightweight DL models and compression techniques should take into consideration the underlying hardware architecture, enabling hardware-algorithm codesign to achieve more efficient inference. DL developers should prioritize optimization for heterogeneous processors (Guo et al. 2023), expanding support for various types of operators and enhancing single-operation performance. In many scenarios, more powerful CPUs and accelerators, especially GPUs and DSPs, can significantly accelerate inference (Zhang et al. 2023). This encourages DL researchers to design models well-suited for GPU computing, emphasizing operators with high parallelism while minimizing memory-intensive operations that hinder parallelism.

Neural Network Hardware Accelerator. To design a reasonable scheduling mode in a complex and multi-application operating environment, it is vital to consider the relatively primitive driver management compared to CPU and GPU. There is a significant research opportunity to address this gap by introducing flexibility in SoCs to effectively handle and adapt to the evolving requirements of improved DL operations. The addition of flexibility can enhance silicon efficiency and lead to cost-friendly solutions. Consequently, the integration of dynamic reconfigurability into SoCs is expected. However, it is crucial to minimize power consumption and area in SoCs that incorporate extra logic. Therefore, research efforts focused on reducing power consumption and optimizing the area of such SoCs are actively pursued.

DL Library Selection. It is crucial to assess the advantages and disadvantages of various DL libraries and devise a solution that can unify their strengths. Otherwise, the issue of inference performance fragmentation may persist for an extended period, as resolving it requires substantial engineering efforts. Achieving optimal performance in mobile DL applications often necessitates the integration of multiple DL libraries and dynamic loading based on the current model and hardware platform. However, this approach is seldom implemented due to the considerable overhead in terms of software complexity and development efforts. There is a need for a more lightweight system that can harness the superior performance of different DL libraries.

Developing Benchmarks. Proper benchmark standards are crucial for accurately evaluating the inference performance (Ren et al. 2023). To enable meaningful comparisons of DL models, optimization algorithms, and hardware platforms, a universal and comprehensive set of quality metrics specific to inference is essential. Currently, benchmark datasets and models predominantly focus on CNNs evaluated on the ImageNet (Azizi et al. 2023). To ensure a more comprehensive evaluation, it is necessary to develop additional benchmark datasets, libraries, and DL models that cover a wider range of applications and input data types. This will facilitate a more thorough assessment of inference performance.

Explainability in Inference. The adoption of AI in critical domains has raised concerns about transparency and accountability. Explainable AI (XAI) aims to address these concerns by providing transparency in DL algorithms (Adadi & Berrada 2018). However, ensuring explainability in the context of intelligent inference remains a challenging and underexplored research area, particularly considering the trade-off between optimization and accuracy. Resolving this challenge is essential for the responsible deployment of AI.

Complex Audio Processing Models. Most existing research focuses predominantly on image-processing tasks, leaving a notable gap in the exploration of similar methods for audio-processing models. However, we posit that the techniques developed for image processing can also be effectively applied to these audio scenarios. Specifically, when addressing RNN-based models, such as LSTMs (Yu et al. 2019) and GRUs (Jiao et al. 2020), their recurrent nature introduces dependencies between samples that are absent in CNNs. Consequently, this poses a challenge in offloading computations, as the RNNs must be transferred alongside the computation. While the partitioning strategies employed in prior studies demonstrate applicability to various DNN architectures by automatically identifying split-point dependencies, RNNs necessitate specialized treatment. The future of inference systems is expected to encompass a wide range of architectures and use cases, showcasing their versatility and applicability in various domains.

Resource Allocation for Inference. The collaborative DNN inference application scenarios are characterized by dynamic environments where future events are challenging to predict accurately. To effectively handle large-scale tasks, it is crucial to have robust online edge resource coordination and provisioning capabilities (Donta et al. 2023; Adadi & Berrada 2018; Dustdar & Murturi 2020; Alkhabbas et al. 2020; Tsigkanos et al. 2019). Real-time joint optimization of heterogeneous computing, communication, and cache resource allocation, along with customized system parameter configuration based on task requirements, is necessary. Addressing the complexity of algorithm design, an emerging research direction focuses on efficient resource allocation strategies driven by data-driven adaptive learning.

Enhancing Security in Inference. Ensuring the credibility of services in distributed collaborative inference requires the design of a robust distributed security mechanism (Flamis et al. 2021; Sedlak et al. 2022). This mechanism plays a vital role in authenticating subscribers, controlling access to collaborative inference tasks, ensuring model and data security on devices, and facilitating mutual authentication between different devices. Furthermore, ongoing research explores the use of blockchain technology to enhance the security and privacy of devices and data in collaborative inference. This avenue holds promise and warrants further exploration in future collaborative DNN inference, particularly regarding privacy issues.

9.9 Conclusion

This chapter provides a comprehensive review of the current state of DL operating on IoT devices. It discusses various methods for accelerating DL inference across devices, edge servers, and the cloud, highlighting their utilization of the unique structure of DNN models and the geospatial locality of user requests in edge computing. The analysis emphasizes the crucial trade-offs between accuracy, latency, and energy that need to be considered. Despite significant progress, numerous challenges persist, including performance improvements, hardware and software optimization, resource management, benchmarking, and integration with other networking technologies. These challenges can be overcome through technological innovations in algorithms, system design, and hardware accelerations. As DL innovation continues at a rapid pace, it is anticipated that new technical challenges in edge computing will arise, providing further opportunities for innovation. Ultimately, this review aims to stimulate discussion, attract attention to the field of inference, and inspire future research endeavors.

References

Adadi, Amina, and Mohammed Berrada. 2018. Peeking inside the black-box: A survey on explainable artificial intelligence (XAI). *IEEE access* 6: 52138–52160.

Ain, Qurat-ul et al. 2018. IoT operating system based fuzzy inference system for home energy management system in smart buildings. *Sensors* 18 (9): 2802.

Alkhabbas, Fahed, et al. 2020. A goal-driven approach for deploying self-adaptive IoT systems. In *2020 IEEE International Conference on Software Architecture (ICSA)*, 146–156. Piscataway: IEEE.

Almeida, Mario, et al. 2022. Dyno: Dynamic onloading of deep neural networks from cloud to device. *ACM Transactions on Embedded Computing Systems* 21 (6): 1–24.

Azizi, Shekoofeh, et al. 2023. Synthetic data from diffusion models improves imagenet classification. arXiv preprint. arXiv:2304.08466.

Bajrami, Xhevahir, et al. 2018. Face recognition performance using linear discriminant analysis and deep neural networks. *International Journal of Applied Pattern Recognition* 5 (3): 240–250.

Bradski, Gary, Adrian Kaehler, et al. 2000. OpenCV. *Dr. Dobb's Journal of Software Tools* 3 (2): 1–81.

Cheng, Yu, et al. 2015. An exploration of parameter redundancy in deep networks with circulant projections. In *Proceedings of the IEEE International Conference on Computer Vision*, 2857–2865.

Choudhary, Tejalal, et al. 2020. A comprehensive survey on model compression and acceleration. *Artificial Intelligence Review* 53: 5113–5155.

Courville, Vanessa, and Vahid Partovi Nia. 2019. Deep learning inference frameworks for ARM CPU. *Journal of Computational Vision and Imaging Systems* 5 (1): 3–3.

Deng, Yunbin. 2019. Deep learning on mobile devices: A review. In *Mobile Multimedia/Image Processing, Security, and Applications 2019*. Vol. 10993, 52–66. Bellingham: SPIE.

Developers, TensorFlow. 2022. TensorFlow. In *Zenodo*.

Donta, Praveen Kumar, and Schahram Dustdar. 2022. The promising role of representation learning for distributed computing continuum systems. In *2022 IEEE International Conference on Service-Oriented System Engineering (SOSE)*, 126–132. Piscataway: IEEE.

Donta, Praveen Kumar, Boris Sedlak, et al. 2023. Governance and sustainability of distributed continuum systems: A big data approach. *Journal of Big Data* 10 (1): 1–31.

Dustdar, Schahram, and Ilir Murturi. 2020. Towards distributed edge-based systems. In *2020 IEEE Second International Conference on Cognitive Machine Intelligence (CogMI)*, 1–9. Piscataway: IEEE.

Dustdar, Schahram, and Ilir Murturi. 2021. Towards IoT processes on the edge. In *Next-Gen Digital Services. A Retrospective and Roadmap for Service Computing of the Future: Essays Dedicated to Michael Papazoglou on the Occasion of His 65th Birthday and His Retirement*, 167–178.

Flamis, Georgios, et al. 2021. Best practices for the deployment of edge inference: The conclusions to start designing. *Electronics* 10 (16): 1912.

Girshick, Ross, et al. 2015. Region-based convolutional networks for accurate object detection and segmentation. *IEEE Transactions on Pattern Analysis and Machine Intelligence* 38 (1): 142–158.

Guo, Anqi, et al. 2023. Software-hardware co-design of heterogeneous SmartNIC system for recommendation models inference and training. In *Proceedings of the 37th International Conference on Supercomputing*, 336–347.

Guo, Peizhen, and Wenjun Hu. 2018. Potluck: Cross-application approximate deduplication for computation-intensive mobile applications. In *Proceedings of the Twenty-Third International Conference on Architectural Support for Programming Languages and Operating Systems*, 271–284.

Han, Seungyeop, et al. 2016. MCDNN: An approximation-based execution framework for deep stream processing under resource constraints. In *Proceedings of the 14th Annual International Conference on Mobile Systems, Applications, and Services*, 123–136.

Haris, Jude, Gibson, Perry, Cano, José, Agostini, Nicolas Bohm and Kaeli, David. 2022. Hardware/Software Co-Design of Edge DNN Accelerators with TFLite. 107 (8): 1–4.

Hu, Chuang, et al. 2019. Dynamic adaptive DNN surgery for inference acceleration on the edge. In *IEEE INFOCOM 2019-IEEE Conference on Computer Communications*, 1423–1431. Piscataway: IEEE.

Huang, Jin, et al. 2020. Clio: Enabling automatic compilation of deep learning pipelines across iot and cloud. In *Proceedings of the 26th Annual International Conference on Mobile Computing and Networking*, 1–12.

Huynh, Loc N., et al. 2017. DeepMon: Mobile GPU-based deep learning framework for continuous vision applications. In *Proceedings of the 15th Annual International Conference on Mobile Systems, Applications, and Services*, 82–95.

Iandola, Forrest N., et al. 2016. SqueezeNet: AlexNet-level accuracy with 50x fewer parameters and $<$ 0.5 MB model size. arXiv preprint. arXiv:1602.07360.

Jayakodi, Nitthilan Kanappan, Janardhan Rao Doppa, et al. 2020. SETGAN: Scale and energy trade-off gans for image applications on mobile platforms. In *Proceedings of the 39th International Conference on Computer-Aided Design*, 1–9.

Jayakodi, Nitthilan Kanappan, Syrine Belakaria, et al. 2020. Design and optimization of energy-accuracy tradeoff networks for mobile platforms via pretrained deep models. *ACM Transactions on Embedded Computing Systems (TECS)* 19 (1): 1–24.

Jeong, Hyuk-Jin, et al. 2018. IONN: Incremental offloading of neural network computations from mobile devices to edge servers. In *Proceedings of the ACM Symposium on Cloud Computing*, 401–411.

Jiang, Xiaotang, et al. 2020. MNN: A universal and efficient inference engine. In *Proceedings of Machine Learning and Systems*. Vol. 2, 1–13.

Jiao, Meng, et al. 2020. A GRU-RNN based momentum optimized algorithm for SOC estimation. *Journal of Power Sources* 459: 228051.

Kang, Yiping, et al. 2017. Neurosurgeon: Collaborative intelligence between the cloud and mobile edge. *ACM SIGARCH Computer Architecture News* 45 (1): 615–629.

Kounoudes, Alexia Dini et al. 2021. User-centred privacy inference detection for smart home devices. *2021 IEEE SmartWorld, Ubiquitous Intelligence & Computing, Advanced & Trusted Computing, Scalable Computing & Communications, Internet of People and Smart City Innovation (SmartWorld/SCALCOM/UIC/ATC/IOP/SCI)*, 210–218. Piscataway: IEEE.

Kouris, Alexandros, et al. 2022. Multi-exit semantic segmentation networks. In *Computer Vision–ECCV 2022: 17th European Conference, Tel Aviv, Israel, October 23–27, 2022, Proceedings, Part XXI*, 330–349. Berlin: Springer.

Laskaridis, Stefanos, Stylianos I. Venieris, Hyeji Kim, et al. 2020. HAPI: Hardware-aware progressive inference. In *Proceedings of the 39th International Conference on Computer-Aided Design*, 1–9.

Laskaridis, Stefanos, Stylianos I. Venieris, Mario Almeida, et al. 2020. SPINN: Synergistic progressive inference of neural networks over device and cloud. In *Proceedings of the 26th Annual International Conference on Mobile Computing and Networking*, 1–15.

Lebedev, Mikhail, and Pavel Belecky. 2021. A survey of open-source tools for FPGA-based inference of artificial neural networks. In *2021 Ivannikov Memorial Workshop (IVMEM)*, 50–56. Piscataway: IEEE.

Lebedev, Vadim, and Victor Lempitsky. 2016. Fast convnets using group-wise brain damage. In *Proceedings of the IEEE Conference on Computer Vision and Pattern Recognition*, 2554–2564.

Leiserson, Charles E., et al. 2020. There's plenty of room at the Top: What will drive computer performance after Moore's law? *Science* 368 (6495): eaam9744.

Leon, Vasileios, et al. 2022. Systematic embedded development and implementation techniques on intel myriad VPUs. In *2022 IFIP/IEEE 30th International Conference on Very Large Scale Integration (VLSI-SoC)*, 1–2. Piscataway: IEEE.

Li, En, et al. 2018. Edge intelligence: On-demand deep learning model co-inference with device-edge synergy. In *Proceedings of the 2018 Workshop on Mobile Edge Communications*, 31–36.

Li, Hongshan, et al. 2018. JALAD: Joint accuracy-and latency-aware deep structure decoupling for edge-cloud execution. In *2018 IEEE 24th International Conference on Parallel and Distributed Systems (ICPADS)*, 671–678. Piscataway: IEEE.

Li, Liangzhi, et al. 2018. Deep learning for smart industry: Efficient manufacture inspection system with fog computing. *IEEE Transactions on Industrial Informatics* 14 (10): 4665–4673.

Li, Ying, et al. 2023. Federated domain generalization: A survey. arXiv preprint. arXiv:2306.01334.

LiKamWa, Robert, and Lin Zhong. 2015. Starfish: Efficient concurrency support for computer vision applications. In *Proceedings of the 13th Annual International Conference on Mobile Systems, Applications, and Services*, 213–226.

Liu, Hongye, et al. 2016. Deep relative distance learning: Tell the difference between similar vehicles. In *Proceedings of the IEEE Conference on Computer Vision and Pattern Recognition*, 2167–2175.

Liu, Shaoshan, et al. 2019. Edge computing for autonomous driving: Opportunities and challenges. *Proceedings of the IEEE* 107 (8): 1697–1716.

Lucas, Bruce D., and Takeo Kanade. 1981. An iterative image registration technique with an application to stereo vision. In *IJCAI'81: 7th International Joint Conference on Artificial Intelligence*. Vol. 2, 674–679.

Mao, Jiachen, et al. 2017. MoDNN: Local distributed mobile computing system for deep neural network. In *Design, Automation & Test in Europe Conference & Exhibition (DATE), 2017*, 1396–1401. Piscataway: IEEE.

Mohammadi, Mehdi, and Ala Al-Fuqaha. 2018. Enabling cognitive smart cities using big data and machine learning: Approaches and challenges. *IEEE Communications Magazine* 56 (2): 94–101.

Owens, John D., et al. 2008. GPU computing. In *Proceedings of the IEEE* 96 (5): 879–899.

Panda, Priyadarshini, et al. 2016. Conditional deep learning for energy-efficient and enhanced pattern recognition. In *2016 Design, Automation & Test in Europe Conference & Exhibition (DATE)*, 475–480. Piscataway: IEEE.

Polino, Antonio, et al. 2018. Model compression via distillation and quantization. arXiv preprint. arXiv:1802.05668.

Rastegari, Mohammad, et al. 2016. XNOR-Net: Imagenet classification using binary convolutional neural networks. In *European conference on computer vision*, 525–542. Berlin: Springer.

Ren, Wei-Qing, et al. 2023. A survey on collaborative DNN inference for edge intelligence. In *Machine Intelligence Research*, 1–25.

Romero, Adriana, et al. 2014. Fitnets: Hints for thin deep nets. arXiv preprint. arXiv:1412.6550.

Sedlak, Boris, et al. 2022. Specification and operation of privacy models for data streams on the edge. In *2022 IEEE 6th International Conference on Fog and Edge Computing (ICFEC)*, 78–82. Piscataway: IEEE.

Sengupta, Abhronil, et al. 2019. Going deeper in spiking neural networks: VGG and residual architectures. *Frontiers in Neuroscience* 13: 95.

Soto, José Angel Carvajal, et al. 2016. CEML: Mixing and moving complex event processing and machine learning to the edge of the network for IoT applications. In *Proceedings of the 6th International Conference on the Internet of Things*, 103–110.

Sun, Yi, Chen, Yuheng, Wang, Xiaogang, Tang, Xiaoou. 2014. Deep learning face representation by joint identification-verification. *Advances in Neural Information Processing Systems* 27 (8): 1–8.

Targ, Sasha, et al. 2016. Resnet in resnet: Generalizing residual architectures. arXiv preprint. arXiv:1603.08029.

Teerapittayanon, Surat, et al. 2016. Branchynet: Fast inference via early exiting from deep neural networks. In *2016 23rd International Conference on Pattern Recognition (ICPR)*, 2464–2469. Piscataway: IEEE.

Tsigkanos, Christos, et al. 2019. Dependable resource coordination on the edge at runtime. *Proceedings of the IEEE* 107 (8): 1520–1536.

Viola, Paul, and Michael Jones. 2001. Rapid object detection using a boosted cascade of simple features. In *Proceedings of the 2001 IEEE Computer Society Conference on Computer Vision and Pattern Recognition. CVPR 2001*. Vol. 1, I–I. Piscataway: IEEE.

Wang, Qipeng, et al. 2022. Melon: Breaking the memory wall for resource-efficient on-device machine learning. In *Proceedings of the 20th Annual International Conference on Mobile Systems, Applications and Services*, 450–463.

Wang, Yang, et al. 2017. Effective multi-query expansions: Collaborative deep networks for robust landmark retrieval. *IEEE Transactions on Image Processing* 26 (3): 1393–1404.

Wu, Jiaxiang, et al. 2016. Quantized convolutional neural networks for mobile devices. In *Proceedings of the IEEE Conference on Computer Vision and Pattern Recognition*, 4820–4828.

Xiang, Yecheng, and Hyoseung Kim. 2019. Pipelined data-parallel CPU/GPU scheduling for multi-DNN real-time inference. In *2019 IEEE Real-Time Systems Symposium (RTSS)*, 392–405. Piscataway: IEEE.

Xu, Daliang, et al. 2022. Mandheling: Mixed-precision on-device DNN training with DSP offloading. In *Proceedings of the 28th Annual International Conference on Mobile Computing And Networking*, 214–227.

Xu, Mengwei, Jiawei Liu, et al. 2019. A first look at deep learning apps on smartphones. In *The World Wide Web Conference*, 2125–2136.

Xu, Mengwei, Tiantu Xu, et al. 2021. Video analytics with zero-streaming cameras. In *2021 USENIX Annual Technical Conference (USENIX ATC 21)*, 459–472.

Xu, Mengwei, Xiwen Zhang, et al. 2020. Approximate query service on autonomous iot cameras. In *Proceedings of the 18th International Conference on Mobile Systems, Applications, and Services*, 191–205.

Yim, Junho, et al. 2017. A gift from knowledge distillation: Fast optimization, network minimization and transfer learning. In *Proceedings of the IEEE Conference on Computer Vision and Pattern Recognition*, 4133–4141.

Yu, Yong, et al. 2019. A review of recurrent neural networks: LSTM cells and network architectures. *Neural Computation* 31 (7): 1235–1270.

Zhang, Qiyang, Xiang Li, et al. 2022. A comprehensive benchmark of deep learning libraries on mobile devices. In *Proceedings of the ACM Web Conference 2022*, 3298–3307.

Zhang, Qiyang, Zuo Zhu, et al. 2023. Energy-efficient federated training on mobile device. *IEEE Network* 35 (5): 1–14.

Zhang, Xiangyu, et al. 2018. Shufflenet: An extremely efficient convolutional neural network for mobile devices. In *Proceedings of the IEEE Conference on Computer Vision and Pattern Recognition*, 6848–6856.

Zhao, Zhuoran, et al. 2018. Deepthings: Distributed adaptive deep learning inference on resource-constrained IoT edge clusters. *IEEE Transactions on Computer-Aided Design of Integrated Circuits and Systems* 37 (11): 2348–2359.

Zhou, Kanglei, et al. 2022. TSVMPath: Fast regularization parameter tuning algorithm for twin support vector machine. *Neural Processing Letters* 54 (6): 5457–5482.

Chapter 10
Applications of Deep Learning Models in Diverse Streams of IoT

Atul Srivastava, Haider Daniel Ali Rizvi, Surbhi Bhatia Khan, Aditya Srivastava, and B. Sundaravadivazhagan

10.1 Introduction

The Internet of Things (IoT), automation, and deep learning are three transformational forces that have evolved in the dynamic world of technology to revolutionise the way humans interact with machines, analyse information, and make decisions. These powerful concepts, when combined, are driving innovation and altering industries all over the world.

10.1.1 Internet of Things

At its heart, the IoT is a massive network of interconnected devices, sensors, and systems that communicate and share data via the Internet in real time. This interconnectedness gives common items the potential to gather, analyse, and share data without the need for direct human intervention. IoT applications have

A. Srivastava · A. Srivastava
Department of Computer Science & Engineering, Amity School of Engineering and Technology, Amity University, Lucknow, India

H. D. A. Rizvi
Yogananda School of AI, Shoolini University, Bajhol, India

S. B. Khan (✉)
Department of Data Science, School of Science, Engineering and Environment, University of Salford, Manchester, UK

B. Sundaravadivazhagan
Department of Information and Technology, University of Technology and Applied Sciences, Al Mussana, Muladdah, Oman
e-mail: sundaravadi@act.edu.om

P. K. Donta et al. (eds.), *Learning Techniques for the Internet of Things*,
https://doi.org/10.1007/978-3-031-50514-0_10

permeated multiple areas, from smart homes and cities to industrial processes and healthcare, resulting in increased efficiency, better decision-making, and enhanced user experiences. The ability of IoT to collect real-time data enables the building of smart environments that respond intelligently to changing conditions and user preferences (Mohammadi et al., 2018a; Abhishek et al., 2023).

10.1.2 Automation

Automation, on the other hand, is the act of transferring repetitive and manual jobs to machines or computer systems in order to reduce human interference and increase operational efficiency. Industries can streamline production, reduce errors, and achieve better precision with advancements in robotics, artificial intelligence (AI), and software automation. Automated technologies in manufacturing, logistics, and customer support have considerably increased productivity and cost-effectiveness, freeing up human workers to focus on higher-value activities requiring creativity and problem-solving talents. The convergence of IoT and automation has resulted in the growth of smart factories and self-driving cars, transforming industrial landscapes (Heaton, 2018; Praveen Kumar et al., 2023).

10.1.3 Deep Learning

Deep learning (DL), a subset of AI, is a breakthrough in imitating the neural networks of the human brain in order to process massive amounts of data and learn from patterns. Layered neural networks are trained on vast datasets to recognise complicated patterns and make intelligent decisions. DL algorithms have transformed many industries, including computer vision, natural language processing, and recommendation systems. DL has unleashed a new era of AI capabilities that were once deemed science fiction, ranging from autonomous vehicles that observe their environment to virtual assistants that understand and respond to human speech (Mohammadi et al., 2018a).

Image processing, speech recognition, and pattern identification have all benefitted from DL models. Sensors collect data and information from their surroundings, whereas actuators convert electrical impulses into real actions. Sensors allow physical things and digital networks to communicate with one another. Descriptive analytics is used to assess past accomplishments and track present performance, whereas predictive analytics is used to extract information from raw data to forecast and predict behaviours and events. Prescriptive analytics is more sophisticated than descriptive or predictive analytics.

IoT data is generated rapidly and comes in a variety of formats, including unstructured data, semi-structured data, and structured data. The accuracy and validity of the collected data are critical for the outcomes of analytics. The rapid expansion of big data with IoT has been aided by the development of new software and hardware platforms. Horizontal scaling platforms and vertical scaling platforms are the two types of platforms (Alazab et al., 2021).

Recent advances in predictive analytics for IoT big data have resulted in significant computational and memory requirements, which can be met by specialised, more powerful computational platforms. Cloud computing is a framework that connects servers to various distributed devices via Transmission Control Protocol/Internet Protocol (TCP/IP) networks and allows data to be moved (Praveen Kumar et al., 2022). It also offers a variety of services and applications via the Internet. Fog computing is a technique that brings processing nodes and data analytics closer to end devices as an alternative to cloud computing. Cloud and fog computing paradigms share storage, deployment, and networking capabilities; however, fog is meant primarily for interactive IoT applications that require real-time answers (Vinueza Naranjo et al., 2018).

Edge computing is offered as a new cloud computing paradigm that eliminates and mitigates the shortcomings of cloud computing by processing data at the edge before it is transported to the cloud's core servers (Shi et al., 2016).

AI, machine learning, and DL are all related concepts that describe how computers might learn to accomplish activities that normally require human intelligence. DL is widely employed in a variety of applications, and various open-source frameworks and libraries have been developed to aid with DL research.

TensorFlow is a DL framework written in C++, Torch is a Lua-based open-source framework, and Theano is a Python-based open-source library. Google, Facebook, and Twitter have all used these frameworks to construct their services. Open-source DL frameworks that can be used on GPUs and CPUs include Caffe, Keras, MatConvNet, Deeplearning4j, MXNet, and Chainer. They are appropriate for both convolutional and recurrent networks and have shown promising results in projects such as facial recognition, object identification, and picture classification.

10.1.4 The Synergy

When IoT, automation, and DL come together, they create a tremendous synergy that accelerates technology innovations to previously unimaginable heights. The data-rich environment of IoT provides the essential input for DL algorithms to find patterns and correlations, improving the accuracy and reliability of autonomous decision-making. Automation supplements this partnership by converting DL's intelligent insights into actions, resulting in efficient, adaptive, and self-learning systems. This confluence of technical advances has the potential to transform industries, improve people's lives, and usher in a new era of innovation.

With immense technological capacity, however, comes considerable responsibility to handle issues such as data privacy, cybersecurity, and ethical consequences. As we embark on the IoT, automation, and DL path, it is critical to prioritise ethical considerations and strike a balance between innovation and societal well-being.

Finally, the intersection of the IoT, automation, and DL is an exciting chapter in the growth of technology. This trinity of breakthroughs is causing a paradigm shift in how we perceive and interact with machines, opening up enormous prospects for growth and prosperity. We can construct a future that integrates human brilliance with the revolutionary capabilities of these technologies by wisely utilising their promise, unlocking limitless possibilities for a better and more connected world.

10.2 Applications of DL in IoT Paradigms

10.2.1 Data Analysis

10.2.1.1 Overview of Data Analysis in IoT

In IoT applications, where enormous volumes of data are created from linked devices and sensors, data analysis is essential. In order to gain useful insights, spot patterns, make data-driven decisions, and foster innovation, this data must be thoroughly analysed. Data analysis in the IoT context refers to a variety of methods and strategies used to convert unprocessed data into decision-making information (Mohammadi et al., 2018a).

Importance of Data Analysis in IoT Applications: Data analysis in IoT applications holds great significance due to the following reasons (Mohammadi et al., 2018a):

1. Data analysis supports informed decision-making by revealing patterns, trends, and correlations in IoT data. It gives stakeholders the ability to get useful insights and make timely and correct decisions, resulting in increased operational efficiency and resource optimisation.
2. Predictive Analytics: IoT systems can use data analytic techniques to predict future events, behaviours, and patterns by utilising historical data. Predictive analytics helps with proactive maintenance, risk avoidance, and process optimisation, ultimately improving system performance and dependability.
3. Real-Time Monitoring: Data analysis provides real-time monitoring of IoT devices, allowing for the initial discovery and response to abnormalities or important occurrences. This significantly improves situational awareness, safety, and security in various applications including smart cities, healthcare, and transportation.

Challenges and Opportunities in Analysing IoT Data: Analysing IoT data presents several challenges and opportunities, including:

1. *Volume and Velocity*: The sheer volume and velocity of data created by IoT devices pose storage, processing, and analysis issues. DL models can solve these issues by handling enormous amounts of data effectively and providing real-time processing capabilities.

2. *Data Heterogeneity*: Because IoT data comes in a variety of forms, types, and protocols, it is heterogeneous and difficult to analyse. DL models are capable of handling a wide range of data kinds, including photos, text, and time series, allowing for complete analysis and integration of heterogeneous IoT data.
3. *Data Quality and Dependability*: IoT data might be noisy, incomplete, or error-prone, compromising the accuracy and dependability of analytical findings. DL models may build strong representations from faulty data, reducing the effect of data quality concerns and improving analytical outputs.

DL's Role in Addressing Data Analysis Challenges: DL algorithms have emerged as strong solutions for tackling data analysis difficulties in IoT due to their inherent capabilities:

1. *Feature Extraction*: DL models can automatically build hierarchical representations and extract high-level features from raw IoT data, removing the need for manual feature engineering and enabling effective analysis of complex and unstructured data.
2. *Pattern Recognition*: DL models excel in identifying detailed trends, relationships, and irregularities in IoT data, allowing for accurate classification, clustering, and prediction tasks. They can unearth hidden insights and patterns that traditional analysis tools may find difficult to uncover.
3. *Scalability and Real-Time Processing*: DL models are parallelised and optimised to process huge data in real time. This scalability qualifies them to process the large continuous data produced by IoT devices and allows for quick decision-making.
4. *Adaptability and Generalisation*: DL models can adapt and generalise well to new and previously unknown data, making them appropriate for dynamic IoT contexts where new devices, sensors, or data sources are constantly added. They may learn from a variety of data sources and apply their expertise to new circumstances.

To summarise, data analysis is critical for realising the full potential of IoT data. DL models can be used to analyse IoT data, as well as provide opportunities for key insight extraction, predictive analytics, and improved real-time decision-making across a variety of IoT applications.

10.2.1.2 DL Techniques for IoT Data Analysis

Fundamentals and Architectures of Neural Networks DL has been developed upon neural networks (Heaton, 2018). They are made up of linked layers of artificial neurons that are designed to imitate the function and structure of the human brain. DL makes use of a variety of neural network topologies, including feedforward networks, deep belief networks (DBNs), and deep neural networks (DNNs). These designs enable effective feature extraction and representation learning from IoT data.

Convolutional Neural Networks (CNNs) for Image and Video Data CNNs are frequently used in IoT applications for analysing image and video data. They use convolutional layers to automatically learn spatial feature hierarchies from images. CNNs specialise at object detection, picture classification, and image segmentation, which renders them beneficial for visual analysis in IoT systems.

Sequential Data Recurrent Neural Networks (RNNs) RNNs are developed to handle sequential data in IoT applications. They keep internal memory so as to process inputs sequentially and capture temporal dependencies. RNNs are ideal for IoT sectors such as time series analysis, natural language processing (NLP), and speech recognition.

Time Series Data Long Short-Term Memory (LSTM) Networks LSTM networks are a subset of RNNs that solve the vanishing gradient problem and detect long-term dependencies. They are very useful in analysing time series data in IoT applications. Time series forecasting, anomaly detection, and predictive maintenance are all activities that LSTM networks excel at (Sunil et al., 2021).

Generative Models for Data Production and Augmentation In IoT data analysis, generative models such as generative adversarial networks (GANs) and variational autoencoders (VAEs) are used for data production and augmentation. GANs can produce synthetic data that resembles the true data distribution, allowing restricted datasets to be expanded. VAEs can learn a concise representation of the data and create fresh samples with controlled changes, allowing for improved model training through data augmentation.

IoT Data Analysis Using Transfer Learning and Pre-trained Models Transfer learning makes use of DL models that have already been trained on large-scale datasets such as ImageNet. These models have learnt detailed feature representations that can be fine-tuned or utilised as a starting point for training IoT-related tasks. Transfer learning enables effective model training in circumstances with minimal labelled data, making it an important technique in IoT data processing.

IoT systems can successfully analyse and extract relevant insights from many data kinds, such as photos, videos, sequential data, and time series data, by utilising these DL algorithms. Within IoT applications, these techniques offer efficient representation learning, accurate predictions, and improved decision-making.

10.2.1.3 Predictive Analytics in IoT

In IoT applications, predictive analytics is a critical component of data analysis. Predictive analytics, which employs DL algorithms, enables the extraction of important insights as well as the capacity to generate accurate forecasts about future events, behaviour, or trends. Predictive analytics is critical in the context of IoT for optimising operations, enhancing efficiency, and enabling proactive decision-making. Here are some specific applications of predictive analytics utilising DL models in IoT:

Predictive Modelling with DL for Forecasting Recurrent neural networks (RNNs) and long short-term memory (LSTM) networks are two examples of DL models that are used to create predictive models for forecasting IoT data. These models can accurately predict future values or events by capturing temporal dependencies and patterns in time series data. Applications like energy consumption prediction, demand forecasting, and predictive maintenance scheduling benefit from forecasting utilising DL.

IoT Data Time Series Analysis and Forecasting Time series analysis is a crucial step in the analysis of IoT data. DL models are good at analysing and predicting time series data while taking seasonality, trends, and patterns into account. These models may depict complex relationships in the data, allowing for precise predictions and real-time anomaly identification. The optimisation of industrial process, forecasting the weather, and other IoT areas can all benefit from time series analysis utilising DL.

Outlier Identification and Anomaly Detection DL models are used to identify outliers and detect anomalies in IoT data. These models are capable of accurately identifying anomalies or outliers that depart from the expected patterns because they understand the typical patterns and behaviour of the data. In order to ensure prompt responses and preventive measures in IoT systems, anomaly detection using DL helps identify unexpected events, potential defects, or security breaches in real time.

Real-Time Predictive Analytics for IoT Applications For IoT applications that call for quick decisions or prompt actions, real-time predictive analytics is essential. In order to constantly analyse streaming IoT data and deliver accurate predictions and insights, DL models can be implemented in real-time situations. Applications like smart cities, traffic management, and real-time condition monitoring benefit from real-time predictive analytics.

Predictive Maintenance in IoT Systems In predictive maintenance, the ideal maintenance schedule for IoT systems is determined using predictive analytics. In order to predict probable failures or maintenance requirements, DL models can analyse sensor data, equipment performance, and past maintenance records. These models enable in the optimisation of maintenance schedules, the minimisation of downtime, and cost reduction in sectors including manufacturing, transportation, and healthcare by spotting abnormalities and patterns in the data.

IoT systems can adopt proactive decision-making by applying predictive analytics powered by DL models. These models make use of past and current data to deliver insightful information, precise forecasts, and early alerts. This allows businesses to improve operations, enhance efficiency, and assure the proper operation of IoT equipment.

10.2.1.4 Data Mining and Pattern Recognition

Techniques for data mining and pattern identification are essential for gaining insightful knowledge from IoT data. Powerful feature extraction, unsupervised learning, association rule mining, dimensionality reduction, and anomaly detection abilities are provided by DL models. These methods make it possible to find significant patterns, connections, and hidden structures inside various IoT datasets. We examine DL's uses in data mining and pattern identification for the IoT in this section.

DL for Feature Extraction and Pattern Recognition DL models excel at learning hierarchical representations automatically and extracting useful features from raw IoT data (Heaton, 2018). DL models capture complex patterns and correlations in data by using numerous layers of interconnected neurons. These learnt features are then used in IoT applications for a variety of pattern recognition tasks such as object detection, sentiment analysis, and event categorisation.

Unsupervised Learning Techniques for Clustering IoT Data For clustering IoT data without labelled information, unsupervised learning techniques such as self-organising maps (SOM) and deep clustering algorithms are used. DL models may learn the underlying structure and relationships in data and group similar instances together automatically. Data exploration, anomaly detection, and discovering significant subgroups within big IoT datasets are all aided by clustering approaches.

IoT Data Association Rule Mining The goal of association rule mining is to find interesting correlations or associations between objects in a collection. DL methods, including recurrent neural networks (RNNs) and generative models, are capable of capturing complex connections and uncovering hidden associations in IoT data. In numerous IoT domains, association rule mining can yield significant insights such as co-occurrence patterns, sequential linkages, and item associations.

Techniques for Reducing Dimensionality in High-Dimensional IoT Data IoT data is frequently multidimensional, making it difficult to analyse and visualise. DL techniques, such as autoencoders and variational autoencoders (VAEs), can effectively reduce dimensionality by learning compact representations of data. Dimensionality reduction approaches improve visualisation, anomaly detection, and effective processing of high-dimensional IoT data by compressing the data while keeping its critical properties.

DL-Based Anomaly Detection in IoT Data Detecting anomalies in IoT data is critical for recognising peculiar occurrences, defects, or outliers. DL methods, such as autoencoders, may learn regular trends and correlations in data, allowing deviations from expected behaviour to be detected. DL-based anomaly detection approaches have the benefit of collecting complex patterns and reacting to dynamic changes in IoT systems, strengthening anomaly detection accuracy and dependability.

IoT systems can successfully extract useful insights, discover hidden patterns, and identify abnormal occurrences within varied IoT datasets by utilising DL techniques in data mining and pattern identification. These strategies enable organisations to make educated decisions, optimise operations, spot anomalies, and maximise the value of their IoT data.

10.2.1.5 Visualisation and Interpretability of IoT Data

Understanding and deriving useful insights from complex information depends critically on the visualisation and interpretability of IoT data as well as DL models. Understanding model behaviour, interpreting outputs, and encouraging data exploration are made easier with the use of visual analytics tools, explainable AI methodologies, and DL models. This section looks at the value of visualisation and interpretability in relation to IoT data.

Visualising DL Models and Their Outputs Understanding the internal representations, architectures, and operations of DL models is aided by visualising these models' outputs. Feature visualisations, activation heat maps, and network visualisations are a few examples of techniques that can help you understand how the models analyse and extract features from IoT data. Visualisations can also be used to pinpoint a model's advantages, disadvantages, and future growth areas.

Explainable AI to Interpret DL Models Explainable AI methods seek to explain the judgements made by DL models (Alazab et al., 2021). These methods aid in deciphering and comprehending the logic underlying model results. DL models become more transparent by using techniques like saliency maps, attention mechanisms, and rule extraction, allowing stakeholders to believe and understand their predictions and classifications.

Interpreting and Comprehending Predictions Based on DL To establish trust and comprehend the elements influencing certain outcomes, DL-based predictions must be interpreted (Alazab et al., 2021). Model predictions can be connected to specific input features or patterns using interpretation approaches including feature importance analysis, local explanations, and rule-based reasoning. Stakeholders can learn more and base their decisions on the model's findings by analysing the variables driving the forecasts.

Visual Analytics for IoT Data Exploration and Decision Support To support data exploration, decision-making, and discovery, visual analytics integrates interactive visualisations with analytical approaches (Ahangarha et al., 2020). Visual analytics solutions offer interactive interfaces for examining and analysing IoT datasets, which are frequently large and complicated. These technologies help decision-makers in IoT applications including smart cities, healthcare, and industrial monitoring by enabling stakeholders to spot trends, patterns, and anomalies.

Visualisation and interpretability techniques empower stakeholders to comprehend and gain insights from IoT data and DL models. They enable effective communication, foster trust in the models, and facilitate decision-making processes. By employing visualisations, explainable AI methods, and visual analytics tools, organisations can leverage the full potential of their IoT data, driving innovation and informed decision-making.

10.2.1.6 Case Studies and Applications

DL has been successfully utilised in a variety of IoT domains. In this part, we will look at several notable case studies and applications where DL has made a major contribution:

DL Techniques for Smart Homes and Energy Management In smart home systems, DL techniques are utilised to optimise energy consumption, boost automation, and give a more pleasant user experience. DL models can learn and estimate energy usage patterns by analysing sensor data from appliances, occupancy patterns, and meteorological variables. In smart homes, this enables intelligent energy management, demand response, and personalised energy reduction recommendations.

DL for Healthcare Monitoring and Diagnoses DL has found wide applications in healthcare for monitoring and diagnoses. DL models can aid in early disease detection, diagnosis, and treatment planning by analysing medical sensor data, electronic health records, and medical imaging data. DL models for detecting cancer in medical pictures, predicting disease progression, and personalised monitoring of vital signs for remote patient monitoring are some examples.

Smart City Traffic Prediction and Congestion Management To anticipate traffic and control congestion, smart cities have implemented DL algorithms. DL algorithms can predict traffic congestion, manage traffic flow, and recommend effective routes for automobiles by analysing real-time data from traffic sensors, GPS data, and social media feeds. These schemes shorten commute times and minimise traffic congestion while improving urban transportation efficiency.

Environmental Monitoring and Conservation with DL Models DL models are being used in environmental monitoring and conservation initiatives (Ahangarha et al., 2020). DL systems are capable of identifying and categorising environmental features, identify species in danger, monitor deforestation, and investigate climatic patterns via analysing data from sensor networks, imagery from satellites, and IoT devices. These applications assist in protecting the environment, manage resources, and promote sustainable practises.

Industrial IoT Applications and Process Optimisation with DL DL plays a role in industrial IoT applications because it facilitates process optimisation, predictive maintenance, and quality control (Jiang and Ge, 2021). DL models can predict machine failures, optimise production processes, detect abnormalities, and improve overall operational efficiency through analysing sensor data from industrial machinery. In the industrial sector, these applications result in reduced downtime, higher productivity, and expense savings.

These case studies show the various DL applications in IoT, showing how it has the ability to have an impact on different industries (Jiang and Ge, 2021; Ahangarha et al., 2020). By utilising the abilities of DL models, enterprises may foster creativity, improve efficiency, and make data-driven decisions in the era of linked devices and IoT ecosystems.

10.2.2 Smart Cities and Development

The IoT is the interconnection of billions of smart gadgets that can connect, share information, and coordinate actions. These devices can work in real time, adapt to changing surroundings, and operate without the need for human intervention or supervision.

Researchers have conducted several studies on IoT data analytics, including investigations utilising DL and machine learning techniques. Mohammadi et al. (2018b) have published on the most recent DL approaches for IoT applications such as indoor geolocation, security, privacy, image/speech recognition, and other topics. Mahdavinejad et al. (2018) provided an analysis of the machine learning techniques employed in applications for smart cities to collect and analyse IoT data.

10.2.2.1 DL Techniques for IOT in Smart Cities' Development

Applying DL algorithms for the analysis data from the IoT has resulted in significant transformations in numerous urban areas. Through assimilating insights from previously collected data, these intelligent devices can now make more precise and faster decisions and enact actions. Various applications, such as intelligent transportation, vigilant monitoring, advanced agriculture, and enhanced environmental management, capitalise on IoT devices to enhance urban mobility, diminish crowd congestion, and regulate movement within cities. These applications offer remedies that enhance the flow of traffic. A smart residence denotes a dwelling equipped with web-connected gadgets that interact and share real-time data regarding the home's status. This culminates in a more energy-efficient household. Rashid and colleagues proposed a smart energy monitoring system tailored for household appliances, employing a model with LSTM. This system exhibits the ability to predict energy consumption and accurately forecast bills, achieving an accuracy level surpassing 80%.

Yan et al. (2019) devised a hybrid design rooted in DL (DL) to forecast energy consumption. This approach combines a stationary wavelet transform (SWT) with LSTM neural network. Le et al. (2019) introduced a framework that predicts energy usage by employing two CNNs in conjunction with a bidirectional LSTM. The realm of healthcare can experience enhancements through IoT implementation, empowering healthcare providers to optimise resource utilisation, curtail expenses, and explore novel revenue streams. The prevalence of elderly individuals residing independently in their homes has surged recently. In alternative studies, convolutional neural networks (CNNs) have been harnessed to detect instances of falls in elder care scenarios, achieving remarkable precision. Likewise, a CNN model has demonstrated an exceptional real-time accuracy of 99.98% recognition of occurring falls. Shi et al. proposed a fusion of digital mammograms and a CNN model to predict severe illnesses in ductal carcinoma patients. Early-stage detection of Parkinson's disease has been accomplished utilising a CNN model that applies medical image classification and analysis. Moreover, a CNN model was utilised to pinpoint cardiovascular ailments within mammograms. In the realm of dietary assessment refinement, a system for identifying food images was put forth. This system underwent testing on distinct datasets, yielding favourable outcomes.

Liu et al. (2017) suggested a mobile system for recognising foods that employs edge computing and a CNN model to categorise food photos. To improve traffic and parking management in large cities, smart transportation solutions combine DL models with IoT data. Bura et al. provided and implemented a perimeter computer network, a CNN model, and a smart parking solution.

Amato et al. (2017) introduced a distributed framework for identifying designated spots, which utilises an advanced convolutional neural network (CNN) model to discern between occupied and vacant parking spaces within a parking lot. Yang et al. put forth a DL approach that integrates CNN and LSTM components to anticipate parking occupancy at the level of city blocks. Through the utilisation of deep CNNs coupled with an intelligent car-tracking filter, Cai et al. (2019) developed a real-time video platform for evaluating parking conditions, contributing to the development of smart urban environments. The realm of intelligent surveillance finds application in the management and safeguarding of individuals and objects from issues such as criminal activities and fires. The incidence of vehicular collisions within urban settings has escalated, causing significant damages and considerable losses. Numerous investigations have introduced predictive models for crash risk based on LSTM methodologies.

An application of a convolutional neural network (CNN) was employed to identify plant ailments through leaf images. Employing a multilayer CNN model, the categorisation of diseased mango leaves was conducted, focusing specifically on anthracnose fungal disease. Bu and Wang (2019) devised an intelligent IoT system tailored for agricultural settings, which enhances the food production process. This system leverages a deep reinforcement learning (DRL) architecture with four distinct layers. Häni et al. (2020) harnessed a semi-supervised clustering technique to identify and quantify fruits, alongside developing a CNN-based strategy for intelligent fruit counting. Furthermore, a comprehensive system was proposed amalgamating for the objective of predicting air quality over a 48-hour period, and CNN, long short-term memory (LSTM), and artificial neural networks (ANNs) were used.

A fresh approach rooted in DL has been suggested by Mutis et al. (2020) and also put forth by Markovic et al. (2018) to manage the quality of indoor air. These innovations possess the capacity to perceive external temperature conditions, human activities, and the status of doors being open or closed within office environments.

The recognition of waste materials holds significance within the realm of smart urban developments, and CNN-based DL methodologies have showcased encouraging outcomes when applied to the examination of images collected from urban areas. These techniques excel in identifying the scattering of waste materials and facilitating their disposal.

Zhang et al. (2019) introduced an innovative approach to enhance urban road cleanliness, which involves analysing street photographs to quantify the presence of litter through mobile edge computing. The challenges posed by vast IoT big data encompass issues such as prolonged data storage and analysis necessitating fog (Srirama, 2023) and edge paradigms, scalable cloud computing services, high-performance CPUs, etc. Additionally important is the use of big data analytics tools like Apache Hadoop.

Ensuring these applications to effectively establish a vibrant smart city ecosystem, quality of service is essential. The establishment of smart city services involves the integration of various technologies. A novel learning paradigm called transfer learning, which leverages prior knowledge to address novel challenges, holds relevance within smart city scenarios. These services optimise performance, reduce effort and costs through the utilisation of accumulated insights from past tasks, and bolster accuracy through multi-task learning, all while supporting real-time data analysis.

Microservices technology facilitates the creation of IoT applications using a collection of finely grained, loosely connected, and reusable components. This technology has the potential to enhance IoT big data analytics. By decomposing complex DL services into smaller, reusable microservices, the performance and efficiency of DL applications can be elevated and enhanced.

10.2.3 Home Automation

10.2.3.1 Overview of IOT in Home Automation

The automatic and computerised operation of numerous household appliances is referred to as “home automation”. In plainer language, it implies the capability to effortlessly oversee your home’s functions and services through the Internet, contributing to more practical and safer lifestyle while simultaneously reducing household expenses. Continue reading to gain insights into commonly raised questions about home automation methods, along with some innovative suggestions for integrating automation into your home.

Home automation entails an interconnected arrangement of hardware, communication interfaces, and digital systems that interlink everyday devices through a website. Every device incorporates and sensors possesses wireless connectivity, enabling remote management via a smartphone or tablet, whether you’re within your home or situated at a distance. This empowers you to manipulate elements like lighting, door locks, or even temperature settings from any location. A comprehensive home automation setup comprises three primary components: sensors, controllers, and actuators.

Sensors possess the ability to detect alterations in factors such as sunlight levels, temperature, or movement. These variables, among others, can subsequently be tailored by home automation setups to align with your specific preferences. Controllers encompass the devices responsible for transmitting and receiving updates about the ongoing state of automated functions within your home. These can include desktop computers, laptops, tablets, or mobile phones. Actuators are devices that govern the physical mechanics or operations of an automated home system. These devices can encompass items like motorised valves, motors, or light switches and they are directly wired into the system.

10.2.3.2 Challenges and Opportunities for IOT in Home Automation

IoT Challenges in Smart House Automation: There are certain IoT challenges in smart house automation. They are listed below. Data Security and Latency: In smart home automation, data security and latency are major concerns. IEEE standard protocols are used to enhance the data security. The fog computing is used as a resolution to overcome latency concerns (Chinmaya Kumar et al., 2024).

Mixed Criticality: The use of various systems and functionalities in smart home automation raises various criticalities. Mixed criticality is avoided using distinct low-criticality and high-criticality functions.

Fault Tolerance: Within the smart home automation setup, numerous sensors interact with both hardware and software components. Consequently, identifying the source of a system fault can be challenging. Address this issue by incorporating redundant controllers to mitigate potential system malfunctions.

Functional Safety: Systems with a paramount role in safety, such as those pertaining to fire emergencies, demand prioritisation. These systems must operate consistently. Address this challenge by establishing a dedicated IoT-based emergency system to effectively manage such situations.

IoT-powered smart home automation has the potential to connect previously unconnected objects. People's interactions have transformed as a result of IoT innovation. IoT automation is popular in this day and age.

10.2.3.3 DL Techniques for IOT in Home Automation

Machine learning and DL methods have proven valuable in the context of intelligent home automation. They serve purposes such as monitoring and recognising items, identifying human behaviours, detecting faces, managing smart devices, optimising energy consumption, overseeing household activities, and enhancing safety and security (Mehmood et al., 2019; Liciotti et al., 2020; Jaihar et al., 2020; Popa et al., 2019; Peng et al., 2020; Shah et al., 2018). A machine learning subset known as an algorithm for DL emulates the human brain's structure to perform data analysis through a specific logical framework (Khan et al., 2019). However, various DL strategies have been extensively utilised and shown effectiveness in addressing engineering, classification, and detection complexities within artificial intelligence models for tasks like object classification, detection, and prediction (Khan et al., 2020).

A CNN model represents a type of artificial neural network employed to detect and manipulate images. The CNN model, renowned for its proficiency in containing numerous convolutional layers, is frequently harnessed to address challenges associated with image-related tasks (Sahu and Dash, 2021). The fundamental elements of a home security system rest upon motion detection and surveillance. By processing images captured through security cameras utilising the CNN model, specific areas of interest can be targeted for detection purposes. By employing intelligent detection methods grounded in DL models, the potential of a smart home automation framework can be elevated. This enhancement involves discerning observed movements as either household occupants or intruders, subsequently triggering alarms for the user.

A well-designed home automation system contributes to stress reduction, conserves essential resources like electricity and water, and elevates the overall quality of life. Through the integration of ambient intelligence, elements like lighting, entertainment systems, environmental conditions, and other home devices are all controlled by the intelligent home automation system (Lobaccaro et al., 2016).

Challenges that persist within the domain of smart home automation encompass remote household monitoring over long distances, intelligent system decision-making, preservation, precise motion detection, and the immediate retention of data for subsequent forecasting, assessment, and decision-making purposes. Addressing these challenges necessitates the establishment of an economical, cloud-based smart home automation system that operates in real time, relying on an Android smartphone application.

It's crucial to clarify that when referring to "low cost", it indicates that the proposed system's prototype was realised using budget-friendly IoT hardware, which encompasses components like microcontroller boards, sensors, and cameras. This implementation approach ensures economic feasibility and ease of configuration. An Android smartphone is used to control the house, with the graphical user interface application for smart home automation offering a comprehensive view of the home's various conditions. Real-time sensor data is collected by the mobile application and stored through a platform-as-a-service model, accompanied by graphical representations of environmental readings.

Additionally, the initial IoT hardware deployed in this implementation is both cost-effective and adaptable, ensuring accessibility and scalability.

10.2.4 Energy-Efficient IoT

Energy is so important in the IOT network; thus, it must be analysed depending on the different classifications it has. Effective use of energy protocols has been

Table 10.1 Highlights of the IoT-enabled network classification techniques

Matrices	Li et al. (2019)	Vinueza Naranjo et al. (2018)	Kaur et al. (2019)	Ammad et al. (2020)	Ibrahim et al. (2021)	Kumar et al. (2020)
Specific application support	L	H	H	H	H	L
Technology adopted	Edge	Fog	Data centres	Fog	–	–
Durability of network	M	M	H	M	H	H
Latency	H	H	L	L	L	L
Data aggregation support	N	N	Y	N	Y	N

developed to help save energy and maximise energy use and distribution. The increase of urbanisation and IoT services has resulted in a significant need for energy efficiency systems. The IoT platform offers smart solutions that are attuned to context for tasks like energy provisioning, conservation, harvesting, and administration. Wireless devices play a key role in IoT setups, enabling user engagement, information exchange, and resource accessibility. Ensuring energy efficiency is of paramount importance within IoT platforms, contributing to their durability and sustainability over time.

Li et al. (2019) proposed a smartphone edge offloading with energy awareness method for IoT on diversified networks. Ammad et al. (2020) provided a case study that verifies a multiple-level fog-based environmentally friendly architecture for IoT-enabled smart environments. Using a first-fit decreasing technique, Kaur et al. (2019) developed a software-defined data centre that is an energy-efficient architecture for IoT deployments. Ibrahim et al. (2021) used data aggregation, compression, and forecasting approaches to reduce sending information to the cluster head (CH) and eliminate redundancy in the acquired data. Vinueza Naranjo et al. (2018) offer a new resource management strategy in virtualised network with fog systems that are superior than the current cutting-edge benchmark resource managers. Table 10.1 highlights the IoT-enabled methods for network classification.

Abdul-Qawy et al. (2020) introduced a categorisation framework to analyse papers focused on energy-saving solutions for diverse wireless nodes in the IoT. Ashraf et al. (2019) devised a method to harvest energy from haptic IoT sensors while maintaining substantial data throughput and minimising queuing delays. Tang et al. (2020) presented an approach for IoT fog computing systems with decentralised compute offloading and energy harvesting capabilities. Ozger et al. (2018) proposed a networking model that combines IoT-connected smart grids using energy harvesting and cognitive radio, addressing challenges and outlining future research directions. Nguyen et al. (2018) developed a routing system that promotes energy efficiency in setups for diverse IoT networks that are aware of distributed

Table 10.2 Comparative analysis of the feature

Matrices	Ozger et al. (2018)	Abdul-Qawy et al. (2020)	Tang et al. (2020)	Ashraf et al. (2019)	Nguyen et al. (2018)
Data rate improvement	N	Y	Y	N	N
Energy conversation	Y	Y	Y	Y	Y
Residual energy consideration	Y	Y	N	N	Y
Data queue utilisation	L	L	L	H	H
CPU utility	H	L	H	L	L

Table 10.3 Comparison of the energy use methods in a descriptive way

Matrices	Zhong et al. (2018)	Said et al. (2020)	Tom et al. (2019)	Yu et al. (2018)
Mean square error	L	L	H	L
Cost-effectiveness	N	Y	Y	N
Estimation of power requirement	H	M	M	H

energy harvesting. Table 10.2 offers further insights into the characteristics featured in the aforementioned approaches.

The IoT enhances energy use by utilising traditional and dedicated infrastructures, as well as a smart grid environment. Pawar et al. integrate complicated IoT framework-based smart energy management technologies.

Regarding an IoT fog-enabled electricity distribution system, Tom et al. (2019) incorporated both customers and utilities, offering the potential for intelligent energy management and demand reduction. In a green IoT context, Said et al. (2020) introduced a solution for energy management that significantly outperforms previous methods. Their work introduced a fog-based architecture for the Internet of Energy, assessing the performance of bandwidths and delays in comparison to cloud-based models. To facilitate efficient energy scheduling in smart construction, Zhong et al. (2018) put forth a distributed auction system founded on ADMM principles. In the IoT context, Yu et al. (2018) designed a network of object architecture designed specifically for smart homes and building energy management. Table 10.3 presents a comprehensive comparative overview of the strategies discussed above.

10.2.5 Malware Detection

10.2.5.1 Overview of Malware Detection in IOT

Malware represents software crafted with the intention of causing damage to digital devices. This malicious software assumes different monikers, including worms, Trojan horses, viruses, bots, and botnets, as well as ransomware, adware, and spyware. Its repercussions encompass heightened security vulnerabilities and financial losses. The creators of malware are continually enhancing the intricacy of their code, rendering even robust detection methods less effective. To counter this, a framework has been established employing conventional methodologies for malware identification. This section explores algorithms used to categorise malware, tactics for obscuring malware code, datasets for analysis, techniques involving sandbox environments, and prevalent forms of malware. The discussion extends to encompass strategies for detecting malware on mobile platforms and the application of machine learning in detecting malware, particularly in relation to sandbox-based techniques. The oligomorphic approach encrypts and decrypts using two separate keys: the polymorphism method makes additional copies of the encrypted payload, and the metamorphic method allows for dynamic code modification.

The identification of malware involves a decision-making process that begins with simple static analysis and progresses to effective dynamic analysis. Sandbox environments are used to run malicious malware downloaded from unknown attachments or dubious URLs.

n-Gram Model Static and dynamic features are used to create features, analytical characteristics, and graph model: Graphs are generated from system calls.

In the malware detection classification stage, malware datasets are used. Machine learning algorithms conduct data operations such as regression, categorisation, and grouping.

10.2.5.2 DL Technique for Malware Detection in IOT

Present-day malware is intentionally designed to exhibit characteristics drawn from multiple families, complicating the process of categorising it. In order to enhance data representation and decision-making efficiency, image processing techniques are also integrated with extensive datasets. A profiling model called MalIn-sight, developed by Han et al. (2019), has been utilised for malware detection. By considering structural, low-level, and high-level behavioural aspects of malware, this model achieves an impressive accuracy of 97.21% in identifying novel types of malwares. Kim et al. (2018) employed a multimodal DL approach to identify malware targeting Android smartphones. Their methodology incorporates a range of features, including strings, method APIs, shared library function opcodes, permissions, parts, and external influences.

Vinayakumar et al. (2019) utilised reinforcement learning to establish a strategy for circumventing anti-malware engines, employing deep Q learning. Additionally, they introduced a feature selection architecture based on deep Q learning. A multi-level DL system (MLDLS) has been developed by Zhong et al. to improve the effectiveness of DL-based malware detection systems (MDSs). Several DL models were coordinated by this system using a tree architecture. Vinayakumar et al. (2019) formulated a two-step malware detection framework. This framework amalgamates both conventional machine learning methods and DL frameworks that include image processing techniques as well as dynamic and static methods.

Li et al. (2019) employed the domain generation algorithm (DGA) to create a machine learning framework geared towards malware detection. This framework consists of a two-level model as well as a prediction model, yielding an accuracy of 95.89%. An approach for identifying malware is introduced, employing image texture analysis in conjunction with support vector machine (SVM). The technique leverages image-based data for swift malware detection, offering substantial insights into the structure of malicious software. SigMal constitutes a malware detection framework that relies on similarity and draws from various image-based malware datasets.

When combined with artificial intelligence, machine learning gives rise to DL, a technique involving the construction of hierarchical layers. DL finds application in various domains, including self-driving vehicles, image processing, and natural language understanding. A deep neural network, a multi-layered artificial neural network, encompasses multiple strata positioned between its initial input layer and its final output layer. In comparison to a single-layered neural network, it exhibits superior performance regarding two-class error rates.

Droid-Sec incorporates DL and compares its results with older machine learning models like MLP, SVM, LR, naive Bayes, C4.5, and SVM. In the context of identifying malware within 2D binary code, a deep neural network is deployed. Based on a dataset of four million software binary files, this model achieved an impressive accuracy of 95% detection rate (DR) and a false-positive rate (FPR) of 0.1%.

Convolutional neural networks are used to create deep neural networks (DNNs), which are used to enhance visualisation. It has been used to recognise photos, detect viruses, and create 2D greyscale visuals. CNN-Bi LSTM model does not need specialised subject expertise and depends on data for feature extraction. The model proposed by Kolosnjaji et al. combines CNN and LSTM to achieve the best accuracy performance when compared to CNN.

RNNs are artificial neural network (ANN) types that are utilised in recognition of speech and natural language processing. It is used to solve the gradient problem and has several applications such as speech generation and robot control, speech recognition, translation by machines, and music composition. Malware is categorised using neural networks into different families. RNN is utilised in LSTM before building feature vectors to categorise input into nine categories of malware. Control flow graphs are used for rapid malware detection and to recognise the infection and classify it based on its family (CFGs). To categorise the different malwares, Euclidean distance and multi-layer perceptron are used. The destination registers of runtime traces are used to classify malware. To recognise the functions of each program, complete profile is reviewed, and clustering is used to group the data. Thirty-five static properties of programs are extracted and explained with Shapley additive explanation (SHAP) values to detect the malware. PAIRED, a lightweight Android malware detection solution, is created utilising explainable machine learning.

Image processing techniques can be employed to simplify the malware classification model, and the malware classification accuracy is 0.98 in 56 seconds. Andrew Image, a new malware embedding approach that uses black-and-white graphics produced from hexadecimal-sized instructions in the disassembly file, is offered. In terms of sturdiness and interpretability, Andrew Image outperforms Nataraj Image. Using neural network approaches such as computer vision and picture classification, a framework for detecting anonymous susceptible activities is created. The maximum accuracy is attained when random forest is used. Currently, the signature-based technique is used to protect against malware attacks. However, the signature-based strategy has two key drawbacks: it is a difficult and time-consuming procedure, and it is incompatible with the attackers' obfuscation techniques. Malware coders alter popular software sales platforms such as Google's Play Store and implant dangerous payloads into the app's original code. They distribute fraudulently programmed software to the marketplace, deceiving legitimate consumers who are ignorant of the distinction between legitimate and harmful apps. By monitoring malware-affected consumers, mobile OS providers seek answers to this unending deluge of malware. Google has been given permission to play this role, and it checks each new program that is added to the Play Store. The Mac OS is more secure than other operating systems such as Windows and Android.

The ISAM malware detection model uses Infection and self-propagation via wireless transmission across iPhone devices. Between 2009 and 2015, 36 families of iOS malwares were detected, and a histogram was developed as a result of the analysis of these programs. Finding PHI-UIs and semantic feature sets developed using non-trivial methods in iOS Chameleon apps requires the usage of the Chameleon-Hunter detection tool. Those research papers show the unpredictability of OS X malware. In 2019, the CVE database logged approximately 660 formal cases concerning system security issues. Windows 10 is a vulnerable operating system. Despite assaults and security flaws, Windows 7 retains 30% use. For Windows malware, various detection methods are used. Anonymous malware is classified as genuine or harmful using an active learning malware detection framework. The study (Satrya et al., 2015) creates an 8-bit botnet identification algorithm based on hybrid analysis. A framework is created, more precisely a hybrid multi-filter, to identify runtime environment behaviours to quickly detect malware dynamically (Huda et al., 2018).

It is detailed a malware detection framework that is power-aware and centred on an energy monitor and data analyser. It is tested on an HP iPAQ running Windows OS and achieves 81.3% categorisation accuracy, 64.7% NB accuracy, and 86.8% RF accuracy. The most difficult step is protecting against ransomware.

Because of its invariance in the environment, it is difficult to detect, and it employs evasion methods. Delay execution, self-destruction, polymorphism, and network traffic encryption are some examples. Early warning signs of a ransomware outbreak can be discovered by using Crypto Drop (Scaife et al., 2016), which monitors real-time updates for user data and files. Primary and secondary indicators are important.

To detect malware, Huang and Stokes (2016) used deep belief network (DBN) and recurrent neural network (RNN). The real +ve rate for this model is 82%, whereas the fake +ve rate is 5%. Raw malware byte code is taken into account for malware detection, and after down collection, a DL model can be used. When utilising DL, evasion attempts make malware detection more difficult, and a gradient-based attack is also viable. By merging visualisation and convolutional neural network techniques, a DL-based method for identifying Windows malware is put into practice. The approach achieves 91.41% accuracy for static visualisation and 94.70% accuracy for hybrid visualisation. Malware process behaviours are represented as images in the work (Pascanu et al., 2015) and categorised as malicious or not using CNN. An autonomous vision-based Android malware detection (AMD) method has been proposed by building 16 distinct CNNs based on DL. The AMD model obtains 99.40% accuracy for balanced android samples and 98.05% efficiency for imbalanced android samples. Modelling DNN analyses Windows application binaries, and hidden layered DNN involves malware classification using feature vectors.

Vis Droid is a malware detection system for Android that uses an image-oriented analytic approach. To categorise malware families, it employs five datasets of greyscale photos and six unique ML classifiers. Mobi Tive is intended to improve Android virus detection through the use of customised neural networks in real-time mobile contexts. It computes and analyses five critical criteria in order to overcome the drawbacks of server-side malware detection. Cuckoo sandbox employs uni-grams for tracing malware runtime behaviour and DNN for signature generation. SVM is used in classification. A hybrid analysis for detection of malware (HADM) is created by extracting features both statically and dynamically. There is a suggested CNN-based technique for finding Windows malware, which makes use of runtime executable file behaviour aspects to detect and categorise unknown malware.

A KNN classifier is used to detect IoT malware, and it achieves 98.2% accuracy. The implementation of dynamic analysis of data for IoT malware detection (DAIMD) proposes a well-known and cutting-edge IoT malware detection method. The CNN model is employed for learning, while the augmented CNN model is utilised to identify variants. Cuckoo sandbox employs uni-grams for tracing malware runtime behaviour and DNN for signature generation. SVM is used in classification. Hybrid analysis for detection of malware (HADM) is created by extracting features both statically and dynamically. It is suggested to use CNN to identify Windows malware, which makes use of runtime executable file behaviour aspects to detect and categorise unknown malware. A KNN classifier is used to detect IoT malware, and it achieves 98.2% accuracy.

10.2.6 DL for IOT Healthcare and Telemedicine

10.2.6.1 Overview of DL in Healthcare and Telemedicine

DL emerges as a novel facet of machine learning, rooted in artificial neural networks that encompass multiple strata. This notion of neural networks for data processing draws inspiration from the biological frameworks observed within the brain. Intricate artificial neural networks with several layers are known as deep neural networks. The input layer collects the data, the hidden layers keep the weights they were given, and the result layer produces the outcomes. IoT serves as a conduit linking the tangible and virtual realms, employing communication protocols to interconnect objects. The realm of Medical Internet of Things (MIoT) is dedicated to offering groundbreaking medical services to hospitals, patients, and healthcare practitioners. Presently, IoT devices generate substantial volumes of data, which are then processed using DL algorithms to extract valuable insights. The integration of DL algorithms is becoming increasingly prevalent in medical and healthcare applications, showing continued growth. This section seeks to delve into the most recent research and assessments concerning the amalgamation of DL technologies and IoT within medical and healthcare domains. It also addresses the challenges associated with the utilisation of related applications and explores potential future prospects.

Nweke et al. (2018) explored methodologies for implementing DL in contexts related to mobile and human activities. They delved into techniques like restricted Boltzmann machines, autoencoders, deep mixture models, and sparse coding. The discussion also encompassed a hybrid generative and discriminative model aimed at enhancing feature learning. The realm of Internet of Health Things (IoHT) offers the potential to optimise resource allocation and reduce service disruptions by collecting, processing, and presenting crucial data regarding vital signs within hospital environments. Nevertheless, the principal limitations of this analysis are attributed to the extensive quantity and diverse nature of the gathered data.

10.2.6.2 DL Techniques for IOT in Healthcare and Telemedicine

Islam et al. (2015) explored a range of medical network configurations and platforms for IoT-driven healthcare technologies; however, these devices are integrated with relatively sluggish processors. Tuli et al. (2020) introduced the concept of Health Fog for automated diagnosis of heart ailments by harnessing DL and IoT. This approach embraces the benefits of using edge computing and fog computing as energy-efficient and quick data processing techniques. Sarraf et al. (2016) asserted that DL techniques have notably improved EEG decoding accuracy and facilitated the identification of anomalous health conditions. Nevertheless, acquiring datasets for EEG pathologies poses a challenging endeavour.

Cerebrovascular accidents (CVA) are conditions characterised by the stoppage of specific mental processes as a result of either ischaemia or haemorrhaging. Within an IoT context, convolutional neural networks can be applied to classify strokes by analysing CT scans. A DL model was created by Faust et al. (2018) to detect atrial fibrillation utilising heart rate signals. The model's performance was evaluated on a sample of 20 individuals, utilising labelled HR signal data sourced from the PhysioNet atrial fibrillation database. In the realm of traditional Chinese medicine, the treatment of infectious fevers involves syndrome differentiation. DL stands as a viable approach for facilitating computer-assisted differentiation of infectious fever syndromes through comprehensive integration.

Bray et al. (2018) employ lung tumour identification using deep reinforcement learning models that produce reliable results. However, the challenge lies in developing an effective algorithm to adjust the Q-value for each action. Melanoma, a significant type of skin cancer, is addressed using an IoT-based system aimed at classifying skin lesions. The proposed method employs picture acquisition that uses CNN models trained on the ImageNet dataset, although the limitation lies in requiring Internet access. DL can serve to establish a connection between smartphone sensor data and personal health. This approach adopts deep-stacked autoencoders (SAE), characterised by a straightforward structure, modest computational demands, and impressive performance.

A tooth health IoT system, employing smart hardware, DL, and a mobile terminal, was engineered to identify and categorise dental disorders. Its standout feature is its compact dimensions, measuring 5.5 mm in width and 4 mm in thickness. Sharma et al. (2018) introduced a system dedicated to monitoring physiological signals within the healthcare domain. They put forth a deep neural network (DNN) approach aimed at evaluating the quality of signals from multiple sensors, with the purpose of learning physiological markers from patient data. An essential aspect of fall detection involves constructing a notably precise detection model for cost-effective devices. While DL models possess the potential for elevated detection accuracy, they exhibit limitations in addressing everyday activities.

Reboucas Filho et al. (2017) said that smart patches may track human health problems using IoT sensors, and cloud computing technology assists in sending data elicited and analysed by IoT devices via the Internet using various machine learning, DL, and CNN. A smart health monitoring device named "smart-log" is designed to track calorie intake and expenditure using the technology behind the "Internet of Things". This wearable or gadget serves the purpose of screening and falls within the realm of smart healthcare. The "smart-log" comprises a smart sensor board in conjunction with a smartphone application. The ongoing research centres on integrating IoT into the domain of sports injuries. This involves utilising a mobile terminal to ascertain the presence of muscle injuries, analysing and processing data collected via the Zigbee network, and ultimately presenting the findings.

Implementing sensors to display heart rate as determined by users in a medical setting context comes with several challenges. Learning a semi-supervised sequence approach was adopted to enhance the identification of excessive cholesterol, high blood pressure, and sleep apnoea. In the realm of epilepsy management, a brain-computer interface (BCI) was employed for seizure prediction. The suggested BCI systems utilise cloud computing to perform real-time computations on incoming data, presenting a promising avenue for constructing a patient-centric real-time seizure prediction and localisation system.

Zhao et al. (2019) tackled challenges in recognising human action through the development of a deep residual long short-term memory (LSTM) network that is bidirectional. This technique holds potential for application in complex and extensive human activity recognition scenarios, despite its limitation in terms of available data points. Chang et al. (2019) introduced ST-Med-Box, a smart device built on DL algorithms, designed to aid patients with chronic conditions in accurately administering multiple medications. This device ensures precise medication intake and adherence among patients.

10.2.7 Security and Privacy

In the fast-growing IoT ecosystem, protecting security and privacy is critical. With the exponential expansion of networked devices and the sensitive nature of data being communicated and processed, IoT systems must be protected from cyberthreats. In this chapter, we look at how DL models can be used to improve security and privacy in various IoT streams. DL algorithms offer increased capabilities for detecting intrusions, safeguarding sensitive data, and reinforcing IoT systems against new security threats.

10.2.7.1 The Significance of Security and Privacy in IoT

As IoT devices become an essential part of healthcare systems, critical infrastructure, and personal surroundings, the risks of illicit use, data breaches, and malicious assaults escalate. The potential implications of security and privacy lapses emphasise the necessity for robust security measures for protecting IoT networks.

10.2.7.2 DL for IoT Security

DL models offer workable answers to IoT security problems. The use of DL algorithms to protect IoT systems includes the detection of malware, intrusion, and anomalies. DL models have the ability to detect the normal patterns and behaviours of IoT devices, enabling them to discover irregularities, identify intrusions, and defend against malicious acts (Al-Garadi et al., 2020).

10.2.7.3 DL-Based Intrusion Detection

Recognising and thwarting possible attacks in IoT systems requires intrusion detection. Recurrent neural networks (RNNs) and deep belief networks (DBNs) are two examples of DL models that can analyse network traffic, system logs, and device behaviour to find aberrant behaviours that may be signs of intrusions. In this chapter, we examine the approaches and algorithms used by DL-based intrusion detection systems and how well they perform at spotting different sorts of attacks (Mohammadi et al., 2018a).

10.2.7.4 DL for Privacy Preservation in IoT

Due to the sensitive nature of the gathered and transmitted personal data, privacy preservation in IoT is a major concern. DL models that anonymise, encrypt, or obscure sensitive data can help to protect user privacy. We investigate methods like federated learning for decentralised and privacy-conscious model training and generative adversarial networks (GANs) for privacy-preserving data synthesis (Bharati and Podder, 2022).

10.2.7.5 DL for Authentication and Access Control in the IoT

Mechanisms for authentication and access control are essential for ensuring that only authorised individuals have access to IoT devices and data. DL models that use biometric identification, behaviour analysis, and multi-factor authentication can improve authentication systems. We go over how DL is used in IoT ecosystems for access control, device authentication, and user identification (Sadhu et al., 2022).

10.2.7.6 Secure Communication in IoT Using DL

To ensure data confidentiality and integrity in IoT systems, communication channels must be secured. DL models can be used to detect data exfiltration efforts, identify malicious network activity, and discover anomalies in network traffic. Investigating DL based methods are widely used for securing communication protocols, identifying protocol flaws, and guaranteeing secure data transmission in IoT networks (Sadhu et al., 2022).

10.2.8 Transportation and Autonomous Vehicles

The concept of mobility has completely changed as a result of the application of DL models to the field of transportation and autonomous cars. DL is essential for allowing safer, more effective, and intelligent transportation systems for the development of IoT technology, sensors, and connectivity.

10.2.8.1 Intelligent Transportation Systems and DL

DL models can analyse enormous amounts of data from different sources, such as sensors, cameras, and GPS devices, to provide real-time traffic monitoring, congestion management, and efficient routing. Researchers highlight the possible advantages of using DL techniques in ITS for enhanced safety, efficiency, and sustainability (Atat et al., 2018).

10.2.8.2 DL for Transportation Object Identification and Recognition

DL models, notably convolutional neural networks (CNNs), have revolutionised object identification and recognition in transportation applications. It investigates how CNNs can recognise and classify items including vehicles, people, traffic signs, and traffic signals. By providing a comprehensive perceptual foundation, these models enable advanced driver assistance systems (ADAS) and help to the development of driverless vehicles.

10.2.8.3 Vehicle Localisation and Mapping Based on DL

Accurate vehicle localisation and mapping are critical for autonomous driving. DL models, when used with sensor fusion techniques, can estimate a vehicle's position and orientation and map the surrounding area with high accuracy. We look at DL approaches like simultaneous localisation and mapping (SLAM) for autonomous vehicle navigation and mapping in difficult real-world scenarios (Wang and Ji, 2020).

10.2.8.4 DL for Predictive Behaviour and Trajectory Planning

Understanding and forecasting the behaviour of other vehicles, pedestrians, and cyclists is critical for autonomous driving to be safe and efficient. DL models, such as RNNs and LSTM networks, may learn complicated driving patterns and capture temporal connections. We investigate how these models can forecast and plan autonomous vehicle trajectories, ensuring smooth and collision-free navigation (Grigorescu et al., 2020).

10.2.8.5 DL for ADAS

DL techniques have significantly enhanced ADAS by allowing real-time perception and decision-making. DL models are used in ADAS for applications like lane detection, traffic sign recognition, driver sleepiness detection, and collision avoidance. These modern technologies offer critical safety features and aid human drivers in a variety of transportation settings (Grigorescu et al., 2020).

10.2.8.6 DL for Autonomous Vehicle Control

DL is important in autonomous vehicle control because it enables end-to-end learning methodologies. DL models can transfer sensory information directly to vehicle control signals, avoiding the need for handmade rules or intermediate representations. These models enable autonomous vehicles to learn complex driving behaviours and effectively navigate a variety of road situations (Chi and Mu, 2017).

10.2.9 Environmental Monitoring and Conservation

DL combined with the IoT has ushered in a new era of possibilities in a variety of sectors, including environmental monitoring and conservation. This chapter explores how DL models are changing environmental monitoring by delivering real-time data insights and enabling effective conservation initiatives. Researchers and practitioners may address major environmental concerns and work towards a more sustainable future by using the power of IoT and DL.

10.2.9.1 Environmental Monitoring with DL

Remote Sensing and Image Analysis Convolutional neural networks (CNNs) showcased extraordinary performance in image analysis and remote sensing. To monitor fluctuations in land cover, deforestation, urban development, and numerous other critical environmental indicators, these models can interpret data from satellites, drones, and ground-based sensors (Chen et al., 2017; Hatcher and Yu, 2018).

Air Quality Monitoring IoT devices paired with DL algorithms have enabled real-time, high-resolution air quality monitoring. DL models can analyse air quality sensor data and anticipate pollutant levels, assisting in the understanding of air pollution trends and guiding policy decisions (Xu et al., 2023).

10.2.9.2 Biodiversity Conservation

Species Identification and Monitoring Recurrent neural networks (RNNs) and transformer-based architectures are examples of DL models that have showed promise in the identification and monitoring of species. These models can identify and track endangered animals by examining audio, visual, and sensor data, enabling focused conservation efforts (Incze et al., 2018).

Predictive Modelling for Conversation DL's capacity to handle complicated and diverse data has enabled conservation predictive modelling. Models can predict habitat loss, shifts in species distribution, and potential risks to biodiversity, assisting conservationists in developing proactive interventions (Zhao et al., 2017).

10.2.9.3 Water Resource Management

Water Quality Monitoring DL models have been used successfully in river, lake, and reservoir water quality monitoring. These algorithms can detect anomalies, estimate contamination levels, and assist in water resource management by analysing data from IoT-enabled water quality sensors (Nandi et al., 2023).

Flood Prediction and Management DL models like long short-term memory (LSTM) networks and graph neural networks (GNNs) have been shown to be beneficial in flood prediction and control. These models can deliver timely alerts about floods and optimise immediate measures through integrating data from IoT devices, weather forecasts, and satellite images (Nandi et al., 2023).

10.2.10 Industrial Internet of Things

The Industrial Internet of Things (IIoT) is a game-changing combination of old industrial systems and new digital technologies. Industries may achieve new levels of efficiency, production, and automation by combining IIoT and DL models. This chapter investigates DL applications in IIoT, concentrating on its role in increasing smart manufacturing and optimising various industrial processes. The synergistic interaction between IIoT and DL is paving the path for Industry 4.0, or the Fourth Industrial Revolution.

10.2.10.1 Foundations of IIoT and DL

Understanding IIoT The IIoT is a network of networked devices, sensors, machines, and data analytics systems that collect and exchange data to optimise industrial operations. IIoT deployment offers industries with real-time insights and predictive capabilities, empowering them to make data-driven decisions (Shi et al., 2016).

DL Basics DL is a subset of artificial intelligence (AI) that employs neural networks with multiple layers to analyse and extract patterns from large datasets. Convolutional neural networks (CNNs) and recurrent neural networks (RNNs) are prominent architectures in DL used for various IIoT applications (Heaton, 2018; Schmidhuber, 2015).

10.2.10.2 Applications of DL in IIoT

Predictive Maintenance One of the significant applications of DL in IIoT is predictive maintenance. By analysing sensor data from industrial equipment, DL models can detect early signs of equipment failures and recommend proactive maintenance actions, reducing downtime and optimising maintenance costs (Schwendemann et al., 2021).

Quality Control and Defect Detection DL models have proven effective in quality control processes by inspecting products for defects and anomalies in real time. Through image recognition and classification, IIoT-enabled DL systems can identify imperfections, ensuring high-quality production and minimising waste (Saberironaghi et al., 2023).

10.2.10.3 Energy Optimisation

Optimising energy consumption is crucial for sustainable and cost-effective industrial operations. DL algorithms can analyse historical energy consumption patterns and real-time sensor data to optimise energy usage, leading to reduced costs and environmental impact.

10.2.10.4 Challenges and Future Perspectives

Data Security and Privacy The vast amounts of data generated by IIoT devices pose significant challenges in terms of data security and privacy. Implementing DL models in IIoT systems necessitates robust cybersecurity measures to safeguard sensitive industrial data.

Integration and Scalability The integration of DL models into existing IIoT infrastructure can be complex and resource-intensive. Additionally, ensuring the scalability of DL solutions to handle the ever-growing volume of data requires careful planning and optimisation (Miorandi et al., 2012).

Edge Computing and Real-Time Processing Industrial processes often require real-time data analysis and decision-making. Deploying DL models on the edge, closer to the data source, enables faster processing and reduced latency, addressing the requirements of time-sensitive applications in IIoT.

10.3 Conclusion

Data analysis is critical for gaining important insights from the massive amounts of data generated by IoT systems. DL models are powerful analytical tools for IoT data, offering predictive analytics, pattern detection, and actionable insights. IoT applications can benefit from higher efficiency, improved decision-making, and increased automation by employing DL capabilities. The potential of DL models in many aspects of IoT data processing is highlighted in this chapter, paving the way for advanced and intelligent IoT systems. The convergence of DL algorithms and IoT data has brought about a transformative impact on smart cities, revolutionising transportation, energy efficiency, healthcare, agriculture, and environmental monitoring. By harnessing the power of real-time data analytics and predictive modelling, DL-enabled IoT devices have enhanced traffic flow, reduced energy consumption in smart homes, improved healthcare diagnostics and patient care, optimised agricultural practices, and fostered cleaner and greener urban environments. Challenges in data handling and processing are being addressed through cloud and edge computing, while transfer learning and microservices technology promise to boost the performance and scalability of smart city applications. As these technologies continue to advance, the vision of smarter, more sustainable cities becomes increasingly attainable, promising a future of interconnected urban ecosystems that improve the quality of life for citizens worldwide.

Improved convenient living, a healthy lifestyle, comfortability, and home security are areas of interest and development. The elderly, handicapped, and sick need to reduce daily activities that can stress them and negatively impact their health. To this end, a smart home automation system that can facilitate local and global monitoring, control, and safety of the home was developed. This work contributes to the existing research in home automation with the design and development of a multifunctional Android-based mobile application for the smart home automation domain. We have proposed an approach to enhance home security using the CNN DL model to classify and detect intruders in the home. The detection is based on the identification of motion in the home environment. Using this method shows that users will have enhanced security of their houses while having minimal disturbance from notifications.

This chapter provides an overview of various energy utilisation and IoT strategies. It also discusses the essential role of IoT-based networks in energy optimisation and overall energy management in IoT. The techniques outlined are grouped into energy efficiency, harvesting, and optimisation for IoT networks. Shared traits within each category are presented to offer a quick summary. However, the scope of the discussed methods is restricted; the latest approaches are evaluated for their specific achievements and effects on the IoT. Security gives us lessons about being proactive rather than reactive. DL-based malware detection technology reduces the flaws of both conventional and traditional methods and gives researchers a thorough understanding of malware analysis.

In conclusion, DL in healthcare IoT presents immense potential to transform the medical landscape by enhancing diagnostics, personalised treatments, patient monitoring, and drug development. However, it also requires careful attention to data security, privacy, and the collaboration of multiple stakeholders for successful implementation and advancement in the field. As technology evolves and research progresses, the synergy of DL and IoT is expected to lead to more efficient and effective healthcare solutions. The design and implementation of IoT systems must take security and privacy into account. In order to address security issues, improve privacy protection, and defend IoT networks from emerging threats, DL models offer strong capabilities. Businesses may develop reliable and resilient IoT systems that protect sensitive data and uphold user confidence by utilising DL algorithms for intrusion detection, privacy preservation, authentication, and secure communication. DL model integration with transportation and autonomous cars opens up new avenues for safer, more efficient, and intelligent mobility. Transportation systems can become more flexible, responsive, and capable of handling complex traffic scenarios by employing DL methods for object detection, behaviour prediction, mapping, and decision-making. DL developments in the IoT area move us towards a future with disruptive transportation solutions and broad adoption of self-driving vehicles.

The combination of DL models with IoT devices has pushed environmental monitoring and conservation to new heights. DL's versatility and scalability have revolutionised data analysis and decision-making in the field of environmental sciences, from remote sensing and air quality monitoring to biodiversity conservation and water resource management. To continue investigating the potential of this explosive combination, it is critical to prioritise sustainable practises and embrace creative solutions in order to maintain and protect our planet for future generations. The combination of IIoT and DL is revolutionising industries by ushering in unprecedented levels of automation, efficiency, and optimisation. From predictive maintenance and quality control to energy optimisation, the applications of DL in IIoT are diverse and impactful. However, challenges related to data security, integration, and real-time processing must be addressed to fully unlock the potential of this powerful combination. As industries continue to embrace the possibilities of IIoT and DL, they are well-positioned to drive the Fourth Industrial Revolution and create smarter, more sustainable industrial processes.

References

Abdul-Qawy, Antar Shaddad H., et al. 2020. Classification of energy saving techniques for IoT-based heterogeneous wireless nodes. *Procedia Computer Science* 171: 2590–2599.

Abhishek, Hazra, et al. Feb. 2023. Cooperative transmission scheduling and computation offloading with collaboration of fog and cloud for industrial IoT applications. *IEEE Internet of Things Journal* 10: 3944–3953.

Ahangarha, Marjan, et al. 2020. Deep learning-based change detection method for environmental change monitoring using sentinel-2 datasets. *Environmental Sciences Proceedings* 5 (1): 15.

Alazab, Mamoun, et al. 2021. Deep learning for cyber security applications: A comprehensive survey. https://doi.org/10.36227/techrxiv.16748161.v1.

Al-Garadi, Mohammed Ali, et al. 2020. A survey of machine and deep learning methods for internet of things (IoT) security. *IEEE Communications Surveys & Tutorials* 22 (3): 1646–1685.

Amato, Giuseppe, et al. 2017. Deep learning for decentralized parking lot occupancy detection. *Expert Systems with Applications* 72: 327–334.

Ammad, Muhammad, et al. 2020. A novel fog-based multi-level energy-efficient framework for IoT-enabled smart environments. *IEEE Access* 8: 150010–150026.

Ashraf, Nouman, et al. 2019. Combined data rate and energy management in harvesting enabled tactile IoT sensing devices. *IEEE Transactions on Industrial Informatics* 15 (5): 3006–3015

Atat, Rachad, et al. 2018. Big data meet cyber-physical systems: A panoramic survey. *IEEE Access* 6: 73603–73636.

Bharati, Subrato, and Prajoy Podder. 2022. Machine and deep learning for iot security and privacy: applications, challenges, and future directions. In *Security and Communication Networks* 2022, 1–41.

Bray, F., et al. 2018. GLOBOCAN estimates of incidence and mortality worldwide for 36 cancers in 185 countries: Global cancer statistics. *CA: A Cancer Journal for Clinicians* 68 (6): 394–424.

Bu, Fanyu, and Xin Wang. 2019. A smart agriculture IoT system based on deep reinforcement learning. *Future Generation Computer Systems* 99: 500–507.

Cai, Bill Yang, et al. 2019. Deep learning-based video system for accurate and realtime parking measurement. *IEEE Internet of Things Journal* 6 (5): 7693–7701.

Chang, Wan-Jung, et al. 2019. A deep learning-based intelligent medicine recognition system for chronic patients. *IEEE Access* 7: 44441–44458.

Chen, Liang-Chieh, et al. 2017. Deeplab: Semantic image segmentation with deep convolutional nets, atrous convolution, and fully connected crfs. *IEEE Transactions on Pattern Analysis and Machine Intelligence* 40 (4): 834–848.

Chi, Lu, and Yadong Mu. 2017. Learning end-to-end autonomous steering model from spatial and temporal visual cues. In *Proceedings of the Workshop on Visual Analysis in Smart and Connected Communities*, 9–16.

Chinmaya Kumar, Duhrey, et al. Mar. 2024. Securing clustered edge intelligence with blockchain. *IEEE Consumer Electronics Magazine* 13: 2.

Faust, Oliver, et al. 2018. Automated detection of atrial fibrillation using long short-term memory network with RR interval signals. *Computers in Biology and Medicine* 102: 327–335.

Grigorescu, Sorin, et al. 2020. A survey of deep learning techniques for autonomous driving. *Journal of Field Robotics* 37 (3): 362–386.

Han, Weijie, et al. 2019. MalDAE: Detecting and explaining malware based on correlation and fusion of static and dynamic characteristics. *computers & Security* 83: 208–233.

Häni, Nicolai, et al. 2020. A comparative study of fruit detection and counting methods for yield mapping in apple orchards. *Journal of Field Robotics* 37 (2): 263–282.

Hatcher, William Grant, and Wei Yu. 2018. A survey of deep learning: Platforms, applications and emerging research trends. *IEEE Access* 6: 24411–24432.

Heaton, Jeff. 2018. Ian Goodfellow, Yoshua Bengio, and Aaron Courville: Deep learning: The MIT Press, 2016, 800 pp. ISBN: 0262035618. In *Genetic programming and evolvable machines* 19.1-2, 305–307.

Huang, Wenyi, and Jack W. Stokes. 2016. MtNet: A multi-task neural network for dynamic malware classification. In *Detection of intrusions and malware, and vulnerability assessment*, ed. by Juan Caballero et al., 399–418 Cham: Springer International Publishing.

Huda, Shamsul, et al. 2018. A hybrid-multi filter-wrapper framework to identify run-time behaviour for fast malware detection. *Future Generation Computer Systems* 83: 193–207.

Ibrahim, Marwa, et al. 2021. All-in-one: Toward hybrid data collection and energy saving mechanism in sensing-based IoT applications. *Peer-to-Peer Networking and Applications* 14 (3): 1154–1173.

Incze, Agnes, et al. 2018. Bird sound recognition using a convolutional neural network. In *2018 IEEE 16th International Symposium on Intelligent Systems and Informatics (SISY)*, 000295–000300. IEEE.

Islam, SM Riazul, et al. 2015. The internet of things for health care: a comprehensive survey. *IEEE Access* 3: 678–708.

Jaihar, J., N. Lingayat, P. S. Vijaybhai, G. Ventakesh, and K. P. Upla. 2020. Smart home automation using machine learning algorithms. In *International Conference for Emerging Technology (INCET)*, Belgaum, India.

Jiang, Xiaoyu, and Zhiqiang Ge. 2021. Information fingerprint for secure industrial big data analytics. *IEEE Transactions on Industrial Informatics* 18 (4): 2641–2650.

Kaur, Kuljeet, et al. 2019. A big data-enabled consolidated framework for energy efficient software defined data centers in IoT setups. *IEEE Transactions on Industrial Informatics* 16 (4): 2687–2697.

Khan, Murad, Bilal Jan, et al. 2019. Deep learning methods and applications. In *Deep learning: convergence to big data analytics*, 31–42.

Khan, Murad, Junho Seo, et al. 2020. Towards energy efficient home automation: a deep learning approach. *Sensors* 20 (24): 7187.

Kim, TaeGuen, et al. 2018. A multimodal deep learning method for android malware detection using various features. *IEEE Transactions on Information Forensics and Security* 14 (3): 773–788.

Kumar, Sushil, et al. 2020. Toward energy-oriented optimization for green communication in sensor enabled IoT environments. *IEEE Systems Journal* 14 (4): 4663–4673.

Le, Tuong, et al. 2019. Improving electric energy consumption prediction using CNN and Bi-LSTM. *Applied Sciences* 9 (20): 4237.

Li, Shilin, et al. 2019. Energy-aware mobile edge computation offloading for IoT over heterogenous networks. *IEEE Access* 7: 13092–13105.

Li, Yi, et al. 2019. A machine learning framework for domain generation algorithmbased malware detection. *IEEE Access* 7: 32765–32782.

Liciotti, D., M. Bernardini, L. Romeo, and E. Frontoni. 2020. A sequential deep learning application for recognising human activities in smart homes. *Neurocomputing* 396: 501–513.

Liu, Chang, et al. 2017. A new deep learning-based food recognition system for dietary assessment on an edge computing service infrastructure. *IEEE Transactions on Services Computing* 11 (2): 249–261.

Lobaccaro, Gabriele, et al. 2016. A review of systems and technologies for smart homes and smart grids. *Energies* 9 (5): 348.

Mahdavinejad, Mohammad Saeid, et al. 2018. Machine learning for Internet of Things data analysis: A survey. *Digital Communications and Networks* 4 (3): 161–175.

Markovic, Romana, et al. 2018. Window opening model using deep learning methods. *Building and Environment* 145: 319–329.

Mehmood, F., I. Ullah, S. Ahmad, and D. Kim. 2019. Object detection mechanism based on deep learning algorithm using embedded IoT devices for smart home appliances control in CoT. *Journal of Ambient Intelligence and Humanized Computing* 1–17.

Miorandi, Daniele, et al. 2012. Internet of things: Vision, applications and research challenges. *Ad hoc Networks* 10 (7): 1497–1516. https://doi.org/10.1016/j.adhoc.2012.02.016.

Mohammadi, Mehdi, et al. 2018a. Deep learning for IoT big data and streaming analytics: A survey. *IEEE Communications Surveys & Tutorials* 20 (4): 2923–2960.

Mohammadi, Mehdi, et al. 2018b. Deep learning for IoT big data and streaming analytics: A survey. *IEEE Communications Surveys & Tutorials* 20 (4): 2923–2960.

Mutis, Ivan, et al. 2020. Real-time space occupancy sensing and human motion analysis using deep learning for indoor air quality control. *Automation in Construction* 116: 103237.

Nandi, B.P., et al. 2023. Evolution of neural network to deep learning in prediction of air, water pollution and its Indian context. *International Journal of Environmental Science and Technology*, 1–16.

Nguyen, Thien Duc, et al. 2018. A distributed energy-harvesting-aware routing algorithm for heterogeneous IoT networks. *IEEE Transactions on Green Communications and Networking* 2 (4): 1115–1127.

Nweke, Henry Friday, et al. 2018. Deep learning algorithms for human activity recognition using mobile and wearable sensor networks: State of the art and research challenges. *Expert Systems with Applications* 105: 233–261.

Ozger, Mustafa, et al. 2018. Energy harvesting cognitive radio networking for IoT-enabled smart grid. *Mobile Networks and Applications* 23: 956–966.

Pascanu, Razvan, et al. 2015. Malware classification with recurrent networks. In *2015 IEEE International Conference on Acoustics, Speech and Signal Processing (ICASSP)*, 1916–1920. IEEE.

Peng, Z., X. Li, and F. Yan, 2020. An adaptive deep learning model for smart home autonomous system. In *International Conference on Intelligent Transportation, Big Data and Smart City (ICITBS)*, Vientiane, Laos.

Popa, D., F. Pop, C. Serbanescu, and A. Castiglione. 2019. Deep learning model for home automation and energy reduction in a smart home environment platform. *Neural Computing and Applications* 31(5): 1317–1337.

Praveen Kumar, Donta, Murturi Ilir, et al. 2023. Exploring the potential of distributed computing continuum systems. *Computers* 12 (10): 198.

Praveen Kumar, Donta, Srirama Satsh Narayana, et al. July 2022. Survey on recent advances in IoT application layer protocols and machine learning scope for research directions. *Digital Communications and Networks* 8: 729–746.

Reboucas Filho, Pedro P., et al. 2017. New approach to detect and classify stroke in skull CT images via analysis of brain tissue densities. *Computer Methods and Programs in Biomedicine* 148: 27–43.

Saberironaghi, Alireza, et al. 2023. Defect detection methods for industrial products using deep learning techniques: A review. *Algorithms* 16 (2): 95.

Sadhu, Pintu Kumar, et al. 2022. Internet of things: Security and solutions survey. *Sensors* 22 (19): 7433.

Sahu, Madhusmita, and Rasmita Dash. 2021. A survey on deep learning: convolution neural network (CNN). In *Intelligent and Cloud Computing: Proceedings of ICICC 2019*, vol. 2, pp. 317–325. Springer.

Said, Omar, et al. 2020. EMS: An energy management scheme for green IoT environments. *IEEE Access* 8: 44983–44998.

Sarraf, Saman, et al. 2016. DeepAD: Alzheimer's disease classification via deep convolutional neural networks using MRI and fMRI. In *BioRxiv*, 070441.

Satrya, Gandeva B., et al. 2015. The detection of 8 type malware botnet using hybrid malware analysis in executable file windows operating systems. In *Proceedings of the 17th International Conference on Electronic Commerce 2015*, 1–4.

Scaife, Nolen, et al. 2016. CryptoLock (and drop it): stopping ransomware attacks on user data. In *2016 IEEE 36th International Conference on Distributed Computing Systems (ICDCS)*, 303–312. https://doi.org/10.1109/ICDCS.2016.46.

Schmidhuber, Jürgen. 2015. Deep learning in neural networks: An overview. *Neural Networks* 61: 85–117.

Schwendemann, Sebastian, et al. 2021. A survey of machine-learning techniques for condition monitoring and predictive maintenance of bearings in grinding machines. *Computers in Industry* 125: 103380.

Shah, S. K., Z. Tariq, and Y. Lee, 2018. Audion IoT analytics for home automation safety. In *IEEE International Conference on Big Data (Big Data)*, Seattle, WA, USA.

Sharma, Sagar, et al. 2018. Toward practical privacy-preserving analytics for IoT and cloud-based healthcare systems. *IEEE Internet Computing* 22 (2): 42–51.

Shi, Weisong, et al. 2016. Edge computing: vision and challenges. *IEEE Internet of Things Journal* 3 (5): 637–646.

Srirama, Satish Narayana. 2023. A decade of research in fog computing: Relevance, challenges, and future directions. *Software: Practice and Experience*. https://doi.org/10.1002/spe.3243. eprint: https://onlinelibrary.wiley.com/doi/pdf/10.1002/spe.3243. https://onlinelibrary.wiley.com/doi/abs/10.1002/spe.3243.

Sunil, Kumar, et al. Nov. 2021. Land subsidence prediction using recurrent neural networks. *Stochastic Environmental Research and Risk Assessment* 36: 373–388.

Tang, Qinqin, et al. 2020. Decentralized computation offloading in IoT fog computing system with energy harvesting: A dec-POMDP approach. *IEEE Internet of Things Journal* 7 (6): 4898–4911.

Tom, Rijo Jackson, et al. 2019. Smart energy management and demand reduction by consumers and utilities in an IoT-fog-based power distribution system. *IEEE Internet of Things Journal* 6 (5): 7386–7394.

Tuli, Shreshth, et al. 2020. HealthFog: An ensemble deep learning based Smart Healthcare System for Automatic Diagnosis of Heart Diseases in integrated IoT and fog computing environments. *Future Generation Computer Systems* 104: 187–200.

Vinayakumar, R., et al. 2019. Robust intelligent malware detection using deep learning. *IEEE Access* 7: 46717–46738.

Vinueza Naranjo, Paola G., et al. 2018. Design and energy-efficient resource management of virtualized networked Fog architectures for the real-time support of IoT applications. *The Journal of Supercomputing* 74 (6): 2470–2507.

Wang, Zhengyang, and Shuiwang Ji. 2020. Second-order pooling for graph neural networks. *IEEE Transactions on Pattern Analysis and Machine Intelligence* 45: 6870–6880.

Xu, Rui, et al. 2023. A hybrid deep learning model for air quality prediction based on the time-frequency domain relationship. *Atmosphere* 14 (2): 405.

Yan, Ke, et al. 2019. A hybrid LSTM neural network for energy consumption forecasting of individual households. *IEEE Access* 7: 157633–157642.

Yu, Jaehak, et al. 2018. WISE: web of object architecture on IoT environment for smart home and building energy management. *The Journal of Supercomputing* 74: 4403–4418.

Zhang, Pengcheng et al. 2019. Urban street cleanliness assessment using mobile edge computing and deep learning. *IEEE Access* 7: 63550–63563.

Zhao, Rui, et al. 2019. Deep learning and its applications to machine health monitoring. *Mechanical Systems and Signal Processing* 115: 213–237.

Zhao, Zhao, et al. 2017. Automated bird acoustic event detection and robust species classification. *Ecological Informatics* 39: 99–108.

Zhong, Weifeng, et al. 2018. ADMM-based distributed auction mechanism for energy hub scheduling in smart buildings. *IEEE Access* 6: 45635–45645.

Chapter 11
Quantum Key Distribution in Internet of Things

Somya Rathee

11.1 Introduction

The Internet of things (IoT), which connects common devices and facilitates seamless communication and data sharing, has completely changed the way we interact with the world around us (Donta et al. 2022). Throughout the development and increase in IoT devices, security challenges have emerged as a key area of examination. A possible method to drastically improve the security of IoT networks is quantum key distribution (QKD). Quantum key distribution (QKD) allows for the secure transfer of cryptographic keys while securing private data from intercepting parties. Incorporating QKD with IoT ensures secure device-to-device communication and provides protection against new threats. A new generation of trusted and secure IoT applications is empowered by the combination of IoT and QKD, laying the foundation for a more secure and robust digital future (Campbell and Gear 1995).

QKD creates a secure way of dispersing cryptographic keys by utilizing the ideas of quantum physics. It uses the fundamental properties of quantum mechanics, such as the uncertainty principle, superposition, and the no-cloning theorem, to ensure that cryptographic keys are securely distributed and shielded from potential eavesdroppers, in contrast to conventional encryption techniques that rely on mathematical algorithms. QKD is the ideal solution for the increasingly complex IoT world because it offers a higher level of security by leveraging the unique capabilities of quantum physics (Using quantum key distribution for cryptographic purposes: a survey 2014).

S. Rathee (✉)
Informatics, HTL Spengergasse, Vienna, Austria

P. K. Donta et al. (eds.), *Learning Techniques for the Internet of Things*,
https://doi.org/10.1007/978-3-031-50514-0_11

11.1.1 Cryptography and Involvement of Quantum Physics

The study of secure communication known as cryptography has a vast history that dates back to the beginning of existence. To prevent unauthorized access to sensitive information throughout the ages, numerous traditional encryption techniques were devised. Famous individuals like Claude Shannon and Gilbert Vernam made substantial contributions to the field, establishing the groundwork for current cryptographic ideas. Classical cryptography use schemes that are based on mathematical algorithms, which transform and manipulate text into ciphertext using a secret key. Even if they offered some security, these traditional systems were not impervious to attacks, and developments in computing technology increased the risks to their robustness. However, due to unanticipated contributions from the field of quantum physics, the cryptographic environment has drastically changed over the previous 20 years. In the early 1980s, researchers started looking into how quantum mechanics could fundamentally alter the field of encryption. Quantum key distribution (QKD) was introduced by Scarani et al. (2009).

American mathematician, Peter Shor, created an algorithm for quantum computers in 1994, which has the potential to revolutionize the factoring of big integers, named Shor's factoring algorithm. The reason why it is so popular is the fact that given enough advancements in quantum computation, this algorithm can be used to break encryption. Factoring large integers is a crucial component of many classical cryptographic schemes, such as the widely used RSA (Rivest-Shamir-Adleman) algorithm (Milanov 2009). The security of RSA relies on the difficulty of factoring large numbers into their prime factors. Shor's algorithm, however, showcased that a powerful quantum computer could efficiently factorize large numbers, rendering RSA and other similar classical cryptographic algorithms vulnerable.

This breakthrough discovery sent shockwaves through the field of cryptography, as it posed a significant threat to classical cryptographic systems that have been relied upon for secure communication and data protection. As a result, the importance of exploring quantum-resistant cryptographic solutions and developing quantum-safe cryptographic protocols has become a critical area of research in the quest to secure information in the face of quantum computing advancements.

In this chapter, our primary focus centers on the critical cryptographic process of key distribution, with a particular emphasis on its implementation harnessing the principles of quantum physics. It is essential to acknowledge that a secret key serves various fundamental purposes in cryptography, extending beyond merely facilitating message encryption. One of its vital applications lies in message authentication, enabling the verification of the sender's identity and ensuring the integrity and origin of transmitted information. As the realm of the IoT continues to expand, secure key distribution becomes increasingly crucial for safeguarding communication and data integrity within IoT networks, making quantum-based solutions a compelling avenue to explore.

11.1.2 Security in IOT

While the field of cybersecurity is still relatively young, IoT has developed into a well-defined set of use cases that address urgent business issues and reduce operational and financial costs in a variety of sectors, including healthcare, retail, financial services, utilities, transportation, and manufacturing.

IoT devices now make up 30% of all devices on enterprise networks, which has sparked a shift in business processes, thanks to the technology's quick development and adoption. These devices' extensive data collection yields insightful information that helps drive precise predictive modelling and real-time decision-making. IoT also enables the enterprise's digital transformation and has the potential to increase labor productivity, company efficiency and profitability, as well as the general working environment. Despite the numerous benefits and advances that IoT technology makes possible, the interconnection of smart devices poses a significant challenge to businesses in terms of the serious security threats brought on by unsupervised and unprotected devices linked to the network. As the number of IoT devices increases quickly, maintaining their security and protection becomes a top responsibility for people, businesses, and society at large.

One of the key challenges in IoT security lies in hardware hacking, where attackers exploit common hardware interfaces in microcontroller development. Understanding electronics and utilizing specialized hardware tools become crucial in identifying vulnerabilities in the physical components of IoT devices. Security professionals must be knowledgeable and skilled in hardware security since hackers can access IoT devices by disassembling the hardware and exploiting debug interfaces.

The complexity of providing proper protection is increased by the presence of older equipment, various communication protocols, and the seamless integration of connected devices, sensors, and software applications in an IoT ecosystem. Additionally, the absence of security standards and norms for IoT devices creates the possibility of vulnerabilities. To protect the whole IoT infrastructure, security professionals must be skilled at deciphering and safeguarding this complex web of devices and services.

A major danger to conventional encryption techniques is also posed by the development of quantum computing. The security of IoT connectivity might be endangered if quantum computers were to become a reality and effectively defeat current cryptography techniques. This realization highlights the need for exploring and adopting QKD as a more secure key distribution method. This insight therefore emphasizes the necessity of investigating and implementing QKD as a more secure key distribution technique. QKD uses quantum mechanics to create an unbreakable channel for parties to exchange cryptographic keys. QKD uses quantum phenomena like superposition and entanglement instead of traditional cryptography techniques, which rely on complex mathematical procedures, to attain a higher degree of security. Organizations may ensure that their IoT connectivity is secure by using QKD, making it immune to assaults from even the most potent quantum computers.

Security professionals must face the particular difficulties presented by the IoT landscape in the pursuit of safe IoT systems. To create effective defenses against future cyber threats, they need to combine knowledge of hardware hacking, network security, and cryptographic protocols. The improvement of the security posture of IoT devices and services will be greatly enhanced by the creation of standardized security guidelines, best practices, and industry-wide collaboration. Research and innovation in protecting the IoT ecosystem will also depend on developing multidisciplinary collaboration between professionals from other fields, such as computer science, engineering, and quantum physics.

Overall, the future of IoT security is dependent on the collaborative efforts of security professionals, researchers, regulators, and industry stakeholders to establish a safer and more robust IoT ecosystem. We can construct a secure foundation for the rapidly growing field of IoT and take advantage of the full potential of linked devices while ensuring privacy and integrity by addressing obstacles and implementing cutting-edge technology like QKD.

11.2 Fundamentals of Quantum Key Distribution

This section covers the fundamental setup and concepts of quantum key distribution. As previously stated, QKD is a groundbreaking method for secure communication that takes advantage of the unique elements of quantum physics. QKD's fundamental goal is to build an impenetrable channel for parties to share cryptographic keys. Unlike traditional cryptography approaches that rely on sophisticated mathematical algorithms, QKD uses quantum phenomena such as superposition and entanglement to accomplish this increased degree of security. As we go deeper into the fundamentals of QKD, we will look at its major components, such as quantum entangled particles and quantum channels, which play an important role in the key distribution process and additional applications, as well as cases of eavesdropping.

11.2.1 Quantum and Classical Channels

In the universe of quantum key distribution (QKD), Alice and Bob, two trustworthy partners, set out to construct a secret key even though they are many miles apart. In order to accomplish this, they will require two types of channels to link them. The first is the quantum channel, which allows them to exchange specific quantum messages. The second is the classical channel, which allows them to send regular messages back and forth.

It is critical that the traditional channel is verified, which means that Alice and Bob can verify each other's identities. This assures that no one else, even if they are listening in, may engage in their chat. On the other hand, the quantum channel is not protected in the same way and is open to potential tampering from a third person, like an eavesdropper called Eve.

For Alice and Bob, "security" means they never use a non-secret key. They either successfully create a secret key, which is a list of secret bits known only to them, or they stop the process if they suspect any interference. After they exchange a sequence of symbols, Alice and Bob must figure out how much information about their secret bits might have been leaked to Eve. This is quite tricky in regular communication because eavesdropping usually goes unnoticed.

However, in quantum physics, information leakage is directly related to the degradation of communication. This unique characteristic of quantum channels allows Alice and Bob to quantify the security of their key exchange. The beauty of quantum physics lies in its ability to provide insights into how information is protected and how QKD ensures a secure way of sharing secret keys even in the presence of potential eavesdroppers like Eve.

11.2.2 Quantum Phenomena and Security in QKD

Quantum key distribution (QKD) derives its unyielding security from a series of remarkable phenomena rooted in the principles of quantum physics. These phenomena collectively form the bedrock of QKD's resistance against potential eavesdroppers like Eve (Gisin et al. 2002):

No cloning theorem: The "No cloning theorem" is a fundamental tenet of quantum mechanics that places strict limitations on copying unknown quantum states. In practical terms, this theorem implies that Eve, the potential eavesdropper, cannot secretly make an exact copy of the quantum information being transmitted between Alice and Bob. Any attempt to do so would inevitably disrupt the delicate quantum states, leaving behind clear evidence of tampering. This property is of paramount importance in QKD, as it guarantees that any unauthorized attempt to intercept the quantum communication will be immediately detected, ensuring the security and integrity of the secret key exchange.

Quantum measurement: Any attempt by Eve to gather information from quantum states through measurements inevitably alters the states being observed. In quantum physics, measurement inherently disturbs the system under study, providing a telltale sign of eavesdropping on the quantum channel.

Entanglement: The concept of entanglement is one of the most intriguing aspects of quantum mechanics. When particles become entangled, their properties become interconnected, resulting in correlations that defy classical explanations. These entangled states play a pivotal role in QKD's security by rendering any attempt to establish correlations beforehand futile.

Violation of Bell's inequalities: Quantum correlations obtained through separated measurements on entangled pairs violate Bell's inequalities. These correlations cannot have been predetermined, indicating that the outcomes of measurements did not exist before being observed. Consequently, Eve cannot possess information about these outcomes prior to their measurement.

For instance, consider a scenario where Alice and Bob are communicating using quantum states of light, known as photons. As they exchange these quantum signals through the quantum channel, an eavesdropper named Eve attempts to intercept and measure these photons to gain information about the transmitted quantum states. However, due to the intrinsic nature of quantum mechanics, any attempt by Eve to measure these photons inevitably disturbs their quantum states. This disturbance serves as a clear sign of eavesdropping on the quantum channel, immediately alerting Alice and Bob to Eve's presence.

The beauty of quantum key distribution (QKD) lies in this phenomenon, as the disturbance caused by Eve's measurements ensures the security of the communication. Alice and Bob can detect any deviations from the expected behavior of the quantum states, thus preventing the creation of a compromised secret key.

11.2.3 Light as a Medium

Selecting the appropriate medium for quantum information processing, especially in quantum key distribution (QKD), is crucial for its practical implementation. Although quantum information processing can theoretically be achieved with different physical systems like ions, spins, and atoms, using light as the medium has emerged as the most practical and widely adopted approach. This choice becomes particularly significant when Alice and Bob are located far apart, as it requires a medium that enables quantum states to be transmitted over long distances while experiencing minimal decoherence (Using quantum key distribution for cryptographic purposes: a survey 2014). Light is an ideal choice due to its weak interaction with matter, enabling quantum states of light (photons) to be transmitted to distant locations with little decoherence, meaning that the optical modes experience minimal disturbances during propagation. However, the primary challenge with light lies in the phenomenon of scattering and losses, where photons may not reach their destination due to various factors (Bennett and Brassard 2014). These losses impose practical limitations on the achievable secret key rate and distance in a QKD system.

The impact of losses in quantum key distribution (QKD) is quite complex and has several effects. First, it puts a limit on how fast the secret key can be generated, which means that as the distance between the sender and receiver increases, the key rate may decrease. Second, losses can potentially leak information to eavesdroppers who might be listening in on the quantum communication. The type of quantum signal being used also plays a role in this. For instance, using coherent pulses may make it easier for eavesdroppers to gather information compared to using single photons. Similarly, entangled beams have their own unique characteristics that affect the level of information leakage. Moreover, how the QKD system detects and handles losses is crucial. Some implementations use photon counting, where events with no detected photons are ignored, while others use homodyne detection, which always produces a signal, even if some photons are lost. In the latter case, losses are translated into additional noise, which can impact the overall security of the communication. In summary, losses in QKD have various implications, affecting the key generation rate, information leakage, and the detection process. Researchers and practitioners in the field are continually working on improving and understanding these aspects to ensure secure and efficient quantum communication.

11.3 BB84 Protocol

11.3.1 Introduction

In 1984, Charles H. Bennett and Gilles Brassard introduced the BB84 protocol, which became the first quantum communication (QC) protocol. They presented their work in a lesser-known conference in India, highlighting the interdisciplinary nature of quantum communication and its collaboration between diverse scientific communities.

In the BB84 protocol, Alice generates qubits by using a source of single photons.

11.3.2 Polarization

Photons are fundamental particles of light that carry quantum information. The source of photons is designed to produce single photons, meaning that only one photon is emitted at a time, with well-defined spectral properties. This ensures that each photon has a sharply defined characteristic, leaving only one degree of freedom for Alice to manipulate, which is the polarization of the photon.

Photon polarization is determined by the orientation of its electromagnetic wave. In the BB84 protocol, Alice and Bob agree to use two distinct sets of polarizations as the basis for encoding the quantum information. The first basis is the horizontal/vertical (+) basis, in which photons are polarized either horizontally or vertically. The second basis is the complementary basis of linear polarizations (+45/−45 ×), where photons are polarized at either +45° or −45° from the horizontal axis (Using quantum key distribution for cryptographic purposes: a survey 2014). Hence, both bit values 0 and 1 can be encoded in two possible ways, more accurately in non-orthogonal states described as:

$$|\pm 45\rangle = \frac{1}{\sqrt{2}}(|H\rangle \pm |V\rangle)$$

After defining the basis and the encoding of the quantum states, the involved steps in the BB84 protocol can be described as follows:

11.3.3 QKD Procedure

- **Quantum state preparation**: Alice prepares individual photons generated from a source of single photons in one of the four specified states, which are chosen randomly from the two bases. These states are determined based on the two agreed-upon bases: the horizontal/vertical (+) basis and the complementary basis of linear polarizations (+45/−45 ×)
- **Quantum transmission**: Alice sends the prepared qubits to Bob through the quantum channel. This quantum channel allows the transmission of photons with minimal decoherence, ensuring that the information remains secure during transmission. Decoherence refers to the loss of quantum coherence, where the delicate quantum properties of the photons are disrupted by interactions with the environment, such as noise or interference. In classical communication, such disturbances might only cause errors in the received data. However, in the realm of quantum communication, decoherence poses a significant threat to security since any unauthorized measurement or interception of the qubits could disturb their quantum state, potentially revealing the secret key to an eavesdropper. To counteract these vulnerabilities, the quantum channel must be carefully engineered to minimize the interaction of photons with the environment, ensuring that their quantum states remain intact during transmission. This requires sophisticated techniques, such as using optical fibers with low loss and high purity, or even utilizing quantum repeaters to extend the distance over which quantum communication can be achieved. By employing a quantum channel with minimal decoherence, Alice and Bob can ensure the secure transmission of their qubits, preventing any unauthorized measurement or interception from compromising the confidentiality of their shared cryptographic key. This fundamental aspect of the BB84 protocol highlights the importance of robust and well-designed quantum channels in establishing secure quantum communication systems.
- **Basis measurement**: Upon receiving each qubit, Bob randomly selects one of the two bases, either the horizontal/vertical (+) basis or the complementary basis of linear polarizations (+45/−45 ×). He then measures the incoming photons in the chosen basis. Bob's choice of basis is essential because it determines the type of measurement he will perform on the incoming photons. In the horizontal/vertical (+) basis, Bob aligns his measurement apparatus either horizontally or vertically to measure the polarization of the photons. In the complementary basis (+45/ − 45×), Bob positions his measurement apparatus at +45° or −45° relative to the horizontal axis to measure the photons' polarization.

  The BB84 protocol's random selection of bases is a critical security component since it assures that Bob's measurements are unexpected and prevents any pre-established information from being revealed. This randomness plays an important role for guaranteeing the security of quantum communication and preventing possible adversaries, such as Eve, from getting any relevant knowledge about the encoded qubits. Bob obtains the data related to the specific basis used by Alice to encode the qubits by undertaking random and unbiased basis measurements. This phase is essential in establishing the connection between Alice's encoding and Bob's measurements, allowing them to proceed with the protocol's subsequent steps, such as sifting and error rate estimates, to produce a safe secret key for their communication.

- **Correlation checking**: After the transmission of all qubits through the quantum channel, Alice and Bob engage in the crucial step of correlation checking over the classical channel. This step is required in the BB84 protocol to guarantee that they have the same set of qubits encoded and measured in the same basis. To perform the correlation checking, Alice and Bob communicate the basis information for each qubit they sent and received, respectively. To complete the correlation checking, Alice and Bob exchange the basis information for each qubit sent and received, respectively. By comparing this basis information, they may identify instances when they used distinct bases for encoding and measuring. In these circumstances, they remove the corresponding qubits from their lists since the measurements will be uncorrelated due to the usage of distinct bases. When Alice and Bob utilize the same basis for a certain qubit, that qubit is retained for further processing. These qubits are anticipated to provide correlated measurement results, which are critical to establishing the final secret key.

  The correlation checking process is a crucial step in the BB84 protocol, as it helps eliminate any qubits that may have been vulnerable to eavesdropping or errors during transmission. By only considering qubits encoded and measured in the same basis, Alice and Bob can ensure the security and reliability of their quantum communication. This stage also mitigates the influence of possible noise or disruptions in the quantum channel, increasing the security of the final secret key communicated between them..
- **Sifting**: The "raw key" is a list of about $N/2$ bits, where N is the number of transmitted qubits. Due to changes that were applied in bases, this key may include inaccuracies. Sifting, on the other hand, results in a shorter, error-corrected key known as the "sifted key."

  Alice and Bob sift out the qubits that were not measured on the same basis during the sifting procedure. They already have a set of qubits with correlated measurement results as a consequence of correlation checking. However, because of the random selection of bases during measurement, some of these qubits may still include mistakes, resulting in disparities between sent and received bits.

  To fix these flaws and provide a trustworthy secret key, Alice and Bob use a procedure known as "error correction." During mistake correction, they publicly transmit a random sample of bits from their raw keys. They may discover any inconsistencies caused by mistakes or eavesdropping efforts by comparing the transmitted bits with the received bits. This enables investigators to estimate the error rate in the quantum channel, offering insights into potential communication problems. Alice and Bob can find the exact bits in their raw keys that require change using the knowledge received during error correction. They then use appropriate error correction algorithms to fix these inconsistencies and eliminate as many mistakes as feasible. The error correction procedure seeks to make the final secret key created from the raw key as accurate and safe as possible.

Upon the successful correction of mistakes, Alice and Bob are left with a shorter, error-corrected key known as the "sifted key." This filtered key is an important part of the final secret key since it assures that Alice and Bob's encoded and measured qubits are now perfectly aligned. They reduced the impact of errors and eavesdropping by conducting error correction, resulting in a reliable and secure filtered key, suitable for the last phases of the BB84 protocol.

11.3.4 Eavesdropping

The "intercept-resend" eavesdropping approach shows a possible weakness in the BB84 protocol, demonstrating how an attacker like Eve might attempt to get information about the secret key being transmitted between Alice and Bob. When Eve intercepts a photon from Alice, she faces a decision: whether to measure it in the same basis as Alice's preparation or randomly choose another basis. If she selects the same basis, the photon's state remains intact, allowing her to gain full knowledge of Alice's bit without introducing any errors. However, if she chooses the wrong basis, her measurement outcome becomes uncorrelated with Alice's original bit value, and she inadvertently alters the state of the photon. Consequently, even if Bob measures the photon using the correct basis later on, the measurement outcome might not match Alice's intended bit value due to the modification introduced by Eve. This mismatch occurs approximately half of the time, significantly impacting the integrity of the raw key (Table 11.1).

11.3.4.1 Information Gain, Error Rate, Key Length

To evaluate the extent of Eve's information gain and the impact on the raw key, two crucial parameters are considered: the **information gain** (IE) and the **error rate** (Q). On average, the intercept-resend attack yields Eve full information on half of the bits in the raw key ($IE = 0.5$) while introducing an error rate of $Q = 0.25$. These values are essential in assessing the overall security of the protocol and determining whether a secure key can still be extracted despite the eavesdropping attempt (Using quantum key distribution for cryptographic purposes: a survey 2014), shown in Fig. 11.1. To quantify the length of the final secure key that can be extracted, a critical measure known as **mutual information** ($I(A : B)$) is applied. In the BB84 protocol, the mutual information ($I(A : B)$) is a crucial measure used to assess the correlation between Alice's and Bob's raw keys, which determines how much information they share. Mutual information provides insights into the level of security in the key exchange process. Assuming that both bit values are equally probable, meaning there is no bias toward 0 or 1, the mutual information can be calculated using the formula $I(A : B) = 1 - h(Q)$, where h is the binary entropy function.

The binary entropy function $h(Q)$ quantifies the uncertainty or randomness associated with the error rate (Q) introduced by eavesdropping. A higher error rate

Table 11.1 QKD: key sharing and extraction

ALICE—SENDER		
Random	Bits	Bases
0	1	D
1	1	R
1	0	R
0	0	R
1	0	R
1	1	R
0	1	D
0	0	R
1	1	D
BOB—RECEIVER		
Random	Bits	Bases
0	1	D
1	–	R
0	–	D
–	–	D
0	0	R
1	1	R
–	–	D
0	0	R
1	1	R

Shared key
01001011

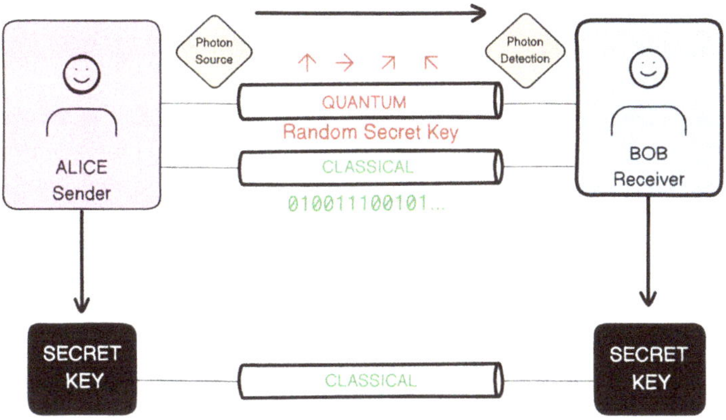

Fig. 11.1 QKD Diagram, Alice and Bob are connected by a quantum channel that Eve can tap into without restrictions, along with an authenticated classical channel that Eve can only listen to

means there is more uncertainty about the original bits exchanged between Alice and Bob. As a result, the mutual information $I(A : B)$ reflects how much information Eve has gained about Alice's secret key compared to what Bob knows.

When the intercept-resend attack numbers are evaluated, it is clear that $I(A : B)$ is smaller than Eve's information gain (IE). In other words, Eve knows more about

Alice's secret key than Bob. This arrangement raises serious security concerns since a secret key should only be communicated between Alice and Bob and should not be revealed to any prospective eavesdropper. If Eve learns more about the raw key, obtaining a secure key becomes difficult since the key's secrecy is jeopardized.

To maintain the security of quantum key distribution, it is critical to minimize the error rate (Q) and maximize the mutual information ($I(A:B)$). This is possible by means of strong security measures and modern cryptographic techniques that protect against eavesdropping attempts and ensure the integrity of the secret key shared between Alice and Bob. The BB84 protocol can provide a secure and reliable quantum key exchange, enabling safe communication in the quantum domain, by maintaining mutual information and minimizing the information acquisition of any prospective attacker.

11.3.4.2 Selective Intercept-Resend Attack

Let us also consider scenarios in which Eve uses a selective intercept-resend attack, intercepting just a**fraction (p)**of the photons transmitted by Alice and leaves the rest alone. In such cases, the error rate (Q) is controlled by the proportion of intercepted photons and is given by $Q = p/4$. Simultaneously, Eve's information gain (IE) is calculated as $IE = p/2$, which is double the error rate.

By evaluating these values, we can observe that if the error rate (Q) exceeds approximately 17%, a secure key cannot be extracted from the BB84 protocol, even when the classical post-processing follows the assumptions outlined in the groundbreaking work by Csiszár and Korner (1978). This highlights the significance of robust security measures in quantum communication protocols to protect against potential eavesdropping attempts and information leakage.

Ensuring the confidentiality and reliability of quantum key distribution is of utmost importance to maintain secure communication in the quantum realm. By minimizing the error rate and carefully managing any potential eavesdropping attempts, quantum communication protocols like BB84 can establish secure and trustworthy channels for transmitting sensitive information. These protocols lay the foundation for quantum cryptography and have the potential to revolutionize secure communication in the digital age. By safeguarding against security threats and leveraging the principles of quantum mechanics, quantum key distribution provides a promising path toward a future of secure, unbreakable communication (Kiktenko et al. 2018).

11.4 Generic QKD Protocols

11.4.1 Classical and Quantum Channels

Classical and quantum channels play crucial roles in quantum key distribution (QKD) protocols, ensuring secure communication between Alice and Bob while defending against eavesdropping attempts by Eve. In the quantum channel, Alice sends quantum signals to Bob, and any interaction with these signals, as per the laws of quantum physics, alters their state. This fundamental feature is the essence of QKD's security. On the other hand, the classical channel enables Alice and Bob to exchange classical messages. Eve can listen in on this channel, but unlike the quantum channel, it must be authenticated to prevent her from altering the messages. Unconditionally secure authentication of the classical channel requires the parties to share an initial secret key or identical random strings. QKD, therefore, expands a short secret key into a longer one, essentially "key-growing," as it cannot create a secret key from scratch using classical means alone. Composability of the secret key is essential to employ it effectively in subsequent tasks (Using quantum key distribution for cryptographic purposes: a survey 2014).

11.4.2 Processing Schemes

In quantum key distribution (QKD), the process begins with Alice transmitting quantum signals to Bob through the quantum channel. The essence of QKD's security lies in the fundamental principle that any interaction with these signals, as governed by the laws of quantum physics, results in a change to their state. This crucial feature ensures that potential eavesdroppers like Eve cannot tap into the quantum channel unnoticed. On the other hand, the classical channel enables Alice and Bob to exchange classical messages back and forth. While this communication remains susceptible to Eve's listening, the classical channel requires authentication to prevent any tampering or alterations to the transmitted messages. This authentication process ensures the integrity and security of the classical communication between Alice and Bob. What is important to note is that QKD does not create a secret key from scratch; instead, it expands an initial short secret key into a longer one. To establish an unconditional secure authentication of the classical channel, Alice and Bob must share an initial secret key or partially secret but identical random strings. This shared initial secret serves as the foundation for further key-growing, making QKD an essential process in secure communication.

The heart of QKD lies in the exchange and measurement of quantum signals on the quantum channel. Alice, in her encoding role, carefully chooses specific quantum states $|\Psi(S_n)\rangle$ to represent a sequence of symbols $S_n = s_1, \ldots, s_n$. Most protocols use quantum states with the tensor product form $|\psi(s_1)\rangle \otimes \ldots \otimes |\psi(s_n)\rangle$. It is crucial that these protocols utilize a set of non-orthogonal states to prevent Eve from easily decoding the information, as a set of orthogonal states could be perfectly cloned, compromising security. Bob plays a critical role in decoding the signals sent by Alice. In addition, he estimates the loss of quantum coherence, which gives valuable insight into Eve's potential knowledge. To achieve this, Bob must employ non-compatible measurements, making it challenging for Eve to gain meaningful information from the signals.

There are two ways to describe QKD protocols: the Prepare-and-Measure (P&M) and entanglement-based (EB) schemes. In the P&M scheme, Alice actively selects the sequence S_n she wants to send, prepares the state $|\Psi(S_n)\rangle$, and sends it to Bob for measurement. In the EB scheme, Alice prepares an entangled state $|\Phi^n\rangle_{AB}$, where the quantum state is entangled between Alice's and Bob's systems. Although both schemes are theoretically equivalent, their practical implementation and feasibility may differ significantly with current technology. The assertion that "entanglement is necessary to extract a secret key" suggests that a secret key cannot be obtained from an entanglement-breaking channel, where the entanglement is lost due to interactions with an adversary. However, this statement does not limit secure implementations to only those based on entanglement. Other secure methods exist as well.

In the context of quantum systems, the term "degree of freedom" refers to a specific property or characteristic that a quantum particle or subsystem can possess. In a quantum system, particles can have multiple degrees of freedom, each corresponding to a distinct aspect of the particle's behavior or state. For example, in the BB84 protocol and other quantum key distribution (QKD) schemes, the quantum system used for communication is typically a qubit. A qubit is a two-level quantum system, and its degrees of freedom represent the possible states it can be in. In the BB84 protocol, Alice prepares qubits in different quantum states to encode her information, and Bob performs measurements on these qubits to decode the transmitted information. The choice of quantum states by Alice and the measurements made by Bob determine the degree of freedom relevant to the key generation process.

In certain scenarios, Alice and Bob may control additional degrees of freedom A' and B', while Eve may not have a purification of the state ρ_{AB} but of $\rho_{AA'BB'}$. This introduces additional complexity, where ρ_{AB} can even be separable, while $\rho_{AA'BB'}$ must be entangled and could even be bound entangled. This consideration arises from the fact that A' and B' can shield meaningful degrees of freedom from Eve's knowledge. Nonetheless, the exploration of QKD with shielding systems remains an ongoing area of research, and practical QKD schemes with shielding systems have yet to be proposed.

11.4.3 Classical Processing

Upon completing the exchange and measurement of numerous signals on the quantum channel, Alice and Bob embark on a critical stage known as classical information processing. During this phase, they communicate through the classical channel, sharing the outcomes of their measurements and conducting data analysis. The primary objective is to extract valuable information about the quantum channel's performance, including essential parameters like decoding error rates, quantum coherence loss, transmission rates, and detection rates. In this classical information processing step, the first task is parameter estimation, where Alice and Bob analyze their observed data to estimate the characteristics of the quantum channel accurately. In some protocols, a preliminary sifting phase may precede parameter estimation, allowing them to discard certain symbols based on decoding errors or other factors.

After parameter estimation and any necessary sifting, Alice and Bob each possess lists of symbols, collectively known as raw keys, with a length of $n \leq N$. However, these raw keys are only partially correlated and contain partial secrecy, making them unsuitable for secure communication. To derive a fully secure key, they employ classical information post-processing techniques, which transform the raw keys into a final secret key denoted as K, with a length $l \leq n$. The length of the secret key K depends on the extent of information that Eve holds regarding the raw keys.

It is essential to emphasize the pivotal role of classical post-processing in ensuring the security of the final key. Through adept application of algorithms and cryptographic methods, Alice and Bob can distill a secure key that withstands potential eavesdropping attempts by Eve. This classical information processing phase plays a central role in the overall success of the QKD protocol, ensuring the confidentiality and reliability of the final secret key for secure communication between Alice and Bob. By extracting a secure key from the initially partially correlated and partially secret raw keys, this process transforms the raw data into a fully secure and usable key, meeting their communication needs with robust security.

11.4.4 Secret Key Rate

In the realm of quantum key distribution (QKD), the secret fraction (r) emerges as a pivotal parameter when dealing with infinitely long keys ($N\infty$). It serves as the linchpin of QKD, meticulously defined in security proofs (II.C.3) to quantify the ratio of the final secret key length (l) to the raw key length (n) as N approaches infinity. The secret fraction (r) essentially represents the portion of the raw key that can be reliably transformed into a secret key, and it plays a crucial role in assessing the effectiveness and security of QKD protocols (Boutros and Soljanin 2023).

However, practical QKD implementations necessitate the consideration of another essential parameter: the raw-key rate (R). This parameter reflects the rate at which raw keys can be generated per unit time and depends on a myriad of factors. The intricacies of the specific QKD protocol and setup intricacies, such as the repetition rate of the source, channel losses, detector efficiency, dead time, and potential duty cycle, all come into play in determining the raw-key rate (R). Achieving a high raw-key rate is vital for efficient and timely key generation in practical QKD systems.

In evaluating the performance of practical QKD systems, a comprehensive view requires the derivation of the secret key rate (K), which is expressed as the product of the raw-key rate (R) and the secret fraction (r):

$$K = Rr$$

The secret key rate (K) serves as a pivotal figure of merit, encapsulating both the efficiency of raw-key generation and the effectiveness of the security measures in transforming the raw key into a reliable secret key. Achieving a high secret key rate is a paramount objective in practical QKD implementations, as it directly impacts the efficiency and scalability of secure key distribution for real-world applications. As we delve into the world of finite-key scenarios, the secret fraction (r) may experience adjustments due to two primary reasons. Firstly, parameter estimation relies on a finite number of samples, obliging us to consider worst-case values to accommodate statistical fluctuations. Finite-key corrections play a crucial role in quantifying the trade-offs between the finite key length and the attainable secret key rate. Secondly, within classical post-processing yields, certain terms persist even in the asymptotic limit, acknowledging the infeasibility of achieving absolute security. Indeed, the probability that Eve gains knowledge of an n-bit key remains strictly positive, at least 2^{-n}. Although finite-key corrections cannot be overlooked, our current focus in this review is on the asymptotic case, wherein the rigorous estimation of finite-key corrections continues to be the subject of ongoing research and exploration.

In summary, the secret fraction and secret key rate are fundamental parameters in QKD, representing the security and efficiency aspects of key generation in both infinite-key and finite-key scenarios. These parameters underpin the foundations of practical QKD implementations, guiding the design and optimization of secure communication systems in the quantum era.

11.5 Types of Protocols

The field of quantum key distribution (QKD) boasts a vast array of explicit protocols, with seemingly infinite possibilities. Remarkably, Bennett demonstrated that even coding a single bit with just two non-orthogonal quantum states can achieve security (Bennett and Brassard 2014). Amidst this multitude of choices, three dominant families have emerged, each distinguished by the detection scheme employed: discrete-variable coding (II.D.2), continuous-variable coding (II.D.3), and the recent distributed-phase-reference coding (II.D.4). The crucial distinction lies in how detection is handled, with discrete-variable and distributed-phase-reference coding relying on photon counting and post-selection of events, while continuous-variable coding leverages homodyne detection (reviewed in Sec. II.G).

11.5.1 Discrete-Variable Coding: The Pioneering Approach

Discrete-variable coding is a fundamental approach used in quantum key distribution (QKD) protocols, which enables the secure exchange of cryptographic keys between two parties, Alice and Bob, over a quantum channel. The term "discrete-variable" refers to the fact that the quantum information is encoded in distinct, discrete states of a quantum system, as opposed to continuous-variable QKD, which uses continuous degrees of freedom.

Technical Details:

Quantum bits (Qubits): The basic unit of quantum information used in discrete-variable QKD is the qubit. A qubit can be realized using various quantum systems, such as photons (polarization or phase encoding), trapped ions, or superconducting circuits. In the context of discrete-variable QKD, photons are commonly used due to their ease of manipulation and transmission over long distances.

Polarization encoding: One common approach in discrete-variable QKD is polarization encoding, especially for free-space implementations. In this scheme, Alice prepares qubits in specific polarization states (e.g., horizontal (H) or vertical (V) polarizations) and sends them to Bob. Bob then measures the received photons' polarizations using appropriate measurement bases (e.g., rectilinear or diagonal basis). The shared key is established based on the measurement results that match the agreed-upon basis.

Phase coding: For fiber-based implementations of discrete-variable QKD, phase coding is often employed. In this technique, Alice prepares qubits in specific phases (e.g., 0 or 180) and sends them through an optical fiber to Bob. The fiber introduces different phase shifts for different states, and Bob measures the relative phases of the received qubits to extract the shared key.

Security analysis: The security of discrete-variable QKD protocols is based on the principles of quantum mechanics. The security analysis involves estimating the level of quantum bit error rate (QBER) and ensuring that the actual eavesdropping attempts do not go undetected. If the QBER is below a certain threshold, the shared key is deemed secure.

Key distillation: After the quantum communication phase, Alice and Bob perform classical post-processing, known as "key distillation," to further enhance the security of the generated key. This process involves error correction and privacy amplification techniques to filter out any residual errors or potential information leakage.

Quantum states and photon sources: The successful implementation of discrete-variable QKD heavily relies on the generation of single photons or entangled photon pairs. Different quantum states can be used, such as single photons, entangled photon pairs, or coherent states, depending on the specific protocol and application.

Quantum channel: The quantum channel is the physical medium through which the qubits are transmitted between Alice and Bob. In discrete-variable QKD, this channel is typically an optical fiber for fiber-based implementations or free-space for free-space setups.

Discrete-variable QKD protocols have been extensively studied and implemented due to their practical advantages and the robustness of the discrete quantum degrees of freedom, such as polarization and phase coding. These protocols continue to be a significant area of research and development in quantum communication and quantum cryptography, paving the way for secure communication in the quantum era.

11.5.2 Continuous-Variable Protocols

Continuous-variable quantum key distribution (QKD) is another essential solution for safe key exchange between Alice and Bob over a quantum channel. Continuous-variable QKD, as opposed to discrete-variable QKD, encodes quantum information using continuous degrees of freedom, such as the amplitude and phase of quantum states. In this technique, quantum information is encoded in continuous variables of quantum systems, allowing for the transmission of an infinite number of states.

Quantum states: Quantum states are the quantum mechanical states of light used to encode information in continuous-variable QKD. Coherent states, which are classical-like states with well-defined amplitude and phase, are the most frequently used type of quantum state. Coherent states are denoted by α, where α is a complex number representing the amplitude and phase of the state.

Homodyne detection: A measuring technique used to identify the amplitude and phase of a quantum state is homodyne detection. A beamsplitter is used to combine the entering quantum state with a local oscillator. The beamsplitter output comprises information about the amplitude and phase of the input state. Alice and Bob may measure the quantum state quadratures, which are the amplitude and phase quadratures, using balanced homodyne detectors.

Gaussian modulation: In continuous-variable QKD, the quantum states are typically modulated using Gaussian distributions. Gaussian modulation is mathematically well-behaved and has a simple mathematical representation. It allows for efficient encoding and decoding of quantum information and can achieve high rates of key generation.

Security analysis: The security analysis in continuous-variable QKD involves evaluating the level of excess noise in the quantum channel and checking for potential eavesdropping attempts. Excess noise can arise from various sources, such as imperfections in the devices or losses in the quantum channel. Continuous-variable QKD is proven to be secure against coherent attacks, where an eavesdropper tries to extract information by performing coherent measurements on the quantum states.

Squeezed states: Squeezed states of light are a special type of quantum state used in some continuous-variable QKD protocols. Squeezed states have reduced noise in one quadrature of the electromagnetic field at the expense of increased noise in the conjugate quadrature. By using squeezed states, continuous-variable QKD protocols can achieve higher levels of security and enhanced tolerance against certain types of collective attacks.

Gaussian post-processing: After the quantum transmission, Alice and Bob perform Gaussian post-processing techniques to extract a final secure key. This involves reconciliation and privacy amplification. Reconciliation is the process of filtering out errors and discrepancies between Alice and Bob's measurement results to obtain an intermediate key. Privacy amplification is a step that further distills the intermediate key to a shorter, final shared key while ensuring that any information leaked to an eavesdropper is negligible.

Quantum channel: The quantum channel is the physical medium through which the continuous-variable quantum states are transmitted between Alice and Bob. It can be an optical fiber or free space. The quantum channel introduces various imperfections, such as losses, excess noise, and phase fluctuations, which need to be considered in the security analysis and post-processing steps.

Continuous-variable QKD protocols offer advantages such as high key rates and compatibility with existing fiber-optic infrastructure. They are a promising avenue for practical quantum communication and cryptography, with ongoing research to address challenges related to security, noise, and error correction. As quantum technologies continue to advance, continuous-variable QKD has the potential to play a significant role in future secure communication networks.

11.5.3 Distribute-Phase-Reference Protocols

Some quantum key distribution (QKD) protocols have been developed by theorists, while certain experimental groups working toward practical QKD systems have devised new protocols that do not fall under the traditional categories. These novel protocols share similarities with discrete-variable protocols in that the raw keys consist of realizations of a discrete variable (a bit), and they are already perfectly correlated in the absence of errors. However, what distinguishes these protocols is the way the quantum channel is monitored using the properties of coherent states, particularly by observing the phase coherence of subsequent pulses. As a result, these protocols have been termed "distributed-phase-reference protocols." In these schemes, the phase coherence of coherent states plays a critical role in encoding and detecting quantum information for secure key distribution. The development and exploration of distributed-phase-reference protocols represent an exciting area of research and innovation in the quest for secure quantum communication (Using quantum key distribution for cryptographic purposes: a survey 2014).

11.5.3.1 Differential-Phase-Shift (DPS) Protocol

The differential-phase-shift (DPS) protocol is a quantum key distribution (QKD) system that uses coherent states with different phases to secure key exchange. In this protocol, Alice creates a series of coherent light states, each with a unique phase that may be changed to 0 or pi. The phase difference between two subsequent coherent states encodes the bits of information. Alice assigns the bit value 0 if the phase difference between two successive states is zero. If the phase difference, on the other hand, is pi, Alice assigns the bit value 1. This phase-based encoding ensures that the raw keys shared between Alice and Bob are already perfectly correlated without any errors (Hatakeyama et al. 2017).

Once Alice encodes the bits, she sends the sequence of coherent states through a quantum channel to Bob, the receiver. The quantum channel can be an optical fiber or free space, through which the coherent states are transmitted. Upon receiving the sequence of coherent states, Bob employs an unbalanced interferometer, a device that splits and recombines light beams, to perform the measurement. The unbalanced interferometer is specifically designed to detect and distinguish between the two phase states of the coherent states.

The interference pattern observed in the unbalanced interferometer allows Bob to unambiguously distinguish between the two phase states, and consequently, he can extract the bit values from the sequence of coherent states. However, one of the key challenges in the DPS protocol is that each pulse in the sequence contributes to the encoding of both the current bit and the subsequent bit. This interdependency between neighboring bits complicates the analysis of the protocol's security and requires careful consideration during the post-processing phase to ensure accurate key extraction and security.

Despite its complexities, several experimental demonstrations of the DPS protocol have been conducted, confirming its feasibility for practical QKD implementations. These experiments have shown that DPS holds promise as a secure quantum key distribution scheme, paving the way for potential applications in secure quantum communication networks.

11.5.3.2 Coherent-One-Way (COW)

The coherent-one-way (COW) protocol, which was proposed in 2004 by Gisin, Ribordy, Tittel, and Zbinden, is another type of distributed-phase-reference quantum key distribution (QKD) scheme used for secure key exchange. The COW protocol was proposed by Gisin, Ribordy, Tittel, and Zbinden in 2004. Similar to the DPS protocol, the COW protocol also employs coherent states of light to encode quantum information (Lavie and Lim 2022).

Each byte of information in the COW protocol is encoded in a series of one non-empty pulse and one empty pulse. By measuring the duration of arrival, these two states, indicating non-empty and empty pulses, may be clearly distinguished. This phase-based encoding enables efficient and consistent bit extraction.

The COW protocol verifies the coherence between two consecutive non-empty pulses to estimate the channel. This check is critical for detecting possible attacks, such as photon number splitting (PNS) assaults, in which an eavesdropper attempts to divide photons from the same pulse in order to measure them individually.

Precision phase control between successive pulses is necessary to enable accurate channel projections and security. This means that the phase between any two successive pulses must be controlled, and thus the entire sequence of pulses must be treated as a single signal. This phase control aspect is similar to the challenge faced in the DPS protocol. A prototype of a full QKD system based on the COW protocol has been reported in recent works, demonstrating progress toward practical implementations of this protocol.

The security analysis of the COW protocol falls into the category of partially secure protocols, where security can be guaranteed under certain assumptions. However, deriving unconditional security bounds for such protocols is complex due to the interdependency between neighboring bits and the need for phase control.

In summary, the coherent-one-way (COW) protocol is a distributed-phase-reference QKD scheme that encodes bits using sequences of coherent states, including non-empty and empty pulses. It utilizes an unbalanced interferometer to discriminate between the two pulse states and performs channel estimation by checking coherence between non-empty pulses. With ongoing research and technological advancements, the COW protocol shows promise for practical applications in quantum communication and cryptography, offering enhanced security and potential benefits in quantum key distribution systems.

Table 11.2 compares different quantum key distribution (QKD) protocols: discrete variable, continuous variable, and distributed-phase-reference. Discrete-variable protocols use single-photon detection and discrete states like polarization or phase coding. Continuous-variable protocols employ homodyne or heterodyne detection with continuous variables. Distributed-phase-reference protocols monitor the quantum channel using coherent states' phase coherence. Each protocol has distinct advantages, and selecting the appropriate one depends on specific application requirements.

Table 11.2 Comparison of quantum key distribution (QKD) protocols

Protocol name	Photon detection scheme	Key encoding method	Basis used	Key generation method	Advantages
Discrete variable	Single-photon detection	Polarization or phase coding	Orthogonal states	Entanglement-based, EPR pairs	Proven security, long-distance transmission capability
Continuous variable	Homodyne or heterodyne detection	Modulation of quadrature amplitudes	Gaussian states	Squeezed-state preparation	Efficient implementation with simple hardware
Distributed-phase-reference	Unbalanced interferometer	Phase difference between coherent states	Coherent states	Coherent-state preparation	Robust against certain attacks, potentially high secret key rates

11.6 Sources

In quantum key distribution (QKD) systems, the generation of quantum states is a critical component that directly impacts the security and performance of the entire communication process. Different types of sources are used to create the required quantum states for transmitting information securely between Alice and Bob. In this section, we explore three key types of sources commonly employed in QKD: **lasers**, **sub-Poissonian sources**, and **sources of entangled photons**.

11.6.1 Lasers

Lasers serve as the most practical and versatile light sources available today, making them the preferred choice for the majority of research groups working in the QKD field. The coherent and stable output of lasers makes them ideal for encoding quantum information and transmitting it through the quantum channel. Lasers are used in both continuous-variable and discrete-variable protocols, depending on the application. However, when lasers are used as attenuated sources for discrete-variable protocols, the need for a phase reference is reduced. The security of laser-based implementations may be affected by photon-number-splitting (PNS) attacks, which must be carefully addressed during security analysis. Photon-number-splitting (PNS) attacks are a type of eavesdropping technique that poses a significant threat to the security of quantum key distribution (QKD) systems using attenuated lasers as discrete-variable sources. PNS attacks exploit the photon bunching property of attenuated lasers, allowing an eavesdropper to split multi-photon pulses into individual photons. This enables the eavesdropper to gain information about the transmitted key without detection at the receiver's end. To counter PNS attacks, QKD protocols may employ "decoy states" and additional security measures to enhance system robustness and protect against potential eavesdropping threats.

11.6.2 Sub-Poissonian Sources

Sub-Poissonian sources, also known as "single-photon sources," are designed to produce light states with reduced probabilities of emitting two or more photons simultaneously. These sources are of paramount importance in many discrete-variable QKD protocols that require single-photon generation. Achieving low values of the second-order correlation function, which measures the probability of detecting two photons at different times, is essential for enhancing the security of QKD systems. Sub-Poissonian sources have been the subject of extensive research and experimentation due to their potential in achieving higher secret fractions and enhancing security against certain attacks.

11.6.3 Sources of Entangled Photons

Sources of entangled photons play a crucial role in entanglement-based QKD protocols and heralded sub-Poissonian sources. The most commonly used method for generating entangled photon pairs is spontaneous parametric down-conversion (SPDC), which involves converting photons from a pump laser beam into pairs of lower-energy entangled photons through nonlinear interaction in an optical crystal. These entangled photons are observed to exhibit quantum correlations in various degrees of freedom, such as time, frequency, polarization, momenta, or orbital angular momenta. The generation of entangled photon pairs is particularly useful in continuous-variable protocols, where the entangled state directly finds application. However, in discrete-variable protocols, the presence of multipair components in entangled photon sources requires careful consideration during security analysis to detect and mitigate potential vulnerabilities (Ma et al. 2007).

The choice of the source type is a critical aspect of QKD system design, as it significantly impacts the security and efficiency of quantum communication protocols. Researchers must carefully analyze the strengths and limitations of different sources to adapt their implementations and enhance the effectiveness of their QKD systems.

11.7 Hacking in QKD

While quantum key distribution (QKD) is theoretically secure against certain eavesdropping attacks, the practical implementation of QKD systems introduces vulnerabilities that adversaries can exploit. These vulnerabilities stem from imperfections in the physical devices and the challenges in achieving perfect quantum states in real-world scenarios. As a result, the security of practical QKD systems relies not only on the underlying principles of quantum mechanics but also on the robustness of the hardware and protocols used.

11.7.1 Trojan Horse Attack

Trojan Horse attacks are a significant class of hacking attacks that pose a serious threat to the security of QKD implementations. In these attacks, Eve, the eavesdropper, seeks to exploit weaknesses in the physical devices utilized by Alice and Bob to gain unauthorized access to sensitive information, particularly the secret key exchanged during the quantum communication process. By carefully probing the devices and analyzing the reflected signals, Eve aims to extract valuable information that could compromise the confidentiality and integrity of the quantum key.

One of the common vulnerabilities targeted by Trojan Horse attacks is the phenomenon known as “afterpulsing” in certain photon counting detectors, such as silicon-based avalanche photo-diodes. When these detectors detect a photon, they may inadvertently emit additional photons at various wavelengths. This afterpulsing phenomenon creates a potential side-channel through which Eve can infer which detector has fired, revealing critical information about the secret key.

To illustrate, consider the scenario where Alice sends a stream of photons to Bob, who uses photon counting detectors to measure incoming photons. Upon detecting a photon, if the detector produces an afterpulse, it provides Eve with valuable information about the measurement outcome. Eve can exploit this knowledge to infer the bit values of the quantum key, potentially compromising the entire key exchange process.

To counteract Trojan Horse attacks and mitigate the afterpulsing vulnerability, various techniques can be employed based on the specific QKD setup:

- Unidirectional light propagation: In setups where the light propagates only unidirectionally, i.e., from Alice’s lab to Bob’s lab, optical isolators can be installed. Optical isolators ensure that any emitted afterpulse from Bob’s detectors remains confined within Bob’s lab and does not escape to the quantum channel, making it inaccessible to Eve.
- Bidirectional light propagation: In more complex setups where light must propagate bidirectionally between Alice and Bob (e.g., Plug and Play configurations), additional monitoring detectors can be introduced. These monitoring detectors serve as sentinels, observing the incoming light for any unexpected signals. If an afterpulse is detected in the monitoring detector, it signals the presence of a potential hacking attempt, and appropriate countermeasures can be implemented to secure the communication.

11.7.2 *Other Hacking Attacks*

Beyond Trojan Horse attacks, adversaries have devised additional strategies to exploit potential weaknesses in specific QKD implementations. Some notable hacking attacks include the following.

11.7.2.1 Faked State Attacks

Faked state attacks involve Eve manipulating the quantum states sent by Alice to Bob. By impersonating the legitimate sender, Eve can introduce errors or extract information from the quantum signals, compromising the security of the system. This attack exploits vulnerabilities in the preparation and measurement stages of QKD (Denny 2011).

11.7.2.2 Phase-Remapping Attacks

Phase-remapping attacks focus on manipulating the phase of the quantum signals during transmission. By introducing phase shifts, Eve can interfere with the correct detection of quantum states, leading to information leakage or disruption of the key generation process (Fung et al. 2007).

11.7.2.3 Time-Shift Attacks

Time-shift attacks involve the malicious adjustment of the timing of quantum signals. By altering the time of arrival of quantum states at the detectors, Eve can cause synchronization issues between Alice and Bob, leading to potential information leakage or even a complete communication breakdown (Qi et al. 2006).

These hacking attacks underscore the need for robust security measures in practical QKD systems. As adversaries continue to develop sophisticated techniques, QKD researchers and engineers must remain vigilant in devising and implementing effective countermeasures to safeguard against these threats. Ensuring the resilience and security of QKD is essential for enabling its widespread adoption in real-world applications where secure communication is paramount. By addressing these hacking challenges, QKD can continue to pave the way for a new era of secure and confidential information exchange in the quantum age.

11.8 The "Uncalibrated-Device Scenario"

In practical QKD experiments, errors and losses can occur both in the quantum channel due to Eve's intervention and within the devices used by Alice and Bob. Specifically, the detectors have finite efficiency (losses) and can produce dark counts (errors). To achieve a meaningful security proof, it becomes crucial to integrate knowledge about these device imperfections into the analysis.

However, incorporating these device imperfections into security proofs is not straightforward. The naive approach of simply removing device imperfections from the parameters used in privacy amplification provides only an upper bound on security, and unconditional security proofs are often only available when attributing all losses and errors to Eve. This assumption, known as the "uncalibrated-device scenario," considers Alice and Bob to have no means of distinguishing the losses and errors of their devices from those originating in the quantum channel.

Despite the challenges, the uncalibrated-device scenario remains a necessary condition to derive lower bounds on the security of practical QKD systems. Researchers are actively exploring this scenario to develop better security proofs and understand the impact of device imperfections on the overall security of QKD.

11.9 Conclusion

This chapter investigates quantum key distribution (QKD) as a critical technology for secure communication in the Internet of things (IoT) era. It discusses the history of cryptography as well as the importance of quantum physics in assuring secure communication. The properties of quantum and classical channels are explained, as well as the fundamental quantum events that make QKD resistant to eavesdroppers.

The use of light as a medium for QKD is highlighted due to its advantages for long-distance quantum state transmission. The BB84 protocol, a groundbreaking quantum communication technique, is thoroughly investigated as well as various QKD protocols, putative quantum information sources, and QKD hacking attacks.

In addition, the chapter discusses the "Uncalibrated-Device Scenario," emphasizing the significance of taking into account calibration uncertainties in QKD systems.

Overall, QKD is a viable path toward a more secure and resilient digital future, protecting communication and data in an interconnected systems and devices. QKD has enormous potential as an integral component of secure communication as quantum technologies progress, leading to greater research and innovation in this cutting-edge sector.

References

Charles H. Bennett, and Gilles Brassard. 2014. Quantum cryptography: Public key distribution and coin tossing. *Theoretical Computer Science* 560 (Part 1): 7–11. ISSN: 0304-3975. https://doi.org/10.1016/j.tcs.2014.05.025. (https://www.sciencedirect.com/science/article/pii/S0304397514004241)

Boutros, Joseph J., and Emina Soljanin. 2023. Time-Entanglement QKD: Secret Key Rates and Information Reconciliation Coding. arXiv: 2301.00486 [cs.IT]

Campbell, S.L., and C.W. Gear. 1995. The index of general nonlinear DAES. *Numerische Mathematik* 72 (2): 173–196.

Csiszár, Imre, and Janos Korner. 1978. Broadcast channels with confidential messages. *IEEE Transactions on Information Theory* 24 (3): 339–348.

Denny, Travis. 2011. Faked states attack and quantum cryptography protocols. arXiv: 1112.2230 [cs.CR].

Donta, Praveen Kumar, et al. 2022. Survey on recent advances in IoT application layer protocols and machine learning scope for research directions. *Digital Communications and Networks* 8 (5): 727–744.

Fung, Chi-Hang Fred, et al. 2007. Phase-remapping attack in practical quantum-key-distribution systems. *Physical Review A* 75 (3). https://doi.org/10.1103/physreva.75.032314. https://doi.org/10.1103%2Fphysreva.75.032314

Gisin, Nicolas, et al. 2002. Quantum cryptography. *Reviews of Modern Physics* 74 (1): 145–195. https://doi.org/10.1103/revmodphys.74.145. https://doi.org/10.1103%2Frevmodphys.74.145

Hatakeyama, Yuki, et al. 2017. Differential-phase-shift quantum-key-distribution protocol with a small number of random delays. *Physical Review A* 95 (4). https://doi.org/10.1103/physreva.95.042301. https://doi.org/10.1103%2Fphysreva.95.042301

Kiktenko, Evgeniy, et al. 2018. Error estimation at the information reconciliation stage of quantum key distribution. *Journal of Russian Laser Research* 39: 558–567. https://doi.org/10.1007/s10946-018-9752-y

Lavie, Emilien, and Charles C.-W. Lim. 2022. Improved coherent one-way quantum key distribution for high-loss channels. *Physical Review Applied* 18 (6): 064053. https://doi.org/10.1103/PhysRevApplied.18.064053. https://link.aps.org/doi/10.1103/PhysRevApplied.18.064053

Ma, Xiongfeng, et al. 2007. Quantum key distribution with entangled photon sources. *Physical Review A* 76 (1): 012307 . https://doi.org/10.1103/physreva.76.012307.https://doi.org/10.1103%2Fphysreva.76.012307.

Milanov, Evgeny. 2009. The RSA algorithm. *RSA Laboratories*: 1–11. https://sites.math.washington.edu/~morrow/336_09/papers/Yevgeny.pdf

Qi, Bing, et al. 2006. Time-shift attack in practical quantum cryptosystems. arXiv: quant-ph/0512080 [quant-ph].

Scarani, Valerio, et al. 2009. The security of practical quantum key distribution. *Reviews of Modern Physics* 81 (3): 1301–1350. https://doi.org/10.1103/revmodphys.81.1301. https://doi.org/10.1103%2Frevmodphys.81.1301

Using quantum key distribution for cryptographic purposes: a survey. 2014. arXiv: quant-ph/0701168 [quant-ph].

Chapter 12
Quantum Internet of Things for Smart Healthcare

Kartick Sutradhar, Ranjitha Venkatesh, and Priyanka Venkatesh

12.1 Introduction

The quantum Internet of things (QIoT) represents the cutting-edge integration of quantum computing and the Internet of things (IoT) (Cheng et al. 2017; Praveen Kumar et al. 2023). It leverages the principles of quantum mechanics to enable revolutionary advancements in communication and computation. In this paradigm, traditional IoT devices are enhanced with quantum-enabled capabilities, allowing them to process and exchange information using quantum bits or qubits. Unlike classical bits, qubits can exist in multiple states simultaneously, leading to exponentially increased computational power and unparalleled levels of security. QIoT has the potential to revolutionize various industries, from finance and healthcare to transportation and manufacturing, by enabling secure, real-time data processing, and precise predictions (Younan et al. 2021). However, significant challenges lie ahead, such as quantum hardware scalability and quantum error correction, which need to be overcome for QIoT to reach its full potential and become a transformative force in the digital era.

Quantum Internet of things holds immense significance in the context of smart healthcare, offering revolutionary opportunities to transform the industry. With the integration of quantum computing and IoT, QIoT brings unprecedented computational power and data processing capabilities. In smart healthcare applications,

K. Sutradhar (✉)
Indian Institute of Information Technology, Sri City, India

R. Venkatesh
Gandhi Institute of Technology and Management, Bengaluru, India
e-mail: rvenkate@gitam.in

P. Venkatesh
Presidency University, Bengaluru, India

P. K. Donta et al. (eds.), *Learning Techniques for the Internet of Things*,
https://doi.org/10.1007/978-3-031-50514-0_12

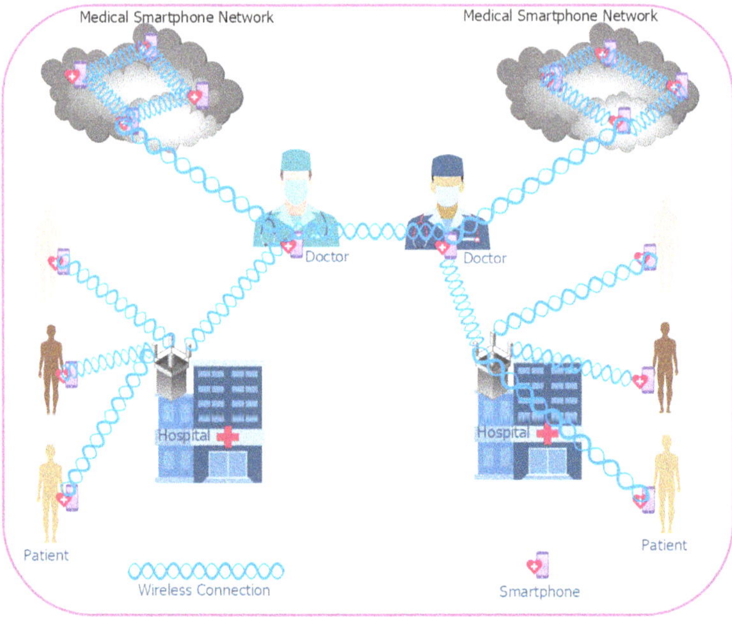

Fig. 12.1 Smart healthcare network

this translates to real-time analysis of vast amounts of medical data, enabling more accurate diagnostics, personalized treatment plans, and drug discovery (Zhu et al. 2019). The ability of quantum devices to handle complex algorithms and perform simulations at an exponential speed opens up new avenues for medical research and advancements. Moreover, QIoT ensures enhanced data security and privacy through quantum encryption and communication protocols, safeguarding sensitive patient information from potential cyber threats. As medical devices and wearables become increasingly interconnected, QIoT's potential to handle the vast streams of data generated by these devices can lead to more efficient remote monitoring, improved patient care, and the potential to predict and prevent health issues proactively. In summary, quantum IoT has the power to revolutionize smart healthcare, leading to better outcomes, reduced costs, and a healthier society overall (Suhail et al. 2020). The smart healthcare network can be shown in Fig. 12.1.

12.2 Quantum IoT: Fundamentals and Components

Quantum Internet of things combines the principles of quantum mechanics with the Internet of things to create a revolutionary paradigm of interconnected devices with quantum-enabled capabilities. There are some fundamentals and components of QIoT.

1. **Quantum computing:** At the core of QIoT lies quantum computing, a type of computing that leverages the principles of quantum mechanics. Unlike classical bits in traditional computing, quantum bits or qubits can exist in multiple states simultaneously, thanks to the phenomena of superposition and entanglement. This property allows quantum computers to perform complex calculations at an exponentially faster rate than classical computers, enabling data processing and analysis on a scale that was previously unimaginable (Steane 1998).
2. **Internet of things:** The Internet of things refers to the network of interconnected physical devices embedded with sensors, software, and other technologies that enable them to collect and exchange data over the Internet (Li et al. 2015). IoT devices can range from simple sensors to complex systems like smart homes, wearable devices, industrial machinery, and more. These devices generate massive amounts of data, which can be utilized for various applications.
3. **Quantum-enabled devices:** In QIoT, traditional IoT devices are enhanced with quantum-enabled capabilities. These quantum-enabled devices use qubits to process and exchange information. For instance, a quantum sensor can provide more precise measurements in medical applications, optimizing patient monitoring and diagnostics. Quantum-enhanced processors can significantly speed up data analysis, enabling real-time insights and predictive maintenance in industrial IoT settings (Wang and Rahman 2022).
4. **Quantum communication:** Quantum communication is a critical component of QIoT that focuses on secure and efficient transmission of quantum information between devices. Quantum communication protocols, like quantum key distribution (QKD), ensure data security through quantum encryption, making it theoretically impossible for hackers to intercept or eavesdrop on the transmitted information (Gisin and Thew 2007).
5. **Quantum entanglement:** Entanglement is a unique quantum phenomenon where two or more qubits become interconnected in a way that the state of one qubit is dependent on the state of another, regardless of the distance between them. Entanglement plays a crucial role in quantum communication and quantum networking, facilitating the creation of secure communication channels and distributed quantum computing (Horodeckiet al. 2009).
6. **Quantum algorithms:** QIoT leverages quantum algorithms specifically designed to take advantage of quantum computing's capabilities. These algorithms can solve complex optimization, simulation, and machine learning problems more efficiently than classical algorithms, offering potential advancements in various applications like drug discovery, personalized medicine, and traffic optimization (Montanaro 2016).

Quantum IoT is a transformative approach that combines quantum computing and IoT to revolutionize various industries. By leveraging quantum capabilities, QIoT offers the potential for faster and more secure data processing, enabling advancements in healthcare, smart cities, logistics, finance, and other domains. However, realizing the full potential of QIoT will require overcoming technical challenges and further advancements in quantum technologies.

12.2.1 Quantum Computing and Its Relevance to Healthcare

Quantum computing holds immense relevance to healthcare, promising to revolutionize the industry in various ways. The ability of quantum computers to process vast amounts of data at unprecedented speeds opens up new possibilities for medical research and drug discovery. Quantum algorithms can efficiently analyze genomic data, protein structures, and interactions, leading to personalized medicine and targeted therapies for individual patients. Quantum computing can also optimize complex healthcare processes, such as hospital scheduling, resource allocation, and supply chain management, leading to improved operational efficiency and reduced costs. Additionally, quantum encryption and secure communication protocols can enhance the privacy and security of patient data, mitigating potential cybersecurity risks in an increasingly interconnected healthcare landscape. While quantum computing is still in its early stages, its potential impact on healthcare is profound, offering a glimpse into a future with more advanced medical treatments, better patient outcomes, and an overall healthier society. As quantum technologies continue to advance, healthcare is expected to be one of the key beneficiaries, ushering in a new era of precision medicine and transformative healthcare solutions (Engelhardt 2017).

12.2.2 Quantum Communication for Secured Healthcare Data Transmission

Quantum communication offers a revolutionary solution for securing healthcare data transmission, addressing one of the most critical challenges in the healthcare industry. With the increasing reliance on interconnected devices and digital platforms for patient data exchange, the risk of data breaches and unauthorized access has become a significant concern. Quantum communication utilizes the principles of quantum mechanics, such as quantum key distribution (QKD), to ensure an unprecedented level of data security. Unlike classical encryption methods, which can be susceptible to hacking through advanced algorithms and processing power, quantum encryption relies on the fundamental properties of quantum mechanics, making it theoretically impossible for eavesdroppers to intercept or tamper with the transmitted information without being detected (Elhoseny et al. 2018).

In a quantum communication system, quantum bits or qubits are used to generate and distribute encryption keys between communicating parties. The process of transmitting the keys involves measuring qubits in various quantum states, and any attempt to intercept or observe these qubits causes a disturbance that can be immediately detected by the legitimate parties. This ensures that the encryption keys remain secure, as any unauthorized attempt to access them will be immediately detected, triggering an alert. By implementing quantum communication for healthcare data transmission, medical institutions can protect sensitive patient information, including medical records, diagnostic data, and treatment plans, from potential cyber threats and data breaches. Moreover, as quantum communication technologies advance, they are expected to provide even stronger security measures, safeguarding healthcare data against future threats posed by quantum computers capable of breaking classical encryption algorithms. While quantum communication for securing healthcare data transmission holds enormous promise, it is still in its early stages of development and practical implementation (Selvarajan and Mouratidis 2023). As researchers continue to advance quantum technologies and overcome current challenges, such as scalability and integration with existing infrastructure, quantum communication is poised to play a pivotal role in creating a safer and more secure healthcare ecosystem, fostering trust and confidence among patients, medical professionals, and healthcare institutions.

12.2.3 Quantum Sensing and Imaging in Healthcare Applications

Quantum sensing and imaging offer promising applications in the healthcare industry, providing advanced tools for diagnostics, monitoring, and treatment. Quantum sensors can measure physical quantities with unparalleled precision, enabling more accurate and sensitive medical devices. The key healthcare applications of quantum sensing and imaging can be shown in Fig. 12.2.

1. **Magnetic resonance imaging (MRI) enhancement:** Quantum sensors based on superconducting qubits or nitrogen-vacancy centers in diamonds can improve the sensitivity of MRI machines. These sensors can detect subtle changes in magnetic fields, leading to higher-resolution images and earlier detection of abnormalities, such as tumors or neurological disorders (Hylton 2006).
2. **Quantum-enhanced imaging modalities:** Quantum sensing techniques, such as quantum illumination and quantum radar, have the potential to enhance

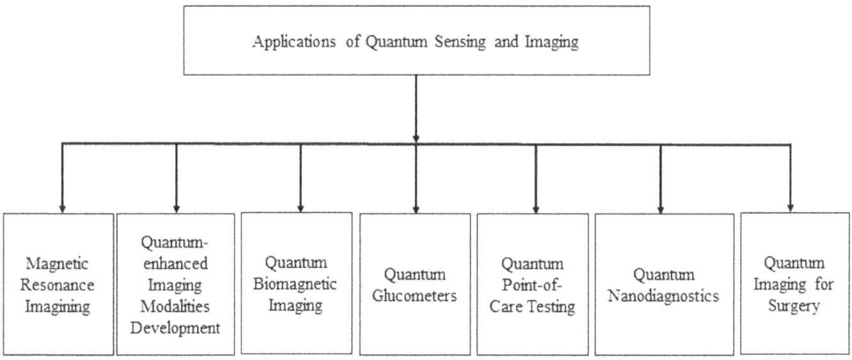

Fig. 12.2 Applications of quantum sensing and imaging

imaging modalities used in medical imaging, such as ultrasound and optical coherence tomography. By utilizing quantum entanglement and squeezing, these techniques can enhance signal-to-noise ratios, allowing for more precise imaging and diagnostics (Ortolano et al. 2019).
3. **Quantum biomagnetic imaging:** Quantum sensors can detect extremely weak magnetic fields generated by the human body, such as the magnetic fields produced by the brain or the heart. These biomagnetic fields can provide valuable information about brain activity, cardiac function, and neurological disorders, leading to improved diagnosis and treatment monitoring (Fong et al. 2004).
4. **Quantum glucometers:** Quantum sensors can be integrated into glucometers to measure blood glucose levels with higher accuracy and sensitivity. This could lead to more effective diabetes management and reduce the need for frequent blood sampling (Tian et al. 2022).
5. **Quantum point-of-care testing:** Quantum sensors could enable highly sensitive and rapid point-of-care diagnostic tests. For example, quantum-enhanced biosensors could detect specific disease markers or pathogens at much lower concentrations, facilitating early disease detection and containment.
6. **Quantum nanodiagnostics:** Quantum dots, nanoparticles with unique quantum properties, can be used as contrast agents in medical imaging. They can target specific biomarkers and provide precise information about cellular structures and functions, aiding in early cancer detection and targeted therapies (Mondal et al. 2012).
7. **Quantum imaging for surgery:** Quantum-enhanced imaging techniques can provide surgeons with real-time, high-resolution images during minimally invasive procedures, enabling more accurate and precise surgical interventions (Li et al. 2012).

Quantum sensing and imaging hold great promise for healthcare applications, but there are still challenges to overcome, such as scaling up quantum technologies for practical use, integrating them with existing medical devices and infrastructure, and ensuring cost-effectiveness. As quantum technologies continue to advance, their integration into healthcare is likely to play a vital role in improving patient outcomes, disease detection, and overall medical diagnostics and treatment.

12.2.4 Integration with Traditional IoT in Healthcare

The integration of quantum sensing and imaging with traditional IoT in healthcare opens up new possibilities for transformative advancements. Traditional IoT devices in healthcare, such as wearable fitness trackers, remote patient monitoring devices, and smart medical equipment, generate vast amounts of data. By incorporating quantum sensors into these devices, healthcare professionals can gain access to more accurate and sensitive measurements. For example, wearable devices with quantum-enhanced sensors can provide more precise health data, allowing for better monitoring of vital signs and early detection of health issues. These quantum-enabled IoT devices can also improve diagnostic imaging, such as MRI machines with quantum sensors that offer higher-resolution images and more detailed insights into medical conditions. Moreover, the combination of quantum-enhanced data analysis and traditional IoT data can lead to more robust predictive analytics and personalized treatment plans (Rejeb et al. 2023). The seamless integration of quantum sensing and imaging with traditional IoT in healthcare holds the potential to revolutionize patient care, improve disease management, and enhance overall healthcare outcomes through advanced data-driven insights and precision medicine. However, to fully realize these benefits, further research and development are needed to address technical challenges and ensure the scalability, security, and compatibility of quantum-enabled IoT devices with existing healthcare infrastructure.

12.3 Smart Healthcare Applications of Quantum IoT

Smart healthcare applications of QIoT present a promising future for the industry, leveraging quantum computing and IoT capabilities to transform patient care and healthcare operations. QIoT can enhance remote patient monitoring by integrating quantum sensors into wearable devices, enabling highly accurate and real-time health data collection. These quantum-enabled devices can detect subtle changes in vital signs, leading to earlier detection of health issues and more proactive interventions. Quantum computing's immense processing power can optimize healthcare logistics, such as hospital scheduling, resource allocation, and supply chain management, streamlining operations and reducing costs. Additionally, QIoT's advanced encryption methods ensure the secure transmission of sensitive patient data, safeguarding against cyber threats and protecting patient privacy. The integration of quantum algorithms with IoT-generated data can lead to more precise predictive analytics, supporting personalized treatment plans and improved disease management. Furthermore, quantum-enhanced imaging technologies can revolutionize medical diagnostics, providing higher-resolution images for accurate and early disease detection (Gardašević et al. 2020). In summary, the application of quantum IoT in smart healthcare has the potential to revolutionize the industry, improving patient outcomes, enhancing operational efficiency, and paving the way for a more interconnected and secure healthcare ecosystem. The smart healthcare applications of quantum IoT can be shown in Fig. 12.3.

12.3.1 Quantum IoT in Diagnostics and Imaging

Quantum IoT offers groundbreaking potential in the field of diagnostics and imaging, revolutionizing the way medical conditions are detected and visualized. By incorporating quantum-enhanced sensors and imaging technologies into traditional

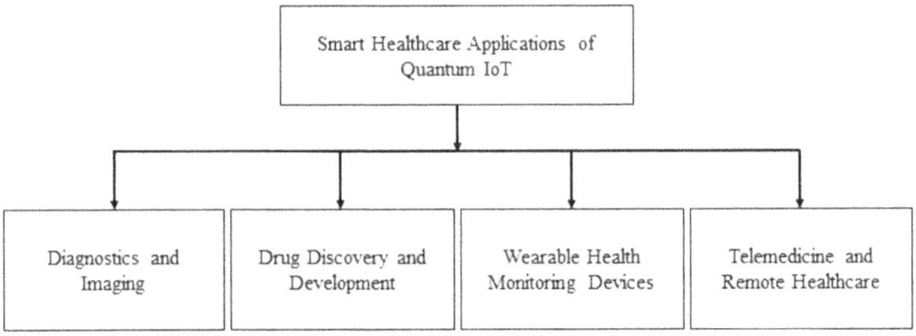

Fig. 12.3 Smart healthcare applications of quantum IoT

medical devices, QIoT can significantly improve the accuracy and sensitivity of diagnostic tools. For instance, quantum sensors integrated into medical imaging devices like MRI machines can provide higher-resolution images, enabling more precise identification of anomalies and early detection of diseases. Quantum-enhanced imaging modalities, such as quantum-enhanced ultrasound and optical coherence tomography, can offer unparalleled visualization of biological tissues and cellular structures, aiding in the early diagnosis of various medical conditions. Additionally, quantum dots and nanoparticles with unique quantum properties can serve as highly sensitive contrast agents, enhancing imaging capabilities and enabling targeted drug delivery in the body. Furthermore, QIoT's ability to process vast amounts of data at extraordinary speeds allows for more advanced analysis and interpretation of medical imaging data, leading to quicker and more accurate diagnoses. As quantum IoT continues to advance, it holds the promise of ushering in a new era of precision medicine, where healthcare professionals can rely on quantum-enhanced diagnostics and imaging technologies to provide personalized and highly effective treatment plans for patients (Elhoseny et al. 2018).

12.3.2 Quantum IoT for Drug Discovery and Development

Quantum IoT offers immense potential in the domain of drug discovery and development, promising to accelerate and optimize the process of identifying novel drugs and therapies. The vast computational power of quantum computing enables the efficient simulation and analysis of complex molecular interactions, which are crucial in understanding the behavior of drugs within the human body. Quantum algorithms can simulate the behavior of molecules at a quantum level, providing more accurate predictions of their interactions with biological targets. This enables researchers to identify potential drug candidates with higher specificity and efficacy, reducing the need for time-consuming and costly experimental trials. Quantum-enhanced simulations can also expedite the screening of vast chemical libraries, narrowing down the search for promising drug candidates. Moreover,

QIoT's quantum encryption capabilities ensure the secure transmission and storage of sensitive pharmaceutical research data, safeguarding intellectual property and proprietary information from potential cyber threats.

Collaboration between researchers and pharmaceutical companies can be streamlined and protected, facilitating advancements in drug development through secure data sharing. Furthermore, quantum sensors can play a vital role in drug manufacturing and quality control. They can precisely monitor various manufacturing processes, ensuring consistency and optimizing production efficiency. Quantum-enhanced sensors can also be employed to assess the purity and quality of pharmaceutical products, ensuring compliance with rigorous regulatory standards. While the integration of QIoT in drug discovery and development is still in its early stages, ongoing research and advancements in quantum technologies hold the promise of transforming the pharmaceutical industry. The synergy between quantum computing's computational prowess and IoT's data-driven insights can significantly expedite the process of bringing innovative drugs to market, addressing medical needs faster and ultimately benefiting patients worldwide.

12.3.3 Quantum IoT-Enabled Wearable Health Monitoring Devices

Quantum IoT-enabled wearable health monitoring devices represent a groundbreaking advancement in the healthcare industry. By integrating quantum sensors into wearable devices, such as smartwatches, fitness trackers, and health patches, these devices can offer unprecedented levels of accuracy and sensitivity in monitoring various health parameters. Quantum sensors can detect and measure physical quantities, such as temperature, pressure, and magnetic fields, with incredible precision, enabling real-time and continuous health monitoring. One of the key advantages of quantum IoT-enabled wearables is their ability to provide more accurate and reliable health data, allowing for better insights into an individual's health status. For example, a quantum-enabled wearable could detect subtle changes in vital signs, such as heart rate and blood pressure, leading to early detection of potential health issues and more timely interventions. Moreover, quantum sensors can enable wearables to measure biomarkers in bodily fluids, like sweat or tears, providing valuable health information without the need for invasive procedures.

Quantum encryption capabilities also ensure the secure transmission of health data between wearables and other healthcare devices, protecting sensitive information from potential cyber threats (Al-Saggaf et al. 2023). This is especially crucial when wearables are used in telemedicine applications or when sharing data with healthcare professionals for remote monitoring and diagnostics. Furthermore, the integration of quantum algorithms with wearable health monitoring devices enables advanced data analytics and personalized health insights. Quantum computing's computational power can process vast amounts of data collected from wearables and generate meaningful patterns and correlations, aiding in disease prevention, diagnosis, and treatment planning. Although the practical implementation of quantum IoT-enabled wearables is still in its early stages, ongoing research and advancements in quantum technologies will likely lead to more sophisticated and widely available devices in the future. These wearable health monitoring devices have the potential to empower individuals to take charge of their health, enable more precise and personalized healthcare delivery, and contribute to overall improvements in public health outcomes.

12.3.4 Quantum-Enhanced Telemedicine and Remote Healthcare

Quantum-enhanced telemedicine and remote healthcare represent a cutting-edge frontier in the healthcare industry, leveraging the power of quantum computing and quantum communication to revolutionize the way medical services are delivered remotely. The integration of quantum technologies in telemedicine has the potential to address some of the key challenges in remote healthcare, including data security, real-time diagnostics, and personalized treatment plans. One of the most significant contributions of quantum-enhanced telemedicine is in data security and privacy. Quantum communication protocols, such as quantum key distribution, offer unbreakable encryption methods, ensuring that sensitive patient data transmitted between remote locations remains secure and immune to eavesdropping or hacking attempts. This level of security is critical when dealing with confidential medical information during virtual consultations, remote monitoring, or the exchange of electronic health records. Moreover, quantum computing's immense computational power enhances remote diagnostics and data analysis (Rasool et al. 2023). Quantum algorithms can efficiently process and analyze vast amounts of patient data from wearables, medical sensors, and imaging devices, providing real-time insights into a patient's health status. This enables healthcare professionals to make quicker and more accurate diagnoses, even from distant locations, leading to faster interventions and improved patient outcomes.

Quantum-enhanced remote healthcare also enables personalized medicine on a broader scale. The ability of quantum computing to analyze complex genetic and molecular data allows for precision medicine approaches tailored to individual patients. Treatment plans can be optimized based on a patient's unique genetic makeup and health history, leading to more effective therapies and reduced adverse effects. Furthermore, quantum-enabled telemedicine can facilitate collaboration among healthcare professionals across the globe. Quantum communication protocols enable secure and instantaneous data sharing, allowing experts to consult and collaborate on complex medical cases in real time, irrespective of their physical locations. While the practical implementation of quantum-enhanced telemedicine is still evolving, ongoing research and advancements in quantum technologies hold the promise of transforming remote healthcare delivery. By addressing issues of data security, accelerating diagnostics, and enabling personalized treatment, quantum-enhanced telemedicine has the potential to improve healthcare accessibility, quality, and outcomes, making healthcare services more efficient and effective, regardless of geographical barriers.

12.4 Advantages and Challenges of Quantum IoT in Smart Healthcare

Quantum IoT in smart healthcare offers several advantages and exciting possibilities, but it also faces notable challenges. The advantages of QIoT in smart healthcare include enhanced data security and privacy through quantum encryption, ensuring that sensitive patient information is protected from cyber threats. Quantum computing's immense computational power enables real-time analysis of vast amounts of medical data, leading to more accurate diagnostics, personalized treatment plans, and drug discovery. Moreover, QIoT can optimize healthcare logistics, resource allocation, and supply chain management, improving operational efficiency and reducing costs. Additionally, quantum sensors and imaging technologies in wearable devices and medical equipment can provide higher-resolution data for remote monitoring and early disease detection. However, QIoT faces significant challenges, such as the scalability and stability of quantum hardware. Quantum technologies are still in their early stages, and integrating them with existing healthcare infrastructure can be complex. Developing quantum algorithms and applications suitable for smart healthcare also requires ongoing research and experimentation. Despite these challenges, the potential benefits of QIoT in smart healthcare are profound, offering the prospect of more secure, efficient, and personalized healthcare services for individuals and communities worldwide. As quantum technologies continue to advance, overcoming these challenges will pave the way for QIoT to become a transformative force in the future of healthcare (Alshehri and Muhammad 2020).

12.4.1 Advantages of Quantum IoT for Healthcare Applications

Quantum IoT offers several significant advantages for healthcare applications, revolutionizing the industry in various ways.

- Enhanced data security: Quantum encryption ensures the highest level of data security, protecting sensitive patient information from potential cyber threats. Quantum communication protocols, such as quantum key distribution, make it practically impossible for hackers to intercept or tamper with transmitted data, ensuring patient privacy and confidentiality.

- Faster and more accurate diagnostics: Quantum computing's immense computational power enables rapid analysis of complex medical data, leading to quicker and more accurate diagnostics. Quantum algorithms can process large datasets, such as genomic information, imaging data, and patient records, providing healthcare professionals with real-time insights for timely interventions and treatment decisions.
- Precision medicine and personalized treatment: QIoT facilitates precision medicine by leveraging quantum computing's ability to analyze molecular interactions and genetic data. This enables personalized treatment plans tailored to an individual's unique genetic makeup and health history, resulting in more effective therapies and better patient outcomes.
- Drug discovery and development: Quantum algorithms can simulate and analyze molecular interactions with unprecedented accuracy, accelerating drug discovery processes. Quantum computing's computational capabilities can efficiently screen potential drug candidates, leading to the identification of promising compounds and reducing the time and cost of bringing new medications to market.
- Remote monitoring and telemedicine: Quantum-enhanced sensors in wearable devices enable remote monitoring of patients' health parameters. This allows healthcare providers to track patients' conditions in real time, improving disease management and reducing the need for frequent in-person visits. Quantum-enabled telemedicine also enables secure and efficient communication between healthcare professionals and patients, regardless of geographical distance.
- Optimized healthcare logistics: Quantum computing can optimize healthcare logistics, resource allocation, and supply chain management. This leads to better utilization of medical resources, improved patient care, and reduced operational costs for healthcare institutions.
- Advanced medical imaging: Quantum-enhanced imaging technologies offer higher-resolution images and enhanced contrast, aiding in the early detection of diseases and providing detailed insights into complex medical conditions.

Quantum IoT is still in its early stages of development, and these advantages highlight its enormous potential to transform healthcare, leading to more secure, efficient, and personalized medical services for individuals and populations worldwide. As quantum technologies continue to advance, the application of QIoT in healthcare is expected to yield even more transformative benefits in the years to come.

12.4.2 Security and Privacy Considerations in Quantum IoT

Security and privacy considerations are of paramount importance in QIoT due to the revolutionary impact of quantum computing on cryptographic methods. While quantum computing offers significant advantages, it also poses unique challenges to data protection. Quantum computing's immense computational power can potentially break conventional encryption algorithms, rendering sensitive data vulnerable to malicious attacks. As such, robust and quantum-resistant encryption techniques, like quantum key distribution, are crucial to ensure the confidentiality and integrity of data in QIoT systems. QKD utilizes the principles of quantum mechanics to securely generate and distribute encryption keys, making it virtually impossible for eavesdroppers to intercept or alter the transmitted data. Moreover, QIoT must address privacy concerns related to the collection and storage of vast amounts of personal health data from wearable devices and medical sensors. Transparent data handling practices, strong data anonymization techniques, and strict adherence to privacy regulations are essential to safeguard individuals' privacy rights. As quantum technologies evolve and QIoT applications become more widespread, collaborative efforts between researchers, industry stakeholders, and policymakers will be vital to establish comprehensive security standards and privacy safeguards, ensuring that the potential benefits of quantum IoT in healthcare are realized without compromising data security and individual privacy (Alshehri and Muhammad 2020).

12.4.3 Technological and Implementation Challenges

Implementing quantum QIoT in healthcare faces several technological and implementation challenges that need to be addressed to fully realize its potential. One of the primary challenges is the scalability of quantum computing hardware. Quantum computers are still in their early stages of development, and building large-scale, stable, and commercially viable quantum systems is a complex task. Overcoming these hardware limitations is essential to handle the vast amounts of data generated by IoT devices in healthcare and to perform complex quantum computations efficiently. Another significant challenge is quantum error correction. Quantum information is highly sensitive to noise and environmental interference, leading to errors in quantum computations. Developing robust error correction codes and fault-tolerant quantum algorithms is critical to ensure reliable and accurate results in QIoT applications. Integration with existing healthcare infrastructure and IoT devices is another obstacle. Adapting quantum-enabled sensors and communication protocols to work seamlessly with traditional IoT devices and systems requires careful consideration of compatibility, data formats, and communication protocols. Additionally, the human factor plays a role in the implementation of QIoT in healthcare. There is a need for skilled quantum scientists, researchers, and engineers who can develop and manage quantum systems and algorithms. Educating and training healthcare professionals on quantum technologies and their applications is also essential to effectively use QIoT in medical settings. Furthermore, QIoT raises ethical and regulatory challenges related to data privacy, security, and ownership. Ensuring compliance with data protection laws and regulations becomes even more critical when handling sensitive medical information using quantum technologies.

Finally, cost considerations are a significant factor in the implementation of QIoT in healthcare. Quantum technologies are still relatively expensive to develop and deploy. The challenge is to balance the potential benefits of QIoT with the associated costs to make it economically viable for healthcare institutions. Addressing these technological and implementation challenges will require collaboration between quantum scientists, healthcare experts, industry stakeholders, and policymakers. Overcoming these hurdles will pave the way for the successful integration of quantum IoT in healthcare, unlocking its transformative potential for improved patient care, better diagnostics, and personalized medicine.

12.4.4 Regulatory and Ethical Implications

The integration of QIoT in healthcare raises several regulatory and ethical implications that require careful consideration. From a regulatory standpoint, QIoT technologies may be subject to new and specific regulations given their potential impact on data security and privacy. Healthcare data is highly sensitive, and the use of quantum encryption and communication methods may necessitate updated legal frameworks to address the unique challenges posed by quantum technologies. Regulatory bodies will need to ensure that QIoT systems comply with data protection laws, maintain patient confidentiality, and establish guidelines for the secure storage and transmission of quantum-encrypted healthcare information. Ethical considerations also come to the forefront when deploying QIoT in healthcare. Transparency in how QIoT technologies function and collect data is essential to maintain patient trust. Patients and healthcare professionals must understand the implications of using quantum-enabled devices and the potential benefits and risks associated with their implementation. Informed consent becomes paramount, especially when dealing with the transmission and sharing of sensitive medical information through quantum networks. Moreover, ethical questions may arise regarding data ownership and usage. Clear policies should define how patient data collected through QIoT devices can be accessed, shared, and used for research purposes.

Ensuring that patients have control over their data and have the option to opt out or revoke consent will be critical in maintaining ethical standards in QIoT-enabled healthcare applications. As with any emerging technology, there is also the risk of potential biases or unintended consequences in the use of QIoT in healthcare. Algorithms used in quantum computing can still be influenced by biases in data, leading to skewed results or unequal treatment. Ensuring fairness, accountability, and transparency in the development and implementation of quantum algorithms becomes crucial to avoid perpetuating existing healthcare disparities. Navigating the regulatory and ethical landscape of QIoT in healthcare requires a collaborative effort involving policymakers, healthcare providers, legal experts, and technology developers. By proactively addressing these implications, we can ensure that the integration of quantum IoT in healthcare is guided by ethical principles, respects patient autonomy, and upholds the highest standards of data security and privacy.

12.5 Current Advances and Case Studies

Several exciting advances and case studies have demonstrated the potential of QIoT in healthcare. Quantum computing companies and research institutions have been making progress in developing quantum algorithms for drug discovery, molecular simulations, and optimization of healthcare logistics. For instance, researchers at IBM and other institutions have been exploring how quantum computing can accelerate the discovery of new drugs by simulating the behavior of molecules and predicting potential drug candidates with higher accuracy. In the field of medical imaging, quantum-enhanced imaging techniques have been investigated to improve the resolution and sensitivity of imaging devices. Quantum sensors integrated into medical imaging devices, such as MRI machines, have shown promising results in providing higher-quality images, leading to more accurate diagnostics.

12.5.1 Research Initiatives and Collaborations

Research initiatives and collaborations in the field of quantum IoT for healthcare have been gaining momentum in recent years. Leading technology companies, research institutions, and healthcare organizations have joined forces to explore the potential applications and benefits of quantum technologies in healthcare. Quantum computing companies, such as IBM, Google, and Microsoft, have been investing in quantum research and collaborating with academic institutions to develop quantum algorithms for medical applications, drug discovery, and optimization of healthcare processes. These initiatives aim to harness the power of quantum computing to solve complex healthcare challenges and accelerate medical advancements. Academic institutions and research centers have been actively involved in exploring the use of quantum-enhanced sensors and imaging technologies in medical devices. Collaborations between quantum physicists and medical researchers have led to innovative approaches for improving medical imaging, remote monitoring, and disease detection. Furthermore, there are initiatives focused on exploring the integration of quantum communication protocols, such as quantum key distribution, in healthcare systems. Research collaborations in this area aim to ensure secure and private transmission of medical data, protecting sensitive patient information from cyber threats. In addition to technology companies and research institutions, collaborations between healthcare providers and quantum experts are emerging. These partnerships aim to bridge the gap between quantum technologies and healthcare needs, with the ultimate goal of translating quantum advancements into real-world healthcare solutions.

Government agencies and funding bodies are also recognizing the potential of QIoT in healthcare and providing financial support for research initiatives. This support fosters collaboration between academia and industry, accelerating the development of practical quantum-enabled healthcare technologies. The growing number of research initiatives and collaborations in quantum IoT for healthcare reflects the increasing interest and potential of this field. By combining expertise from diverse disciplines, these collaborations have the potential to pave the way for groundbreaking advancements in healthcare, enhancing patient outcomes and transforming the delivery of medical services. As quantum technologies continue to evolve, these collaborative efforts will play a crucial role in shaping the future of QIoT in healthcare.

12.5.2 Case Studies of Quantum IoT Applications in Healthcare

Specific case studies of quantum IoT applications in healthcare were relatively limited due to the early stage of quantum technologies. However, there have been notable research initiatives and studies that have demonstrated the potential of QIoT in healthcare:

12.5.2.1 Quantum Encryption for Secure Medical Data Transmission

Some research groups have explored the use of QKD to enhance the security of medical data transmission in telemedicine and remote patient monitoring. QKD ensures the secure exchange of encryption keys, protecting sensitive patient information from potential cyber threats during data transmission.

Zhao et al. (2023) proposed a quantum protocol for secure Internet of things. The integrity and equity of the exchange of medical data are ensured by this study. This work offers an OUCS-based mutual authentication system (BBS-OUC) that is based on the mutual authentication of Blum Blum Shub and Okamoto Uchiyana Cryptosystem (OUCS). Qu (2022) discussed a quantum IoT framework for secure medical information using blockchain. Based on security concerns, this research proposes a new private quantum blockchain network and creates a unique distributed quantum electronic medical record system. The data structure of this quantum blockchain connects the blocks via entangled states. By automatically creating the time stamp by joining quantum blocks with predetermined actions, less storage space is required. The hash value of each block is stored in a single qubit. The quantum electronic medical record protocol goes into great detail on how quantum information is processed. Qu et al. (2023) introduced a quantum blockchain based for the secure Internet of things. In this study, a novel quantum blockchain-based medical data processing system (QB-IMD) is designed. In QB-IMD, a revolutionary electronic medical record algorithm (QEMR) and a quantum blockchain structure are presented to guarantee the validity and impermeability of the processed data.

12.5.2.2 Quantum-Enhanced Imaging for Improved Diagnostics

Studies have investigated the use of quantum sensors in medical imaging devices, such as MRI machines, to enhance image resolution and sensitivity. Quantum-enhanced imaging techniques have the potential to provide higher-quality images, leading to more accurate and earlier disease detection.

Janani and Brindha (2021) proposed a protocol to secure the medical image that can improve the diagnostics process. The privacy-preserving procedure of medical images can be strengthened using the recommended quantum block-based scrambling, and it has been discovered. Additionally, it introduces specialized quantum encryption for ROI-based regional data to guarantee the integrity of medical images. Camphausen et al. (2023) introduced a technique for improve the diagnostics process by improving the medical image. The result is a significant first step toward scaling real-world quantum imaging advantage and could be used for both basic research and biomedical and commercial applications.

12.5.2.3 Quantum Algorithms for Drug Discovery

Quantum computing companies and research institutions have been exploring the use of quantum algorithms to accelerate drug discovery processes. These algorithms can simulate molecular interactions more efficiently, leading to the identification of potential drug candidates with higher precision.

Blunt et al. (2022) discussed a research paper about drug discovery using quantum algorithms. This work presents unique estimates of the quantum computational cost of simulating increasingly bigger embedding sections of a pharmaceutically significant covalent protein-drug complex involving the medication Ibrutinib. They also briefly summarize and compare the scaling features of cutting-edge quantum algorithms. Mustafa et al. (2022) proposed a quantum technique for drug discovery. The variational quantum eigensolver (VQE) and the quantum approximate optimization algorithm (QAOA) are two alternative methods that are used in this chapter to investigate how this problem might be solved utilizing quantum computing and Qiskit Nature.

These case studies represent promising steps in the application of QIoT in healthcare, and it is important to note that the field is still in its early stages of development. The full potential of QIoT in healthcare is yet to be realized, and further research, development, and collaboration are needed to unlock its transformative impact on patient care and medical advancements. As quantum technologies continue to progress, we can expect more comprehensive and impactful case studies of QIoT applications in healthcare to emerge in the coming years.

12.5.3 Implementations and Real-World Deployments

The implementations and real-world deployments of QIoT in healthcare were limited due to the nascent stage of quantum technologies. However, there have been some notable advances and early-stage implementations that show promise:

- Quantum key distribution for secure data transmission: Although not widespread, some research initiatives and collaborations have explored the integration of QKD in healthcare systems for secure data transmission (Scarani et al. 2009). These experiments demonstrated the potential of quantum encryption in safeguarding sensitive medical data during telemedicine consultations and remote patient monitoring.
- Quantum-enhanced imaging technologies: Research groups have started investigating the use of quantum sensors in medical imaging devices to improve image resolution and sensitivity. While not yet commercially available, these quantum-enhanced imaging techniques have shown potential for more accurate diagnostics and early disease detection.
- Quantum algorithms for medical research: Pharmaceutical companies and research institutions have begun exploring the use of quantum algorithms for drug discovery and molecular simulations. These early-stage implementations aim to leverage the computational power of quantum computing to accelerate medical research and identify potential drug candidates more efficiently.
- Quantum-inspired machine learning: Although not true quantum computing, quantum-inspired machine learning algorithms have been applied in healthcare settings. These algorithms can process large and complex datasets to identify patterns and correlations, facilitating predictive analytics and personalized medicine.

While successful implementations of QIoT in healthcare are still limited, ongoing research and advancements in quantum technologies are paving the way for wider real-world deployments in the future. As quantum computing hardware becomes more stable and scalable and quantum algorithms mature, we can expect to see more practical applications of QIoT in healthcare. These implementations have the potential to transform patient care, improve medical diagnostics, and revolutionize drug discovery processes, leading to significant advancements in the healthcare industry.

12.6 Future Directions and Emerging Trends

Quantum IoT is poised to play a transformative role in healthcare, with several emerging trends and directions that hold promise for the industry.

- Advancements in quantum computing hardware: As quantum computing hardware continues to improve, with the development of larger and more stable quantum processors, the computational power of QIoT will increase significantly. This will enable more complex simulations, optimizations, and data analysis, leading to more sophisticated applications in drug discovery, personalized medicine, and medical imaging.

- Quantum algorithms for healthcare: Researchers are actively exploring new quantum algorithms and their applications in healthcare. Quantum machine learning, quantum optimization, and quantum chemistry algorithms are among the areas gaining attention for medical research and healthcare optimization.
- Quantum-enhanced medical imaging: Quantum-enhanced imaging techniques have the potential to revolutionize medical diagnostics. Advancements in quantum sensors and imaging devices will lead to higher-resolution and more sensitive imaging, enabling early detection of diseases and providing deeper insights into biological structures.
- Quantum sensors in wearable devices: Integration of quantum sensors into wearable health monitoring devices will enable more accurate and continuous monitoring of vital signs and biomarkers. Quantum-enabled wearables will facilitate remote patient monitoring and contribute to preventive healthcare.
- Quantum encryption and cybersecurity: The use of quantum encryption methods, like QKD, will become increasingly important to ensure the security and privacy of healthcare data in an era of ever-evolving cyber threats. Quantum-resistant cryptographic standards will be crucial to protect sensitive patient information (Wallden and Kashefi 2019).
- Quantum communications networks: The development of quantum communication networks will facilitate secure and real-time data transmission between healthcare providers and medical devices. Quantum communication protocols will offer unbreakable encryption for telemedicine and remote healthcare applications.
- Public-private collaborations: Collaborations between quantum computing companies, academic institutions, and healthcare providers will accelerate the translation of quantum research into practical healthcare applications. Public-private partnerships will foster innovation, share expertise, and overcome technological and implementation challenges.
- Regulatory frameworks: With the advancement of QIoT in healthcare, regulatory bodies will likely develop specific guidelines to address data security, privacy, and ethical considerations unique to quantum technologies. Clear and standardized regulations will be essential to ensure responsible and ethical implementation of QIoT in healthcare.

As quantum technologies continue to mature and more applications are explored, the future of quantum IoT in healthcare holds tremendous potential for improving patient care, advancing medical research, and addressing complex healthcare challenges. While there are still challenges to overcome, the rapid pace of quantum advancements and the growing interest in QIoT suggest that the healthcare industry is on the brink of a quantum-powered transformation in the years to come.

12.6.1 Roadmap for Quantum IoT in Smart Healthcare

Developing a roadmap for quantum IoT in smart healthcare involves a step-by-step plan to harness the potential of quantum technologies in transforming the healthcare industry. The roadmap should address technological advancements, regulatory considerations, and collaborative efforts to achieve successful implementation.

- Research and development: Focus on advancing quantum computing hardware, quantum algorithms, and quantum-enhanced sensors relevant to healthcare applications. Collaborate with quantum computing companies, research institutions, and healthcare experts to identify specific use cases and challenges in healthcare that can benefit from quantum technologies.
- Proof of concept: Conduct pilot projects and proof-of-concept studies to demonstrate the feasibility and benefits of QIoT in healthcare. Explore quantum encryption for secure data transmission, quantum-enhanced medical imaging, and quantum algorithms for drug discovery and personalized medicine.
- Infrastructure and integration: Invest in building the necessary infrastructure to support QIoT in healthcare. Develop secure quantum communication networks for telemedicine and remote healthcare. Integrate quantum sensors into wearable health monitoring devices and medical imaging equipment (Al-Saggaf et al. 2023).
- Regulatory and ethical frameworks: Engage with regulatory bodies and policymakers to develop appropriate regulatory frameworks for QIoT in healthcare. Address data security, privacy, and ethical considerations unique to quantum technologies. Ensure compliance with existing healthcare regulations and data protection laws.
- Public-private partnerships: Foster collaborations between quantum computing companies, academic institutions, healthcare providers, and government agencies. Encourage public-private partnerships to accelerate research, share expertise, and overcome challenges in implementing QIoT in smart healthcare.
- Scale and integration: Scale up quantum computing hardware and algorithms to handle large and complex healthcare datasets. Integrate QIoT solutions into existing healthcare systems and infrastructure. Develop seamless interfaces between quantum-enabled devices and traditional IoT devices.
- Education and training: Invest in educating and training healthcare professionals, researchers, and IT experts on quantum technologies and their applications in healthcare. Ensure that the workforce is equipped with the necessary skills to leverage QIoT for improved patient care and medical advancements.
- Real-world deployments: Deploy QIoT solutions in real-world healthcare settings. Monitor and evaluate the impact of QIoT on patient outcomes, healthcare efficiency, and cost-effectiveness. Continuously improve and optimize QIoT applications based on feedback and experience.
- Continuous innovation: Embrace continuous innovation and research in quantum technologies to stay at the forefront of QIoT advancements. Collaborate with the quantum computing community to leverage the latest developments and discoveries for healthcare applications.

12.6.2 Potential Impact on the Healthcare Industry

The potential impact of quantum IoT on the healthcare industry is profound, offering transformative advancements that can revolutionize patient care, medical research, and healthcare operations.

- Personalized medicine: QIoT can enable more precise and personalized medicine by leveraging quantum computing's computational power to analyze vast amounts of patient data. This will lead to tailored treatment plans based on individual genetic makeup, health history, and real-time health data from wearable devices.
- Accelerated drug discovery: Quantum algorithms can significantly speed up drug discovery processes by simulating molecular interactions more efficiently. This can lead to the identification of potential drug candidates faster, reducing the time and cost of bringing new medications to market (Bergström and Lindmark 2019).
- Improved medical imaging: Quantum-enhanced imaging technologies can offer higher-resolution and more sensitive medical imaging, providing better visualization of biological structures and earlier disease detection.
- Enhanced data security: Quantum encryption ensures the highest level of data security, protecting sensitive patient information from potential cyber threats during data transmission and storage.
- Remote healthcare and telemedicine: QIoT enables more secure and real-time data transmission, supporting remote patient monitoring and telemedicine consultations. This can expand healthcare access, especially for patients in remote areas.
- Healthcare logistics optimization: Quantum computing can optimize healthcare logistics, resource allocation, and supply chain management, improving operational efficiency and reducing costs.
- Faster and accurate diagnostics: Quantum computing's computational power enables quicker and more accurate diagnostics, leading to timely interventions and better patient outcomes.
- Advancements in medical research: Quantum algorithms can accelerate medical research, leading to breakthroughs in understanding diseases, genomics, and biological processes.
- Collaborative healthcare research: QIoT fosters collaborations between quantum experts and healthcare professionals, driving interdisciplinary research to address complex healthcare challenges.
- Precision healthcare analytics: Quantum-inspired machine learning algorithms can process large datasets, identifying patterns and correlations for more accurate predictive analytics and insights.
- Drug target identification: Quantum algorithms can aid in identifying potential drug targets and understanding the interactions between drugs and biological targets.

- Secure medical data sharing: Quantum communication protocols enable secure and instantaneous data sharing among healthcare providers, supporting collaborative research and diagnostics.

12.6.3 Opportunities for Further Research and Development

Opportunities for further research and development in Quantum IoT (QIoT) for smart healthcare are abundant, offering promising avenues for advancing medical technologies and patient care.

1. **Quantum algorithms for healthcare optimization:** Develop and refine quantum algorithms tailored to specific healthcare optimization tasks, such as resource allocation, supply chain management, and patient scheduling. Optimizing healthcare operations using quantum computing can lead to improved efficiency and cost-effectiveness.
2. **Quantum-enhanced medical imaging:** Continue research into quantum-enhanced imaging technologies to improve resolution, sensitivity, and contrast in medical imaging devices. This can enhance early disease detection and provide more detailed insights into physiological structures.
3. **Quantum-inspired machine learning in healthcare analytics:** Explore the potential of quantum-inspired machine learning algorithms to handle large and complex healthcare datasets. Utilize quantum machine learning for predictive analytics, patient risk stratification, and treatment recommendations.
4. **Quantum encryption and communication protocols:** Further develop quantum communication protocols, like quantum key distribution, to enhance data security and privacy in telemedicine, remote healthcare, and medical data exchange.
5. **Quantum sensors for wearable devices:** Investigate the integration of quantum sensors into wearable health monitoring devices to enable more accurate and continuous health data monitoring. Quantum-enabled wearables can offer precise measurements of vital signs and biomarkers (Kim et al. 2017).
6. **Quantum-enabled drug discovery:** Continue exploring quantum algorithms and simulations for drug discovery to accelerate the identification of potential drug candidates and optimize treatment efficacy.
7. **Quantum computing hardware advancements:** Invest in research to improve the stability, scalability, and error correction capabilities of quantum computing hardware. Advancements in quantum processors will empower more complex and computationally intensive healthcare applications (De Leon et al. 2021).
8. **Real-world deployments and case studies:** Conduct more real-world deployments and case studies to demonstrate the practical benefits and impact of QIoT in smart healthcare. Gathering empirical evidence will drive wider adoption and showcase the transformative potential of quantum technologies in healthcare.

9. **Interdisciplinary research collaborations:** Foster collaborations between quantum computing experts, medical researchers, healthcare providers, and data scientists. Interdisciplinary research can lead to innovative applications and address healthcare challenges from multiple perspectives.
10. **Education and training:** Invest in educational programs to train a quantum-ready workforce in healthcare. Offer training to healthcare professionals, researchers, and IT experts to bridge the gap between quantum technologies and healthcare applications.

By exploring these research opportunities and investing in the development of quantum technologies for smart healthcare, we can unlock groundbreaking advancements that have the potential to revolutionize patient care, improve medical outcomes, and shape the future of the healthcare industry.

12.7 Conclusion

Quantum Internet of things is a convergence of quantum technologies with the Internet of things in the healthcare industry, offering numerous advantages and applications. It enhances data security through quantum encryption, enables faster and more accurate diagnostics, supports personalized medicine, and accelerates drug discovery. QIoT can be applied to medical imaging with quantum-enhanced sensors, secure data transmission through quantum communication protocols, remote patient monitoring using wearable devices with quantum sensors, and more. However, challenges include quantum hardware scalability, interoperability, regulatory considerations, and the need for a quantum-ready workforce. Despite these challenges, the future of QIoT in healthcare looks promising, with opportunities for research in quantum algorithms, machine learning, imaging, and sensors. As QIoT matures, it has the potential to revolutionize healthcare, improving patient outcomes and streamlining healthcare operations through innovative and secure technologies. The future of QIoT in smart healthcare is incredibly promising and transformative. As quantum technologies continue to advance, the integration of quantum computing, quantum communication, and quantum sensing with the Internet of things has the potential to revolutionize the healthcare industry.

References

Al-Saggaf, Alawi A., et al. 2023. Lightweight two-factor-based user authentication protocol for IoT-enabled healthcare ecosystem in quantum computing. *Arabian Journal for Science and Engineering* 48 (2): 2347–2357.

Alshehri, Fatima, and Ghulam Muhammad. 2020. A comprehensive survey of the internet of things (IoT) and AI-based smart healthcare. *IEEE Access* 9: 3660–3678.

Bergström, Fredrik, and Bo Lindmark. 2019. Accelerated drug discovery by rapid candidate drug identification. *Drug Discovery Today* 24 (6): 1237–1241.

Blunt, Nick S., et al. 2022. Perspective on the current state-of-the-art of quantum computing for drug discovery applications. *Journal of Chemical Theory and Computation* 18 (12): 7001–7023.

Camphausen, Robin, et al. 2023. Fast quantum-enhanced imaging with visible wavelength entangled photons. *Optics Express* 31 (4): 6039–6050.

Cheng, Chi, et al. 2017. Securing the Internet of Things in a quantum world. *IEEE Communications Magazine* 55 (2): 116–120.

De Leon, Nathalie P., et al. 2021. Materials challenges and opportunities for quantum computing hardware. *Science* 372 (6539): eabb2823.

Donta, Praveen Kumar, et al. 2023. Exploring the potential of distributed computing continuum systems. *Computers* 12: 198.

Elhoseny, Mohamed, et al. 2018. Secure medical data transmission model for IoT-based healthcare systems. *IEEE Access* 6: 20596–20608.

Engelhardt, Mark A., 2017. Hitching healthcare to the chain: An introduction to blockchain technology in the healthcare sector. *Technology Innovation Management Review* 7 (10): 22–34.

Fong, Luis E., et al. 2004. High-resolution imaging of cardiac biomagnetic fields using a low-transition-temperature superconducting quantum interference device microscope. *Applied Physics Letters* 84 (16): 3190–3192.

Gardašsević, Gordana, et al. 2020. Emerging wireless sensor networks and Internet of Things technologies-foundations of smart healthcare. *Sensors* 20 (13): 3619.

Gisin, Nicolas, and Rob Thew. 2007. Quantum communication. *Nature Photonics* 1 (3): 165–171.

Horodecki, Ryszard, et al. 2009. Quantum entanglement. *Reviews of Modern Physics* 81 (2): 865.

Hylton, Nola. 2006. Dynamic contrast-enhanced magnetic resonance imaging as an imaging biomarker. *Journal of Clinical Oncology* 24 (20): 3293–3298.

Janani, T., and M. Brindha. 2021. A secure medical image transmission scheme aided by quantum representation. *Journal of Information Security and Applications* 59: 102832.

Kim, Jaemin, et al. 2017. Ultrathin quantum dot display integrated with wearable electronics. *Advanced Materials* 29 (38): 1700217.

Li, Yan, et al. 2012. In vivo cancer targeting and imaging-guided surgery with near infrared-emitting quantum dot bioconjugates. *Theranostics* 2 (8): 769.

Li, Shancang, et al. 2015. The internet of things: A survey. *Information Systems Frontiers* 17: 243–259.

Mondal, Somrita, et al. 2012. Physico-chemical aspects of quantum dot–vasodialator interaction: Implications in nanodiagnostics. *The Journal of Physical Chemistry C* 116 (17): 9774–9782.

Montanaro, Ashley. 2016. Quantum algorithms: an overview. *npj Quantum Information* 2 (1): 1–8.

Mustafa, Hasan, et al. 2022. Variational Quantum Algorithms for Chemical Simulation and Drug Discovery. In *2022 International Conference on Trends in Quantum Computing and Emerging Business Technologies (TQCEBT)*, 1–8. Piscataway: IEEE.

Ortolano, Giuseppe, et al. 2019. Quantum enhanced imaging of nonuniform refractive profiles. *International Journal of Quantum Information* 17 (08): 1941010.

Qu, Zhiguo, Zhexi Zhang, et al. 2022. A quantum blockchain-enabled framework for secure private electronic medical records in Internet of Medical Things. *Information Sciences* 612: 942–958.

Qu, Zhiguo, Yunyi Meng, et al. 2023. QB-IMD: A secure medical data processing system with privacy protection based on quantum blockchain for IoMT. *IEEE Internet of Things Journal*.

Rejeb, Abderahman, et al. 2023. The Internet of Things (IoT) in healthcare: Taking stock and moving forward. *Internet of Things* 22: 100721.

Scarani, Valerio, et al. 2009. The security of practical quantum key distribution. *Reviews of Modern Physics* 81 (3): 1301.

Selvarajan, Shitharth, and Haralambos Mouratidis. 2023. A quantum trust and consultative transaction-based blockchain cybersecurity model for healthcare systems. *Scientific Reports* 13 (1): 7107.

Steane, Andrew. 1998. Quantum computing. *Reports on Progress in Physics* 61 (2): 117.
Suhail, Sabah, et al. 2020. On the role of hash-based signatures in quantum-safe internet of things: Current solutions and future directions. *IEEE Internet of Things Journal* 8 (1): 1–17.
Tian, Tao, et al. 2022. Ascorbate oxidase enabling glucometer readout for portable detection of hydrogen peroxide. *Enzyme and Microbial Technology* 160: 110096.
Ur Rasool, Raihan, et al. 2023. Quantum computing for healthcare: A review. *Future Internet* 15 (3): 94.
Wallden, Petros, and Elham Kashefi. 2019. Cyber security in the quantum era. *Communications of the ACM* 62 (4): 120–120.
Wang, Chonggang, and Akbar Rahman. 2022. Quantum-enabled 6G wireless networks: Opportunities and challenges. *IEEE Wireless Communications* 29 (1): 58–69.
Younan, Mina, et al. 2021. Quantum Chain of Things (QCoT): A New Paradigm for Integrating Quantum Computing, Blockchain, and Internet of Things. In *2021 17th International Computer Engineering Conference (ICENCO)*, 101–106. Piscataway: IEEE.
Zhao, Zhenwei, et al. 2023. Secure Internet of Things (IoT) using a novel brooks Iyengar quantum byzantine agreement-centered blockchain networking (BIQBABCN) model in smart healthcare. *Information Sciences* 629: 440–455.
Zhu, Qingyi, et al. 2019. Applications of distributed ledger technologies to the internet of things: A survey. *ACM Computing Surveys (CSUR)* 52 (6): 1–34.

Chapter 13
Enhancing Security in Intelligent Transport Systems: A Blockchain-Based Approach for IoT Data Management

Chinmaya Kumar Dehury and Iwada Eja

13.1 Introduction

Intelligent transportation systems (ITS) have witnessed significant technological innovations recently. Several emerging technologies, such as blockchain (Dehury et al., 2022a), edge computing (Dehury et al., 2022b), and fog computing, can potentially revolutionize how ITS operates (Lin et al., 2017). These technologies offer unique benefits and can be combined to create new solutions and previously impossible applications. ITS is an intelligent system that utilizes advanced technologies like sensors, cameras, and communication networks to collect and process real-time data about transportation infrastructure and vehicles (Qi, 2008; Ravi et al., 2023). This data is then used to optimize traffic flow, enhance safety, reduce congestion, and improve overall transportation efficiency.

Blockchain is a decentralized and tamper-proof digital ledger that allows secure and transparent transactions without intermediaries (Luo et al., 2020). When integrated into ITS, blockchain can provide a robust and secure platform for recording transportation-related transactions and data, ensuring the integrity and privacy of data, which is crucial in a sensitive and interconnected transportation ecosystem (Cocîrlea et al., 2020). Edge computing is another vital technology for ITS, enabling real-time data processing and analysis closer to the source, i.e., the edge devices, such as traffic cameras, sensors, and smart traffic lights. Edge computing reduces latency, enhances responsiveness, and minimizes the amount of data that needs to be transmitted to centralized servers or the cloud (Shi et al., 2016). On the other

C. K. Dehury (✉)
Institute of Computer Science, University of Tartu, Tartu, Estonia
e-mail: chinmaya.dehury@ut.ee

I. Eja
Cloud Platform Team, Finnair, Estonia
e-mail: iwada.eja@finnair.com

P. K. Donta et al. (eds.), *Learning Techniques for the Internet of Things*,
https://doi.org/10.1007/978-3-031-50514-0_13

hand, fog computing builds on edge computing by creating a decentralized network of computing resources that extends the edge capabilities and brings additional computational power closer to the edge devices (Shi et al., 2016; Srirama, 2023). A smart city is a larger concept in which ITS plays a crucial role. It involves integrating various technologies, including ITS, to enhance the quality of life for citizens by improving transportation, energy efficiency, public safety, and overall urban management (Angelidou, 2014).

Security is a critical aspect of intelligent transport systems, considering the sensitive nature of transportation data (Harvey and Kumar, 2020). Securing communication channels, protecting data privacy, and ensuring the trustworthiness of connected devices are paramount. Challenges in ITS implementation include scalability, data management, interoperability, and trust in a distributed and unsecured edge environment (Sumalee and Ho, 2018). The increasing volume of data from various edge devices requires efficient storage, processing, and management techniques (Das et al., 2023).

Integrating blockchain technology with edge and cloud computing can address some of these challenges (Luo et al., 2020). Blockchain's decentralized and tamper-proof nature enhances security and trust in ITS systems. It provides a reliable ledger for recording transactions and storing critical transportation data securely. However, the scalability of blockchain networks remains a concern due to slower transaction speeds and higher costs compared to traditional centralized systems (Luo et al., 2020). Storing all the extensive data from edge devices directly on the blockchain network might need to be more efficient (Das et al., 2023). Thus, a hybrid approach like E2C-Block (edge-to-cloud-blockchain) architecture is proposed.

The E2C-Block architecture efficiently collects, securely stores, and processes IoT data from various ITS sensors. It combines blockchain's security and immutability features with offshore data storage for optimized data management. Edge and cloud computing are leveraged to handle data heterogeneity, trust, and efficient data processing within the ITS infrastructure (Table 13.1).

13.1.1 Problem

Integrating blockchain, fog computing, and edge computing in intelligent transportation systems (ITS) poses several challenges that require careful consideration to unlock their potential benefits fully. Scalability emerges as a significant obstacle (Das et al., 2023), leading to slower transaction speeds and increased costs compared to traditional centralized systems. Additionally, the burgeoning volume of data edge devices generates calls for innovative solutions for efficient processing and analysis (Das et al., 2023). The diverse nature of edge devices also challenges interoperability and data standardization within the ITS ecosystem (Liu et al., 2023).

Security assumes paramount importance, particularly in edge computing, where devices often operate in unsecured environments, raising concerns about safeguarding sensitive transportation data (Liu et al., 2023). Moreover, trust becomes a vital

Table 13.1 List of acronyms

Acronyms	Description
ITS	Intelligent transportation system
HLF	Hyperledger fabric
CBN	Cloud blockchain network
FBN	Fog blockchain network
E2C-Block	Blockchain for edge to cloud continuum
API	Application programming interface
SDK	Software development kit
TPS	Transaction per second
IoT	Internet of things
JSON	JavaScript object notation
XML	Extensible Markup Language
HTTPS	Hypertext transfer protocol secure
RDBMS	Relational database management system
AWS	Amazon web services
CA	Certificate authority

concern in edge computing environments, necessitating the establishment of trust between devices, nodes, and systems. Efficiently managing the vast data generated by edge devices further becomes a critical challenge, prompting the exploration of optimized storage, processing, and management techniques. This book chapter seeks to address the following research questions:

1. How can integrating blockchain, fog computing, and edge computing overcome scalability challenges and reduce storage costs in ITS?
2. How can blockchain technology be harnessed to ensure data integrity and immutability in fog and edge computing environments within the context of ITS?

13.1.2 Motivation

The motivation behind this research lies in the tremendous potential of intelligent transportation systems (ITS) to revolutionize urban mobility and create smarter, more efficient cities. As cities grow and face mounting transportation challenges, ITS offers a promising solution to enhance traffic management, reduce congestion, and improve overall transportation efficiency.

Integrating cutting-edge technologies such as blockchain, fog computing, and edge computing in ITS is key to unlocking new possibilities. However, to fully exploit their benefits, several challenges must be addressed. Scalability remains a crucial concern, as the decentralized nature of blockchain networks can hinder real-time data processing and lead to increased costs. We aim to pave the way for more efficient and cost-effective ITS implementations by investigating how these technologies can collaborate to overcome scalability challenges.

Data integrity and immutability are paramount, especially in fog and edge computing environments, where sensitive transportation data is vulnerable to security breaches. By exploring how blockchain technology can guarantee the trustworthiness of data in such environments, we strive to instill confidence in the reliability of ITS systems.

Solving these challenges will propel ITS toward creating safer, smarter, and more sustainable transportation ecosystems. By addressing these research questions and finding innovative solutions, we aspire to contribute to advancing smart cities, fostering a seamless and interconnected urban mobility experience for citizens and visitors alike.

13.1.3 Outline

The rest of this book chapter is organized as follows:

- Sect. 13.1 introduces fundamental concepts, outlines research questions, and presents our motivation for this work.
- Sect. 13.2 explores intelligent transportation systems, edge, fog, cloud computing, and blockchain technologies, including a comparison of Corda and Hyperledger Fabric (HLF) and the reasons for choosing HLF as our reference implementation.
- Sect. 13.3 delves into the design of our proposed architecture, discussing the options considered and providing the rationale for final choices.
- Sect. 13.4 offers insights into the implementation, covering technical aspects like programming languages, frameworks, and tools.
- Sect. 13.5 presents the experiments' outcomes on the deployed proposed architecture.
- Sect. 13.6 concludes by summarizing the findings from our studies.

13.2 Background

This section introduces concepts crucial for applying blockchain to intelligent transportation systems (ITS), including ITS, edge, fog, and cloud computing, and the benefits of blockchain's decentralized and secure nature.

13.2.1 Intelligent Transport System (ITS)

Intelligent transportation systems (ITS) is an interdisciplinary field within transportation engineering and information technology. It encompasses a range of advanced technologies and communication infrastructure to optimize and manage transportation systems to improve safety, efficiency, and sustainability (Sumalee and Ho, 2018; Lin et al., 2017). ITS leverages cutting-edge technologies such as sensors, cameras, communication networks, data analytics, artificial intelligence, and machine learning to collect, process, and disseminate real-time information about various aspects of transportation (Sumalee and Ho, 2018).

The core objective of ITS is to address the complexities and challenges faced by modern transportation systems, including traffic congestion, safety concerns, environmental impact, and resource inefficiencies (Lin et al., 2017). By integrating intelligent solutions, ITS aims to enhance mobility, reduce travel time, mitigate accidents, optimize traffic flow, and minimize environmental footprints. Moreover, it endeavors to improve the overall transportation experience for individuals, communities, and businesses.

One crucial aspect of ITS is its ability to gather and analyze vast amounts of data from different sources, such as vehicles, infrastructure, and weather conditions (Qi, 2008). ITS can extract meaningful insights from this data through sophisticated data analytics and machine learning algorithms to facilitate informed decision-making, proactive management, and predictive capabilities (Cocîrlea et al., 2020). Real-time traffic information, dynamic route guidance, and adaptive traffic control are examples of ITS applications enabled by data-driven decision support.

ITS is also vital in promoting sustainable transportation practices and reducing environmental impacts. By optimizing transportation operations, encouraging public transit usage, promoting ride-sharing and carpooling, and facilitating electric vehicle adoption, ITS contributes to reducing greenhouse gas emissions and overall carbon footprint (Shit, 2020).

Academic research in ITS covers various topics, including communication protocols, data integration, traffic flow modeling (Meena et al., 2020), optimization algorithms (Kaffash et al., 2021), human-machine interaction (Wang et al., 2021), intelligent control systems, and policy analysis (Lv and Shang, 2023). Scholars in transportation engineering, computer science, electrical engineering, and urban planning collaborate to develop novel solutions, conduct simulations, and evaluate the performance of ITS applications under real-world conditions.

Despite its significant potential benefits, implementing ITS entails system complexity, data privacy, cybersecurity, standardization, and cost-effectiveness challenges. Researchers and practitioners continually strive to address these challenges through rigorous investigation, interdisciplinary collaborations, and innovative approaches to ensure the successful deployment and integration of ITS solutions (Fig. 13.1).

13.2.2 Edge, Fog, and Cloud Computing

Edge computing and fog computing hold significant potential to revolutionize data processing and decision-making in real-time transportation applications. Edge

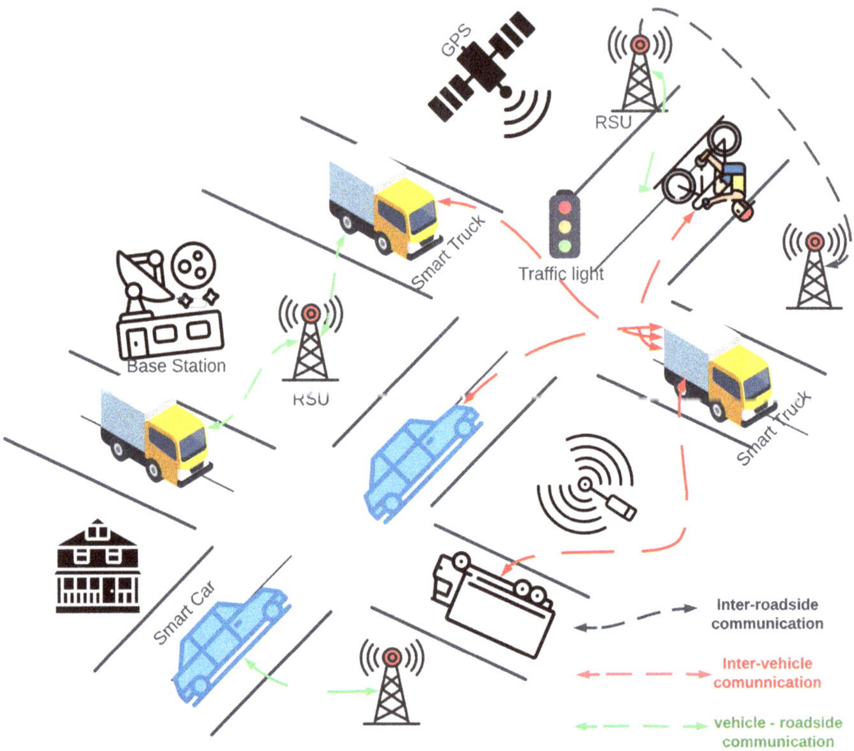

Fig. 13.1 Representation of components in a typical intelligent transportation system

computing, a distributed computing paradigm, brings computing resources closer to the point of demand, such as vehicles and roadside infrastructure, minimizing latency and speeding up response times for critical ITS operations (Fazeldehkordi and Grønli, 2022; Praveen Kumar et al., 2023). By deploying computing capabilities at the network's edge, edge computing enables efficient processing, analysis, and response to data, making it crucial for applications like autonomous vehicles and automated industrial systems (Luo et al., 2020).

Similarly, fog computing, which extends cloud computing to the network's edge, allocates computing, storage, and networking resources between cloud data centers and edge-connected hardware, such as IoT devices and sensors. This approach optimizes performance by processing data closer to the data source, reducing latency, preserving network capacity, and managing significant volumes of data generated by IoT devices (Bonomi et al., 2012).

Nonetheless, integrating edge and fog computing in ITS presents challenges, including ensuring data security and privacy in a distributed system and managing and sustaining distributed computing resources. Careful design and implementation are essential to achieve optimal performance and efficiency in this context. Though

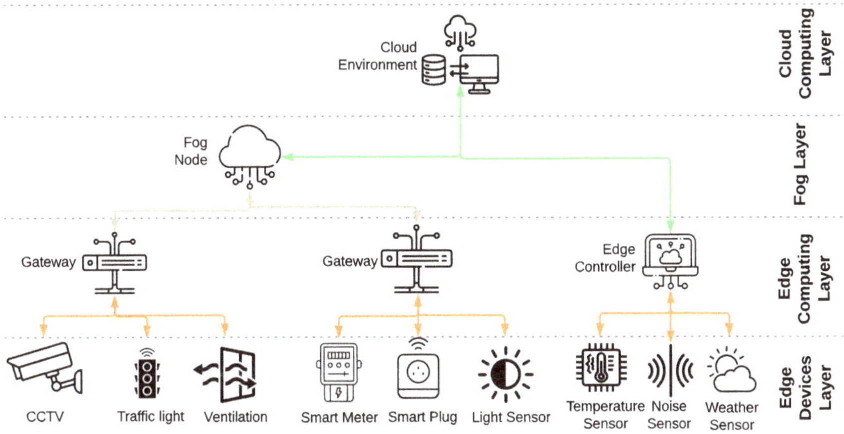

Fig. 13.2 Representation of edge, fog, and cloud computing environments in a hierarchical manner

similar in bringing computing power closer to the data source, edge computing and fog computing serve distinct use cases. While edge computing is suitable for local data processing without cloud services, fog computing is more appropriate when a hybrid cloud and edge architecture are necessary, and greater processing power is required for data processing (Krishnaraj et al., 2022).

Cloud computing, a popular computing model delivering resources over the Internet, including servers, storage, and software, has become integral to modern IT infrastructure due to its scalability, accessibility, and cost savings (Armbrust et al., 2010). It allows organizations to scale computing resources based on demand without incurring additional costs associated with owning and managing hardware, enabling quick responses to changing business requirements. For ITS, fog computing is a complementary technology to cloud computing, providing additional resources and processing power for applications requiring real-time processing of large volumes of transportation data (Lin et al., 2017). By reducing data transmission to cloud data centers, fog computing enhances data privacy, security, and application performance and effectively addresses the low-latency requirements of IoT-based ITS applications (Bonomi et al., 2012). The integration of edge, fog, and cloud computing environments in the ITS domain represents a hierarchical relationship, with cloud computing at the top, fog computing in the middle, and edge computing at the bottom, interconnecting to form a comprehensive computing infrastructure to cater to various ITS use cases and applications (Fig. 13.2). This interconnected ecosystem can revolutionize transportation operations, making ITS more efficient, secure, and responsive to real-time demands.

13.2.3 Blockchain

The popularity of blockchain technology has grown significantly in recent years due to its decentralized approach to securing and verifying data. A blockchain is a distributed database where multiple parties can securely and permanently record transactions. It was first introduced in 2008 through the white paper "Bitcoin: A Peer-to-Peer Electronic Cash System" by Cocîrlea et al. (2020).

A significant advantage of blockchain technology is its decentralized nature, meaning that any single entity or organization does not control it. Instead, it relies on a network of computers (nodes) to validate and record transactions, ensuring higher security and transparency. The data on the chain is difficult to alter or tamper with, providing trust and transparency to digital transactions (Zheng et al., 2017).

One promising area where blockchain technology can bring revolutionary changes is the Internet of things (IoT). As IoT devices become more prevalent (Zheng et al., 2017; Cocîrlea et al., 2020), blockchain can play a vital role in securing the data generated by these devices. Using blockchain to store and verify IoT data can ensure authenticity, integrity, and protection against unauthorized access or manipulation. Industries such as intelligent transport systems, healthcare, smart grid (Hatamian et al., 2023), and finance, which heavily rely on data security and privacy, stand to benefit significantly from this application.

The intelligent transport system (ITS) is a sector that can significantly benefit from blockchain technology. The need for secure and tamper-proof data becomes crucial with the increasing use of IoT devices in transportation, such as connected vehicles and traffic sensors. Blockchain can play a vital role in securing and managing the data generated by these devices. Authenticity, integrity, and protection against unauthorized access or manipulation can be ensured by utilizing blockchain to store and verify IoT data in the ITS (Cocîrlea et al., 2020). Figure 13.3 illustrates the standard data flow, starting from its reception until its storage on a blockchain network.

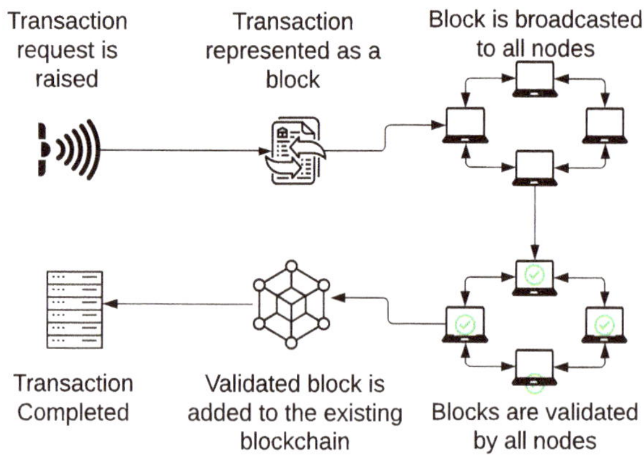

Fig. 13.3 Flow on how data is stored in Blockchain

There are three types of blockchains: private, public, and permissioned (Zheng et al., 2017):

1. **Private blockchains** are restricted to a specific group of participants, controlled by a single entity or organization, and used for securely sharing data among trusted parties. An example is Corda, developed by R3, designed for a consortium of banks.
2. **Public blockchains** are open to anyone for participation, access, validation, and recording of transactions. Bitcoin and Ethereum are well-known examples of public blockchains where anyone can join, validate transactions, and access data.
3. **Permissioned blockchains** sits between private and public blockchains, allowing anyone to join the network, but access to data is restricted to approved participants. It offers a compromise between data security and openness for collaboration among trusted parties.

13.2.4 Hyperledger Fabric (HLF)

Hyperledger Fabric (HLF), an open-source blockchain platform developed by the Linux Foundation, is designed to support the creation and deployment of distributed ledger applications (Ucbas et al., 2023). Its modular, scalable, and secure architecture makes it well suited for various use cases, such as supply chain management, financial services, and healthcare (Honar Pajooh et al., 2021). Demonstrating impressive performance, HLF achieves an end-to-end throughput of over 2980 transactions per second and scales effectively with more than 100 peers (Ucbas et al., 2023).

A prominent feature of HLF is its modular architecture, empowering developers to incorporate diverse components like consensus algorithms, membership services, and data stores to create tailored blockchain solutions for different industries and applications. Another crucial aspect of HLF is its support for smart contracts, self-executing contracts with agreement terms encoded directly into lines of code (Honar Pajooh et al., 2021). These smart contracts enable the automation of processes and enforce specific conditions, facilitating, verifying, and executing contract-related tasks. HLF accommodates smart contracts developed in various programming languages deployed as chain codes on the network.

HLF offers APIs and software development kits (SDKs) in multiple programming languages, including JavaScript, Python, and Java, to facilitate seamless interaction and smart contract deployment. This accessibility streamlines application development on the platform. HLF's architecture comprises several robust and secure components, including the membership service provider, ordering service, peer nodes, and smart contracts (packaged as chain codes). This cohesive design ensures a reliable and efficient blockchain network. Several components make up a HLF network. Among them are peers and orderers. Peers hold and manage the ledgers and smart contracts while executing a consensus protocol

Table 13.2 Components of a hyperledger fabric network

Components	Description
Peer	Non-ordering node that stores and manages copies of ledgers and smart contracts (chain codes)
Orderer	Responsible for managing the ordering service, ensuring proper transaction ordering and block distribution
Channel	Private communication pathway for specific network participants, containing private ledgers and chain codes
Ledger	A record of all transactions in the network, maintained by each peer
Chaincode (smart contract)	Business logic that defines rules and actions for handling transactions on the ledger
Membership Service Provider (MSP)	Manages identities and access control for network participants
Certificate authority (CA)	Issues cryptographic certificates to authenticate network participants
Consensus mechanism	Protocol governing how transactions are agreed upon and added to the ledger
Client application	Interface for users to interact with the network, submitting transactions, and querying the ledger
Endorsing peer	Peer responsible for endorsing and validating transactions before they are committed to the ledger
Anchor peer	Peer that participates in multiple channels and acts as a communication entry point for other peers

for transaction validation and block creation. Working alongside orderers, peers ensure the network's ledgers remain consistent and up to date. Additionally, they provide flexibility and redundancy, offering APIs through the HLF Gateway Service for seamless interaction with client applications. We also have orderer nodes. These are responsible for managing the ordering service, ensuring transactions are appropriately ordered and packaged into blocks. They diligently distribute these blocks to all network participants, ensuring a tamper-proof and reliable ledger. The ordering service in HLF boasts three distinct implementations, providing modularity and a configurable consensus system tailored to specific needs (Zheng et al., 2017). Table 13.2 gives a quick overview of the various components and a brief description of these components. In Fig. 13.4, we present an example of the major components of a HLF network.

13.2.5 Corda

Corda is a versatile and scalable platform that seamlessly integrates with existing enterprise technology in the financial services industry. It operates as a permissioned ledger, asset modeling tool, and workflow routing engine, enabling solutions that

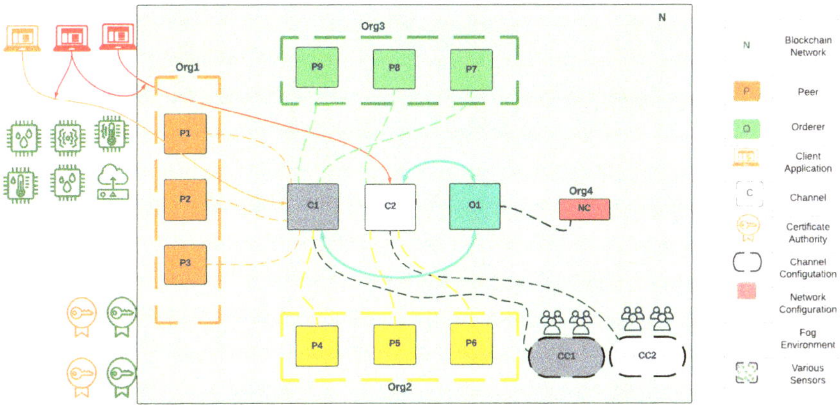

Fig. 13.4 Possible hyperledger fabric components

decentralize assets while ensuring privacy and regulatory compliance (R3, 2023). The primary objective of Corda is to empower businesses to create and manage contracts automatically executed through smart contracts. The platform's identity and privacy management features suit financial use cases well (R3, 2023).

Unlike public blockchains, Corda is not a cryptocurrency platform; instead, it serves as a tool for managing financial agreements. Notably, Corda strongly emphasizes privacy and is intended for use within specific business networks, granting businesses control over data access. Transactions are only visible to the relevant parties involved (Honar Pajooh et al., 2021), making Corda an effective solution for handling sensitive financial information. The platform also incorporates tools for managing identity, enabling businesses to create their identity and access management policies and verify and share identity data. Additionally, Corda includes features to manage legal agreements and ensure compliance with regulatory requirements.

Corda's modular architecture and privacy-oriented approach make it highly adaptable and customizable, catering to the unique needs of diverse industries and use cases. Being open source, Corda fosters a transparent and collaborative environment for building distributed ledger solutions. Figure 13.5 shows an example of the Corda blockchain architecture.

13.2.6 Hyperledger Fabric vs Corda

HLF and Corda are open-source distributed ledger technology platforms designed with distinct architectural differences and intended use cases. Corda's primary focus lies in financial services, prioritizing privacy, and control over data access (Monrat et al., 2020), whereas HLF offers a modular and flexible architecture that caters to a broader range of industries (Saraf and Sabadra, 2018).

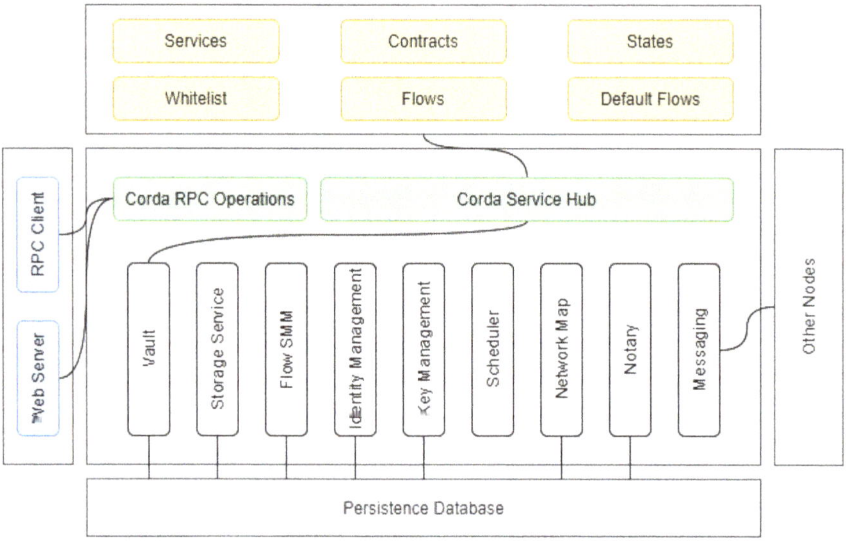

Fig. 13.5 Corda architecture (R3, 2023)

The HLF consensus process involves nodes with different roles (clients, peers, and endorsers) to ensure error-free message delivery (Honar Pajooh et al., 2021). A pluggable algorithm allows for the use of various consensus methods. On the other hand, Corda achieves consensus at the transaction level, involving only relevant parties, with notary nodes used to establish consensus over uniqueness (Honar Pajooh et al., 2021).

Regarding smart contracts, HLF implements self-executing contracts that model contractual logic in the real world. However, the legal validity of these contracts may require further clarification. In contrast, Corda allows smart contracts to include legal prose, with smart legal contracts embodying legal prose expressed and implemented within the smart contract code, granting legitimacy rooted in the associated legal prose (Honar Pajooh et al., 2021).

HLF is a versatile DLT platform suitable for diverse use cases, while Corda is tailored explicitly for financial applications such as trade finance, insurance, and capital markets. We present a comparison between Corda and Hyperledger Fabric in Table 13.3.

13.2.7 Why Hyperledger Fabric?

When considering Hyperledger Fabric and Corda as the reference implementation for our proposed architecture, Hyperledger Fabric emerges as the preferred choice for several compelling reasons. Firstly, Hyperledger Fabric supports a broader range

Table 13.3 Comparison of Corda 4.9 and HLF 2.4

Features	Corda	HLF
Type of platform	Permissioned	Permissioned
Smart contract language	Kotlin	Go, Java, TypeScript
Consensus algorithm	RAFT, BFT-SMaRt	Kafka, SBFT
Privacy	High	Moderate
Scalability	High	Moderate
Transaction throughput	200–300 tps	1000–3000 tps
Participation	Only required parties	All network nodes
Governance	R3	Linux Foundation
Hosting	Self-hosted or cloud	Self-hosted or cloud
License	Apache 2.0	Apache 2.0

of programming languages for smart contract development, including Typescript, which provides greater flexibility and familiarity to software developers working on the project. Additionally, Hyperledger Fabric boasts a more extensive set of resources and documentation, facilitating smoother implementation and reducing potential roadblocks during development.

Moreover, Hyperledger Fabric's interoperability with various open-source benchmarking tools offers an advantage in validating and optimizing the performance of the ITS project. This capability becomes crucial when dealing with the substantial data generation rates inherent in the IoT environment of ITS.

Furthermore, one of the critical criteria for an ITS project is a high transaction per second (TPS) rate to handle the dynamic nature of transportation data (Ucbas et al., 2023). Hyperledger Fabric has demonstrated superior TPS performance compared to Corda in white papers and real-world use cases, making it better suited to handle an ITS-related project's complex and demanding requirements.

Lastly, Hyperledger Fabric's flexibility in accommodating diverse use cases beyond finance aligns well with the comprehensive needs of ITS projects, providing a solid foundation for incorporating various aspects of transportation, safety, and efficiency.

13.3 E2C-Block in ITS Usecase

We propose E2C-Block (blockchain for edge to cloud continuum) as an architecture that integrates blockchain with IoT sensors in an intelligent transportation system. In the following section, we discuss this proposed architecture and elaborate on its design, specifically focusing on its application in the intelligent transport system (ITS). As mentioned, E2C-Block is an innovative model architecture aiming to effectively manage and secure data generated by IoT sensors in a distributed environment, ensuring secure transmission and tamper-proof storage.

The E2C-Block architecture comprises two distinct blockchain networks and an offshore data storage. The first blockchain network, fog blockchain network (FBN), is in a fog computing environment, close to the IoT sensors generating the data. This fog computing environment facilitates secure communication between IoT devices and the cloud infrastructure. Within the FBN, peers process and transmit the data to a second blockchain network in a cloud computing environment called the Cloud Blockchain Network (CBN). The CBN is the larger of the two blockchain networks and is responsible for receiving IoT sensor data from the FBN, hashing and storing the hash of this data on the ledger of its peers, and subsequently forwarding the original IoT sensor data to an offshore data storage. This hashing process becomes crucial for ensuring the integrity of the IoT sensor data, as discussed later.

As data generated by IoT sensors and other participants of an intelligent transport system is enormous, the E2C-Block architecture leverages an offsite data store to handle this enormous data volume. This offshore data storage provides an optimal storage solution, ensuring secure and scalable data storage while preserving data privacy and security. The repository solely receives IoT sensor data from the CBN and stores the unhashed, original versions of the data.

In essence, the E2C-Block process is summarized as follows:

1. Multiple sensors and participants in an ITS System generate data sent to the FBN.
2. The FBN receives sensor data and transmits it to the CBN.
3. The CBN only accepts data from the FBN. It hashes and stores the IoT sensor data hash, forwarding the received initial sensor data to an offshore data storage.
4. The data repository is a storage location for all sensor data. It exclusively receives data from the CBN and stores it as is.

Figure 13.6 provides a high-level overview of E2C-Block, illustrating the flow of sensor data through various computing environments. The diagram shows that the sensor data starts at the FBN in the fog environment, moves to the CBN, and finally reaches the offshore data storage, both situated in a cloud environment. We

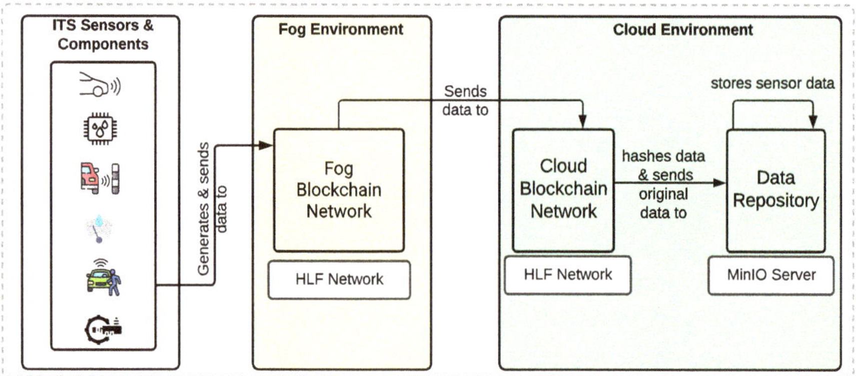

Fig. 13.6 Overview of proposed EC2-Block architecture

shall delve into more comprehensive details of the functioning of each component, including sensors and other components of ITS, the fog blockchain network, the cloud blockchain network, and the offshore data repository, within the E2C-Block architecture in the coming sections.

13.3.1 Intelligent Transport System (ITS)

The data flow in the E2C-Block system starts with IoT sensors used in intelligent transport systems. These sensors capture and transmit data to the Sensors Blockchain Network. Each sensor is assigned a unique ID linking it to its organization. E2C-Block enables administrators to add or remove sensors without compromising security.

Authenticated sensors and other components of the intelligent transport system continuously generate and transmit data to the network. Python scripts were used for data generation due to easy reconfigurability (Uniqueld, send Interval, and sensor data type). JSON format for lightweight and easy consumption by smart contracts on the FBN. These Python scripts execute from the command line, providing a constant stream of data (Code Listing 13.1 shows a sample payload).

Listing 13.1 Sample payload from sensor

```
payload = {
  temperature: 12.22,
  timestamp: '2023-04-02T12:00:00.000Z',
  org: 'Tartucitycouncil',
  device: 'ITSSensorOne',
  id: 'a1b2c3d4-e5f6-4b7c-8d9e-0123456789ab'
};
```

13.3.2 Fog Blockchain Network

The FBN, located in the fog computing environment near the IoT Sensors, is the smaller of the two blockchains in E2C-Block. Its primary role is to act as an intermediary between the IoT sensors and the CBN, facilitating the transmission of sensor data to the CBN. It exclusively communicates with the CBN and does not store data on its peers' ledger. Instead, it constantly listens for the sensor data generated by the IoT sensors. Additionally, the FBN handles authentication and registration for all sensors before data transmission. Communication between the IoT sensors and the FBN occurs through the HTTPS protocol. The FBN modifies the received sensor payload to further optimize the process by adding arrivalTime and departTimeFromFogNode attributes. These attributes benchmark the time a single sensor data point takes to move to the offshore data storage. The decision to use a blockchain network, rather than a single or server cluster, was driven by its inherent benefits, such as ensuring agreement among network peers before authenticating or registering sensors. This mitigates the risk of rogue and compromised proxy servers allowing unauthorized sensor data transmission. Separating the two blockchain networks allows the FBN to be strategically placed in the fog computing environment, closer to the IoT sensor devices. Besides its authentication and registration functions, the FBN initiates host-level IP blocking for sensors that repeatedly fail authentication. This action prevents potential security threats and unnecessary overhead for the CBN.

13.3.3 Cloud Blockchain Network

The CBN is the more significant blockchain within E2C-Block, responsible for receiving IoT sensor data from the FBN. Before storing the received sensor data payload on its peers' ledger, the CBN enhances the payload by adding two additional attributes: arrivalTimeFromFognode, indicating the time it arrived from the FBN, and departureTimeFromPrimaryBlockchain, indicating the time it left the CBN for the offshore data repository. This modified payload is then hashed using the SHA-256 algorithm, producing a fixed-size 256-bit output to ensure data integrity and authenticity.

The hashed data is transmitted to the offshore data storage while the unhashed sensor payload is retained. Figure 13.7 illustrates the data flow from the sensors to the offshore data storage, showing the authentication process, data streaming, and storage on the offshore data storage. The CBN utilizes the HLF (Hyperledger Fabric) as the reference implementation.

13.3.4 Offshore Data Store

The E2C-Block's final component is the offshore data storage, which receives IoT sensor data from the CBN and stores it unhashed. Individual buckets are created for each organization's sensors to ensure well-organized data, and this repository serves as the primary point for querying sensor data. Data authenticity can be periodically verified by querying the CBN using an HTTP request with the hashed value of the sensor's ID payload stored in the repository. By comparing the new hash with the previously stored hash on the CBN, any data tampering since storage in the offshore data storage can be detected.

For offshore data storage, we assessed three potential candidates to use as offshore data storage, MinIO,[1] Amazon S3,[2] and Ceph.[3] MinIO was chosen as the preferred option for the external data repository due to its excellent handling of large data quantities, making it a dependable and efficient choice. It offers robust capabilities for managing extensive data and can be easily deployed on-premises or in the cloud. Its straightforward yet powerful interface for object storage comes with essential features such as access control, versioning, and life cycle policies, ensuring effective data management. Additionally, being open-source, MinIO can be customized to suit specific requirements, providing a flexible solution.

MinIO is a popular open-source object storage system compatible with Amazon S3 cloud storage service. It efficiently stores unstructured data, such as images, videos, and documents, and is designed to run on commodity hardware, optimizing its performance for rapid and efficient data access.

Another option, Amazon S3, is a cloud-based storage service provided by Amazon Web Services (AWS). AWS, a well-established cloud computing platform, includes various services, with S3 being highly scalable, durable, and secure, making it reliable for handling substantial data. S3 also offers features like versioning, access control, and life cycle policies for effective data management. However, in our case, using S3 was not feasible as it would require relying on AWS.

A third alternative considered was Ceph, an open-source distributed storage system. Ceph provides highly scalable and fault-tolerant storage, making it suitable for storing large amounts of data. It offers a range of storage options, including object storage, block storage, and file storage, providing flexibility tailored to specific requirements. Nevertheless, setting up and configuring Ceph can be complex, demanding significant time and resources for deployment and maintenance.

13.4 Implementation of E2C-Block in ITS

In this fictional intelligent transport system (ITS) use case, the Tartu City Council aims to enhance transportation infrastructure by gathering and analyzing data from diverse Internet of things (IoT) sensors. To achieve this, they plan to create a collaborative sensor network with contributions from various entities, including the Tartu Transport Service, Tartu Solar Panel Center, Tartu Meteorological Centre, and Tartu Temperature Monitoring Center. This approach allows all involved parties access a comprehensive data pool for advanced analytics and improved transportation services.

Security is a significant concern due to the sensitive nature of the IoT sensor data. The council requires each organization to access only authorized data to ensure

[1] https://min.io/.

[2] https://aws.amazon.com/s3/.

[3] www.ceph.io.

privacy and confidentiality. They propose integrating blockchain technology into the ITS to establish a secure system that resists tampering and unauthorized access. Blockchain's features ensure data integrity and user privacy, instilling public trust in the ITS.

Without blockchain, the data faces tampering and unauthorized manipulation risks, leading to inaccurate analytics and privacy violations. The secure and immutable ledger provided by blockchain is crucial for the trustworthiness and success of the intelligent transport system.

13.4.1 Registration and Authentication in ITS

Securing and guaranteeing the reliability of data transmitted through E2C-Block relies on the crucial steps of authenticating and registering sensors on the FBN. To begin this process, the ITS sensors must send a POST authentication request to the designated endpoint known as TOKEN_ENDPOINT, which resides within the FBN. This request includes the necessary credentials, specifically the USERID and USER_SECRET values associated with each sensor.

Upon receiving the authentication POST request from the sensors, the TOKEN_ENDPOINT on the FBN verifies the provided credentials to ensure that the requesting sensor is authorized to transmit data over the network. The sensor is granted permission to communicate with the FBN if the credentials are valid. However, suppose the authentication fails and multiple consecutive failed attempts occur, in that case, the FBN takes precautionary action by adding the originating IP address of these failed requests to the UFW firewall Table. Subsequent authentication requests from this sensor are blocked at the OS level, preventing access to the blockchain network. This critical authentication process is facilitated by the Fabric Certificate Authority (CA) server, a vital component of the HLF blockchain framework

Another essential aspect of the authentication process is an authentication window within the FBN. This window is a customizable feature, typically set to a default value of 10 minutes. Within this timeframe, the sensor can interact with the FBN without requiring additional authentication. However, once the authentication window elapses, the sensor must undergo reauthentication to continue communication with the FBN. This serves as an extra security measure, ensuring that only authorized sensors can access and interact with the FBN, bolstering the overall security and integrity of the system.

During the initial authentication request, sensors also provide their SENSOR_ID and SENSOR_ORG values, which verify that the sensor has been added to the FBN's ledger. This step ensures that only authorized sensors can transmit data over the network.

While the sensor initiates the authentication process, the registration of the sensor is simultaneously initiated on the FBN. Sensor registration entails an authorized network administrator adding the SENSOR_ID to the FBN, and this is a one-time process. Additionally, network administrators can de-register a sensor at any point if necessary. For more in-depth information on the sensor registration process, please refer to Sect. 13.4.2.

13.4.2 Fog Blockchain Network

The FBN (Fog Blockchain Network) plays a crucial role in E2C-Block, providing added security and authenticity to data collected by IoT sensors. It is an intermediary between sensors and the CBN (Cloud Blockchain Network), authenticating and registering sensors to allow only authorized data transmission. The FBN verifies sensor data before forwarding it to the CBN, reducing network load and ensuring validated data storage.

Using a blockchain network in fog computing leverages distributed consensus, enabling multiple peers to validate each sensor and data point, reducing the risk of a single point of failure.

Sensor authentication and registration on the FBN are essential for network integrity. The HLF Certificate Authority (CA) server manages digital certificates, allowing secure communication with the network. Sensor enrollment involves sending a certificate signing request to the HLF CA server, simplifying the process using the Hyperledger Fablo Rest API.

Sensor registration is a one-time process requiring administrative privileges, ensuring data immutability and preventing unauthorized access. Authenticated and registered IoT sensors send data to the FBN, which forwards it to the CBN after verification.

Before connecting to the CBN, the FBN undergoes authentication using the HLF CA server. A 10-minute authentication window requires re-authentication. Modified payload is sent to the CBN via a smart contract's POST request.

Overall, the FBN enhances data security and reliability in E2C-Block. Figure 13.7 depicts how data flows from the ITS components through the two blockchain networks to the offshore data repository—a MinIO Storage server.

13.4.3 Cloud Blockchain Network

The CBN, the larger of the two blockchain networks in E2C-Block, consists of ten peers contributed by five participating organizations, each contributing two. It also includes a Solo Orderer and six channels, private sub-networks facilitating secure peer communication. Five of these channels have peers from the same organization, while the remaining channel includes all participating peers, enabling secure communication and data sharing across organizations. One of the primary functions of the CBN is to receive and store the hashed value of sensor data from the FBN.

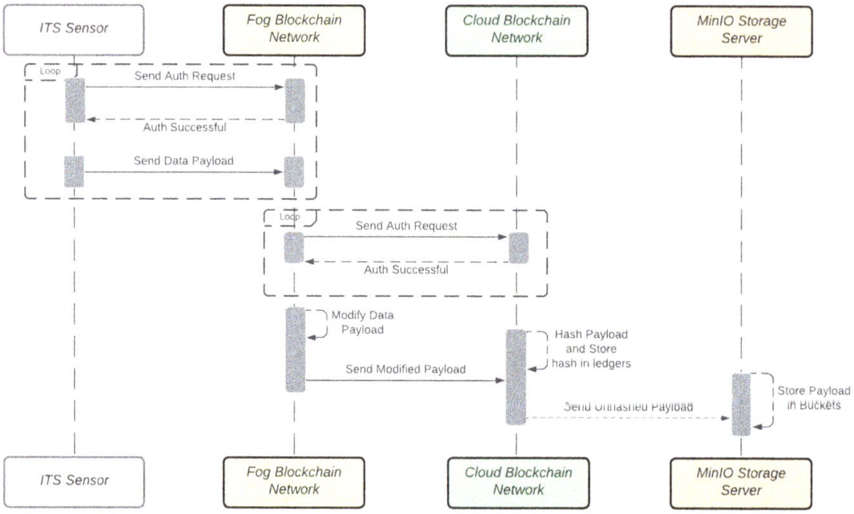

Fig. 13.7 Data flow from ITS sensors to MinIO

Upon receiving the data, a smart contract on the CBN adds two extra attributes to the payload: arrivalTimeToBlockchain and departureTimeFromPrimaryBlockchain. The payload is hashed using an SHA-256 hashing function from the Node.js built-in crypto module. This hashed data is stored in the ledgers of the CBN peers, ensuring data integrity and tamper-evidence while reducing payload size. Subsequently, the CBN forwards the data to an offshore data storage, the MinIO storage server, for offsite storage.

The communication between the CBN and the MinIO storage server is established using MinIO's Javascript SDK, ensuring reliable and secure data transmission. The data transmission process is continuous and asynchronous, transmitting data without interruption or delay, maintaining up-to-date and accurate information. This asynchronous transmission allows the CBN to continue processing transactions and other tasks while data transmission is ongoing.

The CBN also includes a smart contract that verifies the authenticity of previously stored sensor data. It hashes the received sensor payload from the MinIO server with the original hash function and compares the two hashes. If they are equal, it confirms that the data has not been tampered with since storage. This verification process is performed by the MinIO storage server, which serves as the primary query point for all data, not the CBN itself.

13.4.4 Offshore Data Repository

The offshore data storage plays a crucial role in E2C-Block as the centralized repository for all generated IoT sensor data. It receives continuous sensor data from the CBN and stores it as unhashed JSON objects in specific buckets. Each organization's sensors have a unique bucket for efficient data management and access.

The cloud blockchain network is the sole data source for offshore data storage, using its MINIO_ACCESS_KEY and MINIO_SECRET to communicate. Due to its reliability and scalability, we have chosen MinIO Storage Server as the Data Repository in E2C-Block. MinIO's ability to store data as-is, without modifications, ensures data integrity and authenticity throughout the storage process. The MinIO Storage Server operates in a cloud environment and runs on Ubuntu 22.04. All requests to read sensor data are directed to this server, making it the primary query point.

To optimize the MinIO Storage Server, we implemented several enhancements. Firstly, we increased the cache size to reduce disk I/O operations, improving response times and reducing hardware load. Secondly, we configured the server to use Direct I/O instead of Buffered I/O, reducing memory footprint and enhancing overall performance. Lastly, we enabled compression to minimize storage space requirements, which is particularly beneficial when dealing with large amounts of data. This optimization significantly reduced storage costs.

13.4.5 How Is Stored Data Queried?

We have developed an interface that allows users to browse and query the stored ITS Sensor data on the MinIO Storage server. Figure 13.8 illustrates the flow of requests for querying sensor data. The interface displays all available buckets, each corresponding to a sensor owned by different organizations. It serves as a read-only platform, preventing any modifications to the stored sensor data. When users click on a specific sensor, the interface provides detailed readings, including the last verified timestamp of the payload from the CBN, to ensure its authenticity.

Moreover, the interface includes a verification button that allows users to instantly confirm if the sensor payload matches the data on the CBN. When this button clicks, a request is sent to the CBN with the sensor data, which is hashed using the same function employed for the original data hashing. If the hashes match, it indicates that the data remains unaltered.

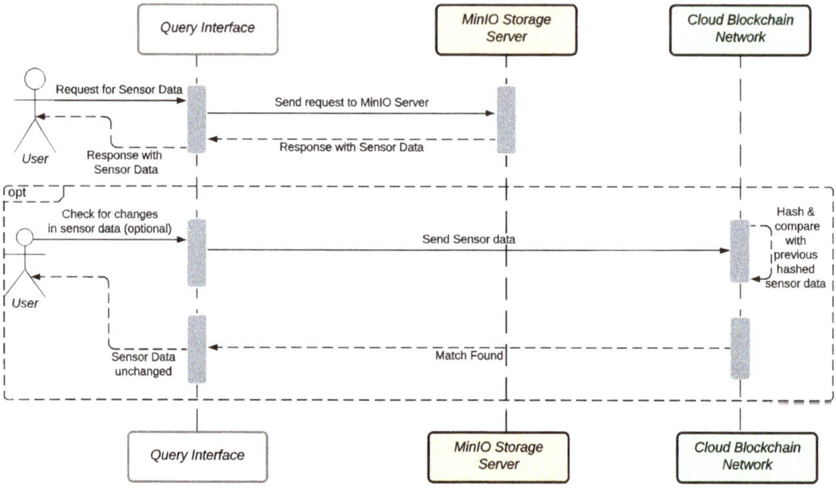

Fig. 13.8 Flow for querying ITS sensor data from MinIO storage

13.4.6 E2C-Block Deployment

We utilized Ansible,[4] a powerful automation tool, to streamline the server provisioning and software package installation setup, making large-scale configuration management more efficient.

We employed Hyperledger Fablo,[5] a specialized open-source tool for deploying blockchain networks. Hyperledger Fablo simplifies the setup, deployment, and management of blockchain networks. By defining network characteristics in JSON files, we could easily create the desired network topology, encompassing multiple organizations, channels, chain codes, and private data collections.

Fablo translated the configuration file into a functioning Hyperledger Fabric (HLF) blockchain network. All components of the Fabric network run as docker containers, allowing easy management and scalability. Fablo supports various consensus protocols like TLS, RAFT, and solo, exposing a REST API to interact with the deployed components of the HLF Network. We utilized this REST API extensively to communicate with the multiple components of the blockchain network in E2C-Block.

[4] www.ansible.com.

[5] https://github.com/hyperledger-labs/fablo.

13.5 Experiments

In this section, we provide an overview of the experiments conducted to benchmark the performance of E2C-Block. Our primary focus was on assessing the performance of the cloud blockchain network, as it constituted the larger of the two networks, and any performance impacts on it would reverberate throughout the entire architecture. To accomplish this, we utilized Caliper as a benchmarking tool to generate the required load for the blockchain system. Concurrently, we collected performance metrics using Grafana and Prometheus.

13.5.1 Experiment Setup

13.5.1.1 Benchmarking Tool

In assessing the performance of E2C-Block's CBN, we utilized Hyperledger Caliper,[6] an open-source benchmarking tool explicitly designed for measuring the efficiency and capabilities of blockchain networks.

13.5.1.2 Network Configuration

The E2C-Block's CBN has ten peers and a solo orderer. A solo orderer is a single-node consensus mechanism in blockchain networks like Hyperledger Fabric. It directly orders transactions as they are received, but its simplicity means it represents a single point of failure. It is commonly used in development or testing environments for its straightforward setup, while more robust consensus mechanisms are preferred in production environments; as such, Transport Layer Security (TLS) support was not available during the experiments. A channel containing the ten peers was created for testing purposes, and the installed chain code was tested. The chaincode's function was to receive data from Hyperledger Caliper, hash the data, store the hash on the CBN peers' ledger, and send the unhashed data to the MinIO Storage server. The batch size was set at 20 MB. These details were defined in a network.yml file.

13.5.1.3 Workload Generation

For benchmarking the CBN, we utilized the workload module of Caliper, which involved defining the smart contract in the *readAsset.js* file. This workload module facilitated the interaction with the deployed smart contract during the benchmark

[6] https://hyperledger.github.io/caliper/.

round. The module extends the Caliper class *WorkloadModuleBase* from caliper-core and includes three overrides:

1. *initializeWorkloadModule*—this function initializes any necessary elements for the benchmark.
2. *submitTransaction*—during the monitored phase of the benchmark, this function interacts with the smart contract method.
3. *cleanupWorkloadModule*—this function performs the necessary cleanup tasks after completing the benchmark.

13.5.1.4 Hardware and Software Specification

The blockchain network, simulation tool, and Caliper were hosted on virtual machines running Ubuntu 20.04 LTS. Each node was configured with eight vCPUs, 64 GB of RAM, and 50 GB of available storage. The nodes were equipped with HLF v2.4 and Caliper v0.5. These computing resources, essential for hosting E2C-Block and conducting the experiments, were generously provided by the HPC Center (University of Tartu, 2018) at the University of Tartu.[7]

13.5.2 Performance Metrics

We conducted three major experiments as follows:

1. The first experiment examined how the block size affects the overall network performance. Transactions were sent at various rates (varying between 10 and 50 transactions), and performance metrics such as throughput, transaction latency, response time, block propagation time, and consensus time were collected for each block size.
2. In the second experiment, the focus was on understanding how the transaction rate impacts the performance of the CBN. We varied the transaction rate from 100 to 3000 transactions per second.
3. The third experiment investigated the influence of the number of participating nodes in the CBN on network performance, particularly on consensus time. Consensus is the process of validating and adding new blocks to the blockchain, and it involves a certain number of nodes agreeing on the validity of a new block. The number of nodes was varied from 10 to 100, and the same performance metrics as in the previous experiments were collected.

The aim of these experiments was to measure the following performance metrics:

[7] https://ut.ee/en.

Table 13.4 Experimental parameters

Parameters	Values
Transactions sending rate	10, 20, 30, ..., 100, ..., 500 (tps)
Number of peers	5, 10, 20, 30
Block size	10, 50

1. *Transaction throughput*: This metric gauges the rate at which the blockchain network successfully commits valid transactions within a specific time frame. It provides valuable insights into the network's efficiency and capacity to process and validate transactions effectively.
2. *Transaction latency*: Transaction latency represents the time the blockchain network takes to confirm and finalize a transaction, from its submission to when it becomes accessible across the entire network. It measures the delay between initiating a transaction and its successful validation and processing.
3. *Block size*: The block size refers to the maximum number of transactions a single block can accommodate. In Hyperledger Fabric (HLF), the block size can be configured by adjusting the maximum block size setting in the network configuration file. This parameter impacts the network's overall throughput, latency, and time required for validating and propagating new blocks.
4. *Block propagation time*: Block propagation time measures how quickly a newly created block disseminates across the network and gets committed to the ledger by all participating nodes. HLF utilizes a gossip protocol for block dissemination, enabling nodes to communicate with a subset of other nodes, exchanging block information efficiently.
5. *Consensus time*: Consensus time refers to the duration it takes for a blockchain network to reach an agreement on a new block and add it to the blockchain. It reflects the time it takes for nodes in the network to collectively agree on the validity of a recent transaction and incorporate it into the blockchain.

13.5.3 Impact of Block Size

Figure 13.9 shows the average throughput across block sizes at varying transaction sending rates. Also, Fig. 13.10 displays the average latency for the same transaction rates. Throughout the experiment, transaction sending rates ranged from 10 tps to 500 tps, with multiple parameters, including transaction sending rate, block size, and the number of peers, listed in Table 13.4.

Figure 13.10 demonstrates that the average latency remained consistently below 1 second during the experiments until it approached approximately 100 tps. As the transaction sending rate increased, the system's throughput grew linearly, eventually stabilizing at around 100 tps, indicating the highest usable rate. Beyond this point, the system's performance deteriorated as the workload increased.

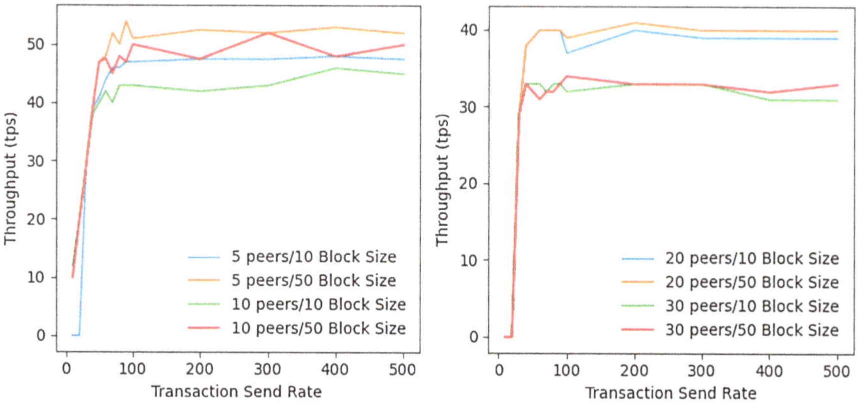

Fig. 13.9 Impact of transaction send rate on throughput under different block sizes and number of peers

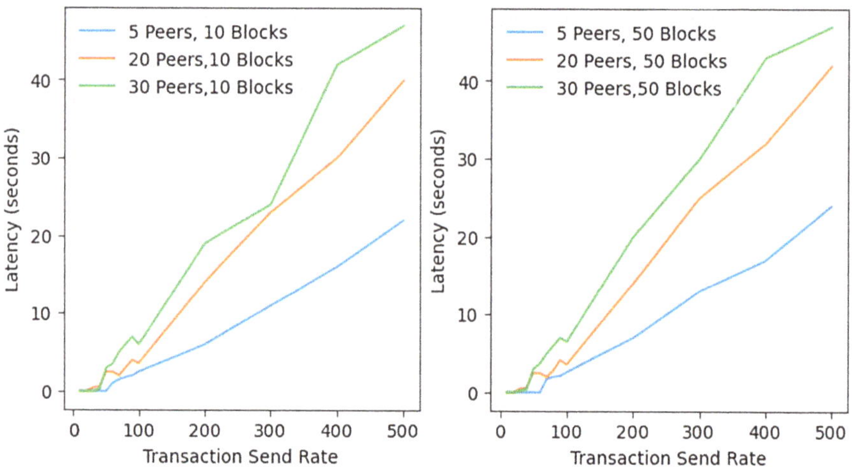

Fig. 13.10 Impact of transaction send rate on latency under different block sizes and number of peers

It is important to note that the performance of the blockchain system is heavily dependent on the hardware capabilities of the system under test and the number of peers involved, both of which can contribute to increased latency. Hyperledger Fabric (HLF) is structured on a Docker-based architecture, where each component of the Hyperledger network operates within its container and communicates through a dedicated network.

To achieve lower transaction latency in real-world applications such as IoT, it is recommended to use smaller block sizes with lower transaction rates. On the other hand, higher transaction rates necessitate larger block sizes to achieve higher throughput while maintaining lower transaction latency.

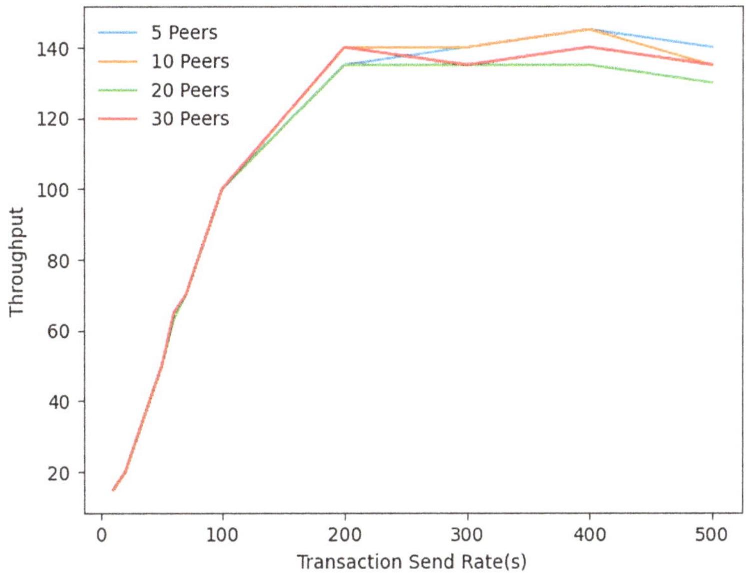

Fig. 13.11 Transaction send rate(s) vs throughput

13.5.4 Impact of Transaction Rates

Figure 13.11 depicts the relationship between transaction rates and throughput for different peer sizes. The graph shows that an increase in peers leads to a decrease in throughput. For instance, at a transaction send rate of 10, the throughput is 15 for five peers and remains unchanged (15) for 30 peers. Similarly, at a transaction send rate of 50, the throughput is 50 for five peers and again remains the same (50) for 30 peers. However, at higher transaction send rates, the difference in throughput between various peer sizes becomes more pronounced.

Another noteworthy observation is that, in some cases, throughput increases with an increase in the transaction send rate. For instance, at a transaction send rate of 300, ten peers' throughput is higher than five peers. However, at a transaction send rate of 400, the throughput for five peers becomes higher than that for ten peers. This suggests the existence of an optimal transaction send rate for a specific peer size, maximizing the network's throughput. In conclusion, the experiment suggests that simply increasing the number of peers in a network may not always result in higher throughput; instead, there may be an optimal transaction send rate that optimizes the network's throughput for a given peer size.

On the other hand, Fig. 13.12 illustrates the relationship between transaction rate and latency for various peer configurations. The figure indicates that latency remains relatively low when only a few peers exist. For example, with only five peers, the latency ranges from 0.1 to 0.5 seconds. However, as the number of peers increases,

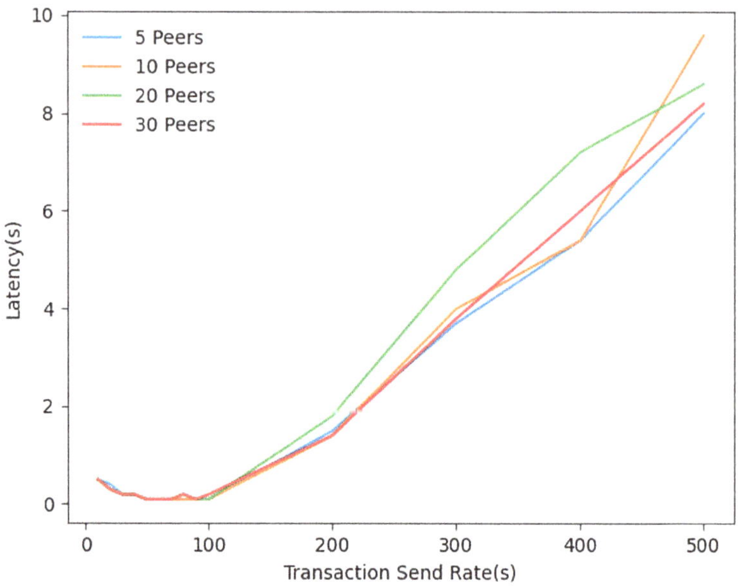

Fig. 13.12 Transaction send rate vs latency

the latency rises significantly. For instance, with 30 peers, the latency can reach 1.4 seconds for a transaction send rate of 200.

Furthermore, the latency continues to increase as the number of peers grows, reaching up to 9.6 seconds with ten peers for a transaction send rate of 500. The figure also highlights the substantial impact of the transaction send rate on latency, mainly when there are numerous peers. For example, with 30 peers, the latency increases from 1.4 to 6 seconds as the transaction send rate escalates from 200 to 400.

However, with the transaction rate increasing from 5 seconds to 30 seconds, the response time also increases from 10 seconds to 50 seconds, the block propagation time increases from 0.5 seconds to 10 seconds, and the consensus time increases from 1 minute to 10 minutes. These findings suggest that while elevating the transaction rate can improve network throughput, it may also adversely affect transaction latency, block propagation time, and consensus time.

13.5.5 Impact of Number of Participating Peers

Figures 13.13 and 13.14 present the findings regarding the blockchain system's performance about latency, throughput, block propagation time, and consensus time while varying the number of peers from 10 to 100.

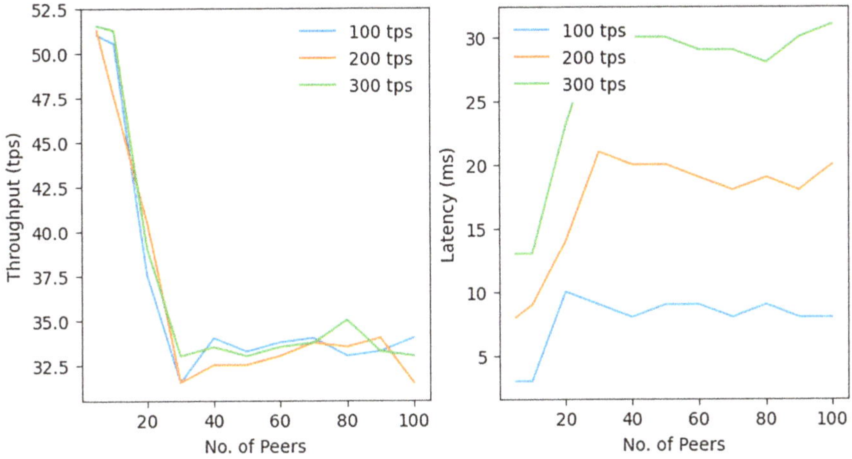

Fig. 13.13 Number of nodes vs. latency/throughput

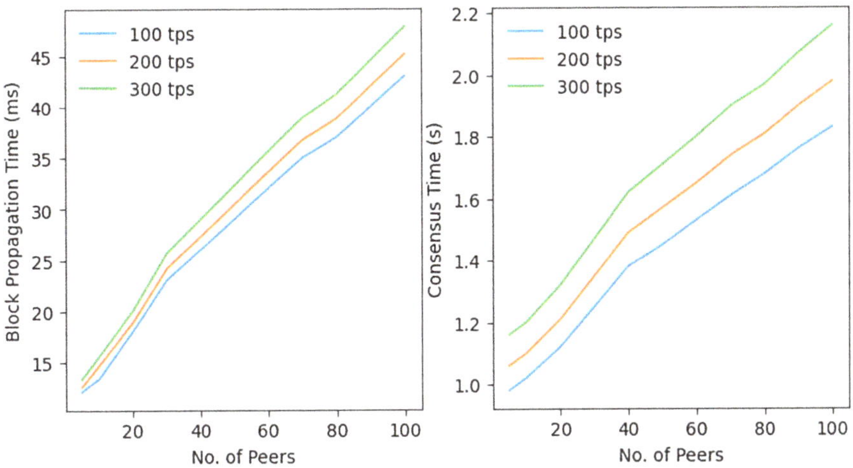

Fig. 13.14 Number of nodes vs. block propagation/consensus time

The experiments encompassed diverse transaction rates and peer configurations, leading to insightful results. At a transaction rate of 100 transactions per second (tps), the throughput declined from 51.00 tps with five peers to 34.00 tps with 100 peers (Fig. 13.13). Similarly, at 200 tps, the throughput decreased from 51.25 tps to 31.50 tps as the number of peers increased from 5 to 100. Although the system reached its peak throughput of 51.50 tps at 300 tps with five peers, this figure dropped to 33.00 tps with 100 peers.

Analyzing Fig. 13.13 reveals that latency increased with the number of peers. At 100 tps, the latency rose from 3 ms with five peers to 8 ms with 40 peers, remaining consistent at 8–9 ms for peer sizes exceeding 40. At 200 tps, the latency increased from 8 ms with five peers to 20 ms with 40 peers, maintaining a range of 18–21 ms for larger peer configurations. Similarly, at 300 tps, the latency escalated from 13 ms with five peers to 31 ms with 40 peers, showing consistent values of 28–31 ms for higher peer sizes.

Figure 13.14 demonstrates that the block propagation time also extended as the number of peers increased. For instance, at 100 tps, the block propagation time increased from 12.1 ms with five peers to 43 ms with 100 peers. Similarly, at 200 tps, the block propagation time increased from 12.6 ms with five peers to 45.15 ms with 100 peers. At 300 tps, the block propagation time increased from 13.356 ms with five peers to 47.829 ms with 100 peers.

Lastly, Fig. 13.14 showcases that the consensus time also rose with the number of peers. At 100 tps, the consensus time increased from 0.98 s with five peers to 1.83 s with 100 peers. Similarly, at 200 tps, the consensus time increased from 1.06 s with five peers to 1.98 s with 100 peers. At 300 tps, the consensus time increased from 1.16 s with five peers to 2.16 s with 100 peers.

13.6 Conclusion

This chapter addressed two critical research questions focusing on securing edge computing environments, utilizing blockchain for data integrity and immutability, and integrating fog computing and edge computing to enhance scalability and reduce storage costs. The study yielded several noteworthy findings. It was established that safeguarding edge computing environments is essential due to their susceptibility to attacks, and this vulnerability can be mitigated by employing the FBN in the fog computing environment. Hashing sensor data on the CBN and storing data on MinIO were identified as practical approaches to reduce storage costs and improve scalability.

Moreover, the experiment demonstrated that increasing the transaction rate can enhance network throughput but may adversely affect other performance metrics. Similarly, augmenting the number of peers can also negatively impact network performance. In conclusion, the study provides valuable insights into leveraging blockchain and related technologies to bolster edge computing environments' security, scalability, and performance. Overall, E2C-Block offers an effective solution for managing and securing IoT sensor data in intelligent transportation systems.

Acknowledgments We thank the HPC center at the University of Tartu for generously offering the computational resources essential for evaluating this architecture and conducting extensive experiments.

References

Angelidou, Margarita. July 2014. Smart city policies: A spatial approach. *Cities* 41: S3–S11. https://doi.org/10.1016/j.cities.2014.06.007.

Armbrust, Michael, et al. 2010. A view of cloud computing. *Communications of the ACM* 53 (4): 50–58. https://doi.org/10.1145/1721654.1721672. https://www.scopus.com/inward/record.uri?eid=2-s2.0-77950347409&doi=10.1145%2f1721654.1721672&partnerID=40&md5=6164f58679f057d5164d1f8f31d3f125.

Bonomi, Flavio, et al. 2012. Fog computing and its role in the internet of things. In *Fog Computing and Its Role in the Internet of Things*, 13–15. https://doi.org/10.1145/2342509.2342513. https://www.scopus.com/inward/record.uri?eid=2-s2.0-84866627419&doi=10.1145%2f2342509.2342513&partnerID=40&md5=3efecb1babc7eefa13ac5001795ea3b8.

Cocîrlea, Dragoş, et al. Oct. 2020. Blockchain in intelligent transportation systems. *Electronics* 9 (10): 1682. https://doi.org/10.3390/electronics9101682.

Das, Debashis, et al. 2023. Blockchain for intelligent transportation systems: applications, challenges, and opportunities. *IEEE Internet of Things Journal*, 1–1. https://doi.org/10.1109/jiot.2023.3277923.

Dehury, Chinmaya, et al. 2022a. Securing clustered edge intelligence with blockchain. *IEEE Consumer Electronics Magazine*, 1–1. https://doi.org/10.1109/MCE.2022.3164529.

Dehury, Chinmaya Kumar, et al. 2022b. CCEI-IoT: clustered and cohesive edge intelligence in Internet of Things. In *2022 IEEE International Conference on Edge Computing and Communications (EDGE)*, 33–40. https://doi.org/10.1109/EDGE55608.2022.00017.

Fazeldehkordi, Elahe, and Tor-Morten Grønli. 2022. A survey of security architectures for edge computing-based IoT. *IoT* 3 (3): 332–365. https://doi.org/10.3390/iot3030019. https://www.scopus.com/inward/record.uri?eid=2-s2.0-85145919146&doi=10.3390%2fiot3030019&partnerID=40&md5=0634d21894a30bd9b828bdce3eba9ce2.

Harvey, Julie, and Sathish Kumar. May 2020. A survey of intelligent transportation systems security: challenges and solutions. In *2020 IEEE 6th Intl Conference on Big Data Security on Cloud (BigDataSecurity), IEEE Intl Conference on High Performance and Smart Computing, (HPSC) and IEEE Intl Conference on Intelligent Data and Security (IDS)*. IEEE. https://doi.org/10.1109/bigdatasecurity-hpsc-ids49724.2020.00055.

Hatamian, Mehdi, et al. 2023. Location-aware green energy availability forecasting for multiple time frames in smart buildings: The case of Estonia. *Measurement: Sensors* 25: 100644. ISSN: 2665-9174. https://doi.org/10.1016/j.measen.2022.100644. https://www.sciencedirect.com/science/article/pii/S2665917422002781.

Honar Pajooh, Houshyar, et al. Jan. 2021. Hyperledger fabric blockchain for securing the edge Internet of Things. *Sensors* 21 (2): 359. https://doi.org/10.3390/s21020359.

Kaffash, Sepideh, et al. Jan. 2021. Big data algorithms and applications in intelligent transportation system: A review and bibliometric analysis. *International Journal of Production Economics* 231: 107868. https://doi.org/10.1016/j.ijpe.2020.107868.

Krishnaraj, N, et al. 2022. EDGE/FOG computing paradigm: Concept, platforms and toolchains. *Advances in Computers* 127: 413–436. https://doi.org/10.1016/bs.adcom.2022.02.012. https://www.scopus.com/inward/record.uri?eid=2-s2.0-85128201315&doi=10.1016%2fbs.adcom.2022.02.012&partnerID=40&md5=63e53e5d93b2c5f11c2ff3612ea9fd9f.

Lin, Yangxin, et al. May 2017. Intelligent transportation system (ITS): concept, challenge, and opportunity. In *2017 IEEE 3rd International Conference on Big Data Security on Cloud (BigDataSecurity), IEEE International Conference on High Performance and Smart Computing (HPSC), and IEEE International Conference on Intelligent Data and Security (IDS)*. https://doi.org/10.1109/bigdatasecurity.2017.50.

Liu, Peng, et al. 2023. Consortium blockchain-based security and efficient resource trading in V2V-assisted intelligent transport systems. *IEEE Transactions on Intelligent Transportation Systems*, 1–12. https://doi.org/10.1109/tits.2023.3285418.

Luo, Chuanwen, et al. 2020. Edge computing integrated with blockchain technologies. In *Lecture Notes in Computer Science (including subseries Lecture Notes in Artificial Intelligence and Lecture Notes in Bioinformatics)* 12000 LNCS, 268–288. https://doi.org/10.1007/978-3-030-41672-0_17.
Lv, Zhihan, and Wenlong Shang. Jan. 2023. Impacts of intelligent transportation systems on energy conservation and emission reduction of transport systems: A comprehensive review. *Green Technologies and Sustainability* 1 (1): 100002. https://doi.org/10.1016/j.grets.2022.100002.
Meena, Gaurav, et al. Feb. 2020. Traffic prediction for intelligent transportation system using machine learning. In *2020 3rd International Conference on Emerging Technologies in Computer Engineering: Machine Learning and Internet of Things (ICETCE)*. IEEE. https://doi.org/10.1109/icetce48199.2020.9091758.
Monrat, Ahmed Afif, et al. Dec. 2020. Performance evaluation of permissioned blockchain platforms. In *2020 IEEE Asia-Pacific Conference on Computer Science and Data Engineering (CSDE)*. IEEE. https://doi.org/10.1109/csde50874.2020.9411380.
Praveen Kumar, Donta, et al. 2023. Exploring the potential of distributed computing continuum systems. *Computers* 12 (10): 198.
Qi, Luo. Aug. 2008. Research on intelligent transportation system technologies and applications. In *2008 Workshop on Power Electronics and Intelligent Transportation System*. IEEE. https://doi.org/10.1109/peits.2008.124.
R3 (2023). *Corda*. https://corda.net/.22.04.2023.
Ravi, Banoth, et al. 2023. Stochastic modeling for intelligent software-defined vehicular networks: A survey. *Computers* 12 (8): 162.
Saraf, Chinmay, and Siddharth Sabadra. May 2018. Blockchain platforms: A compendium. In *2018 IEEE International Conference on Innovative Research and Development (ICIRD)*. IEEE. https://doi.org/10.1109/icird.2018.8376323.
Shi, Weisong, et al. 2016. Edge computing: vision and challenges. *IEEE Internet of Things Journal* 3 (5): 637–646. https://doi.org/10.1109/JIOT.2016.2579198. https://www.scopus.com/inward/record.uri?eid=2-s2.0-84987842183&doi=10.1109%2fJIOT.2016.2579198&partnerID=40&md5=a975bcadfcbc0402da6444a53205ecde.
Shit, Rathin Chandra. Apr. 2020. Crowd intelligence for sustainable futuristic intelligent transportation system: a review. *IET Intelligent Transport Systems* 14 (6): 480–494. https://doi.org/10.1049/iet-its.2019.0321.
Srirama, Satish Narayana. 2023. A decade of research in fog computing: Relevance, challenges, and future directions. *Software: Practice and Experience*. https://doi.org/10.1002/spe.3243. eprint: https://onlinelibrary.wiley.com/doi/pdf/10.1002/spe.3243. https://onlinelibrary.wiley.com/doi/abs/10.1002/spe.3243.
Sumalee, Agachai, and Hung Wai Ho. July 2018. Smarter and more connected: future intelligent transportation system. *IATSS Research* 42 (2): 67–71. https://doi.org/10.1016/j.iatssr.2018.05.005.
Ucbas, Yusuf, et al. 2023. Performance and scalability analysis of ethereum and hyperledger fabric. *IEEE Access* 11: 67156–67167. https://doi.org/10.1109/access.2023.3291618.
University of Tartu. 2018. *UT Rocket*. https://doi.org/10.23673/ph6n-0144.
Wang, Xumeng. et al. Jan. 2021. Visual human–computer interactions for intelligent vehicles and intelligent transportation systems: the state of the art and future directions. *IEEE Transactions on Systems, Man, and Cybernetics: Systems* 51 (1): 253–265. https://doi.org/10.1109/tsmc.2020.3040262.
Zheng, Zibin, et al. 2017. An overview of blockchain technology: architecture, consensus, and future trends. In *An Overview of Blockchain Technology: Architecture, Consensus, and Future Trends*, 557–564. https://doi.org/10.1109/BigDataCongress.2017.85. https://www.scopus.com/inward/record.uri?eid=2-s2.0-85019665012&doi=10.1109%2fBigDataCongress.2017.85&partnerID=40&md5=c2de287d12169cfed5a162648f1a266e.

Index

P. K. Donta et al. (eds.), *Learning Techniques for the Internet of Things*,
https://doi.org/10.1007/978-3-031-50514-0

www.ingramcontent.com/pod-product-compliance
Ingram Content Group UK Ltd.
Pitfield, Milton Keynes, MK11 3LW, UK
UKHW022214051026
15624UKWH00001B/160
* 9 7 8 3 0 3 1 5 0 5 1 3 3 *